Deutsche Gesellschaft für Kristallographie

24. Jahrestagung

14. bis 17. März 2016

Stuttgart

D G K

Abstracts

Table of content

Plenary Talks

PL01
Structural mechanism of sensing and signalling cytosolic DNA

K.- P. Hopfner[1]
[1]*Ludwig-Maximilians-University Munich, Gene Center and Department of Biochemistry, Munich, Germany*

Our immune system detects the infection with pathogens by sensing pathogen or danger associated molecular patterns (PAMPs or DAMPs). Cytosolic DNA and cyclic-di nucleotides (CDNs) are powerful DAMPs and PAMPs that lead to an innate immune host defence through the production of type I interferons and inflammatory cytokines. In this lecture, I will discuss the recent crystallographic and biochemical advances in the emerging field of CDN and DNA signalling in the mammalian immune system. In particular I will focus on the structural biology of the cGAS-STING axis in detecting and signalling cytosolic DNA through the generation of a metazoan CDN. Our recent results show that the DNA sensor cGAS belongs to the family of nucleotidyl-transferases with similarity to e.g. the cytosolic RNA sensor OAS1. cGAS is conformationally activated by DNA and generates cyclic-GMP-AMP in a two catalytic reaction. The produced cyclic-GMP-AMP, or bacterial CDNs activate STING, leading to downstream signalling to produce type I interferons. I will also discuss how STING can be activated by synthetic small molecule ligands, aiding the development of novel approaches to target cancer.

PL02
120 Years of Powder Diffraction

B. David[1,2]
[1]*Rutherford Appleton Laboratory, ISIS Facility, Harwell, Great Britain*
[2]*University of Oxford, Inorganic Chemistry Laboratory, Oxford, Great Britain*

2016 is a significant anniversary year in the history of powder diffraction. Not only is it the centenary of its independent discovery by Paul Scherrer and Peter Debye in Göttingen, and Albert Hull in Schenectady, New York but it is the fiftieth anniversary of the first presentation by Hugo Rietveld of his eponymous method at the IUCr Congress in Moscow in 1966. This talk begins with a timeline of the ingenious pioneers of powder diffraction followed by a brief survey of the current huge impact of powder diffraction across the materials domain. As well as being a retrospective, this talk will also look towards the future and consider the role and place of powder diffraction within the experimental and computational material sciences over the next twenty years.

PL05
Complex Self-Assembly and Mirror Symmetry Breaking in Liquid Crystals

C. Tschierske,[1] M. Prehm,[1] C. Dressel,[1] S. Poppe,[1] X. Zeng,[2] G. Ungar[2], F. Liu[3]
1 Institute of Chemistry, Martin-Luther University Halle, Halle/Saale, Germany
2 Department of Materials Science and Engineering, University of Sheffield, Sheffield, UK
3 Department of Materials Science and Engineering, Xi'an Jiaotong University, Xi'an Shaanxi, P.R. China

Liquid crystals (LCs) represent a state of matter combining the mobility of fluids with long range order, i.e. there is long range order though the individual molecules do not have fixed positions. Previously reported LCs form quite simple phase structures, as for example nematic, smectic and columnar phases. However, in the recent decade significant progress was achieved in increasing the structural complexity of LC phases [1], especially by using different types of polyphilic molecules [2], combining rod-like aromatic units with different types of terminal and lateral chains. This provides access to a wide variety of ordered fluids with periodic multi-compartment structures. Besides different honeycomb tiling patterns, ranging from triangular to giant decagonal, also multicolour tilings, comprising up to seven distinct compartments (colours), tiling patterns combining cells with very different size and shape and even liquid quasicrystals with dodecagonal symmetry were obtained.[2] Expansion of the lipophilic chain volume gives rise to rod-bundle phases where aromatic cores self-assemble into skeletons forming the structural basis of new series of skeletal cubic phases with different space groups and structural similarities to metal organic frameworks.

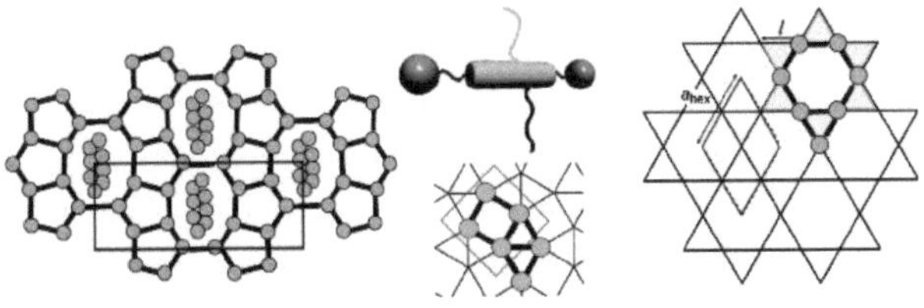

Figure 1. Polyohilic molecule and selected tiling patterns of liquid crystalline honeycombs.

Bicontinuous cubic LC phases are also in the focus of the second part of the talk, especially those formed by dithiophene based polycatenar molecules, representing rod-like aromatic compound with a number of terminal flexible chains. It tuned out that tricontinuous cubic phases of achiral molecules with an *Im*3*m* lattice can be chiral (thus becoming *I*432) [3]. Even more interestingly, isotropic liquid phases of such achiral molecules, occurring besides chiral or achiral bicontinuous cubic phases, were found to spontaneously segregate into conglomerates of chiral domains. Chirality synchronization of molecular conformers and formation of local helical twisted domains are considered as responsible for these symmetry breaking events.[4]

References:
[1] C. Tschierske *Angew. Chem. Int. Ed.* **52**, 8828-8878 (2013).
[2] X. B. Zeng, R. Kieffer, B. Glettner, C. Nürnberger, F. Liu, K. Pelz, M. Prehm, U. Baumeister, H. Hahn, H. Lang, G. A. Gehring, C. H. M. Weber, J. K. Hobbs, C. Tschierske, G. Ungar, *Science*, **331**: 1302-1306 (2011); F. Liu, R. Kieffer, X.-B. Zeng, K. Pelz, M. Prehm, G. Ungar, C. Tschierske, *Nature. Commun.* **3**, 1104 (2012). S. Poppe, A. Lehmann, A. Scholte, M. Prehm, X. B. Zeng, G. Ungar, C. Tschierske, *Nature Commun.* 6, 8637 (2015).
[3] C. Dressel, F.Liu, X. B. Zeng, G. Ungar, C. Tschierske *Angew. Chem. Int. Ed.* **53**, 13115-13120.
[4] C. Dressel, T. Reppe, M. Prehm, M. Brautzsch, C. Tschierske, *Nature Chem.* **6**, 971-977 (1914); C. Tschierske, G. Ungar, *ChemPhysChem* **17**, 9-26 (1016).

PL06

Can we have both, Intercalation and Structural Order in Clays?

J. Breu[1]

[1]*Inorganic Chemistry I, University of Bayreuth, Universitätsstrasse 30, 95440 Bayreuth, Germany, josef.breu@uni-bayreuth.de*

The intrinsically anisotropic bonding in layered materials on one side enables a vivid intracrystalline reactivity that allows for post synthetic functionalization via ion exchange or intercalation and drives the hype on two-dimensional (2D) materials in general. On the the other side, anisotropic bonding is responsible for frequent planar defects that hamper structural characterization and allow for a great deal of hand-waving in the 2D community. The situation is even worsened for insulator materials like clay minerals where a permanent charge arises from isomorphous substitution. Any deviations from an ideal solid solution situation will then consequently lead to heterogeneities in charge density that in turn gives rise to a non-uniform intracrystalline reactivity and random interstratification because cooperativity of intracrystalline reactions is essentially limited to two dimensions. In summary very little structural information is available for clay minerals and their intercalation compounds.

Reconciliation of intracrystalline reactivity and structural order requires control over compositional homogeneity and a subtle balance of the interaction between the host layers and the interlayer species that allows for sufficient directional bonding assuring a fixed phase relationship. Syntheses from the melt at temperatures well above 1,000 K gives rise to hectorites ($[Na_x]^{inter}[Mg_{3-x}Li_x]^{oct}[Si_4]^{tet}O_{10}F_2$) with tuneable and homogeneous intracrystalline reactivities. With these materials defects can be reduced to a level where a more detailed look at structural features becomes available even of hydrated phases and (microporous) intercalates. With structures at hand, insight in the characteristics of host-guest-interactions is gained.

The series of discrete hydration steps with 3D order (crystalline swelling) observed at low water concentrations under certain circumstances is followed by osmotic swelling, in which continuum electrostatic forces govern the interactions between adjacent layers. The separation of individual layers is driven by the electrostatic repulsion and a highly ordered lamellar liquid crystalline phase (Wigner crystal) is formed with basal distances from 1 – 30 nm depending only on water/clay ratio.

1. Kalo, H., Möller, M., Ziadeh, M., Dolejs, D., Breu, J., *Large scale melt synthesis in an open crucible of Na-fluorohectorite with superb charge homogeneity and particle size*, Appl. Clay Sci., **2010,** 48, 39-45.
2. Kalo, H., Milius, W., Brau, M., Breu, J., *Synthesis and single crystal structure refinement of the one-layer hydrate of sodium brittle mica*, J. Solid State Chem., **2013**, 198, 57-64.
3. Kalo, H., Milius, W., Breu, J., *Single crystal structure refinement of one- and two-layer hydrates of sodium fluorohectorite*, RSC Advances, 2012, 2, 8452-8459.
4. Stöter, M., Kunz, D.A., Schmidt, M., Hirsemann, D., Kalo, H., Putz, B., Senker, J., Breu, J., *Nanoplatelets of Sodium Hectorite Showing Aspect Ratios of approximate to 20 000 and Superior Purity*, Langmuir, **2013**, 29, 1280-1285.
5. Herling, M.M., Kalo, H., Seibt, S., Schobert, R., Breu, J., *Tailoring the Pore Sizes of Microporous Pillared Interlayered Clays through Layer Charge Reduction*, Langmuir, **2012**, 28, 14713-14719.
6. Baumgartner, A., Sattler, K., Thun, J., Breu, J., *A route to microporous materials through oxidative pillaring of micas*, Angew. Chem. Int. Ed., **2008,** 47, 1640-1644
7. Stöcker, M., Seidl, W., Seyfarth, L., Senker, J., Breu, J., *Realisation of truly microporous pillared clays*, Chem. Commun., **2008,** 629-631.
8. Herling,M., Breu, J., The Largely Unknown Class of Microporous Hybrid Materials: Clays Pillared by Molecules, Z. Anorg. Allgem. Chem. **2014**, 640, 547–560.
9. Stöter, M., Rosenfeldt, Breu, J., *Tunable Exfoliation of Synthetic Clays,* Annu. Rev. Mater. Res. **2015**, 45, 129–51.

Talks

MS01 - Biostructures I: Challenges in macromolecular crystallography

MS01-02

Crystal structure of alcohol oxidase from *Pichia pastoris*

P. Neumann[1], C. Koch[2], O. Valerius[3], I. Feussner[2], R. Ficner[1]
[1]*Georg-August-Universität Göttingen, Molekulare Strukturbiologie, Göttingen, Germany*
[2]*Georg-August-University Goettingen, Department of Plant Biochemistry, Albrecht-von-Haller-Institute, Göttingen, Germany*
[3]*Georg-August-University Goettingen, Molecular Microbiology and Genetics, Institute for Microbiology und Genetics, Göttingen, Germany*

Introduction

FAD-dependent alcohol oxidases (AOX) are key enzymes of methylotrophic organisms which can utilize lower primary alcohols as sole source of carbon and energy. This metabolic pathway is initiated by the oxidation of methanol to formaldehyde catalyzed by the FAD-dependent alcohol oxidase, which belongs to the family of glucose-methanol-choline (GMC) oxidoreductases. Although crystals of *Pichia pastoris* AOX1 were reported, no atomic structure has been determined so far, probably due to lattice defects in the crystals and/or difficulties in model rebuilding.

Objectives

Solve the phase problem using advanced Molecular Replacement protocols. Perform the crystal structure analysis of the methanol oxidase AOX1 from *Pichia pastoris* based on its three dimensional structure.

Materials & Methods

The crystallographic phase problem was solved by means of Molecular Replacement in combination with initial structure rebuilding using Rosetta model completion and relaxation against an averaged electron density map.

Results

The subunit arrangement of the homo-octameric AOX1 differs from that of octameric vanillyl alcohol oxidase and other dimeric or tetrameric alcohol oxidases, due to the insertion of two large protruding loop regions and an additional C-terminal extension in AOX1. In comparison to other alcohol oxidases the active site cavity of AOX1 is significantly reduced in size explaining the observed preference for methanol as substrate. All AOX1 subunits harbor a modified flavin adenine dinucleotide, which contains an arabityl chain instead of a ribityl chain attached to the isoalloxazine ring.

Conclusion

Homo-octamer formation induces structural changes, which were not expected considering the level of sequence conservation between the target structure and potential templates used for Molecular Replacement searches. Thus, in addition to commonly used MR results validation strategies (TFZ score, packing analysis, calculated R factors), identification of the best initial solution could additionally be facilitated by combining refinement and real space rebuilding employing energy and density-guided refinement in Rosetta program.

MS01-03

Sulfur SAD phasing of the multimeric anaphase-promoting complex Cdc23Nterm with a tailored X-ray beam size of 2.69 Å wavelength at EMBL P13 beamline at PETRA III@DESY

M. Cianci[1], G. Bourenkov[1], J. Kallio[1], G. Pompidor[1], S. Fiedler[1], I. Karpics[1], T. Schneider[1]
[1]*EMBL, Hamburg, Hamburg, Germany*

The capability of reaching wavelengths up to λ= 3.0 Å at EMBL P13 beam line, at PETRA III, the 3rd generation synchrotron at DESY in Hamburg, provides the opportunity of exploring very long wavelengths to harness the sulfur anomalous signal for phase determination. For many heavy atom compounds the anomalous signal is enhanced when collecting data at low energies (>2 Å wavelength) compared to standard data collections at 1 Å wavelength. Data collection at λ= 2.69 Å (4.6 keV) allowed the crystal structure determination at 3.1 Å resolution by sulfur SAD phasing of Cdc23Nterm, a subunit of the multimeric anaphase-promoting complex (APC/C). At this energy, Cdc23Nterm has an expected Bijvoet ratio ⟨|Fanom|⟩/⟨F⟩ of 2.2% with 282 residues with six cysteines and five methionines, and two molecules in the asymmetric unit (65.4 kDa, 12 Cys and 10 Met). Selectively illuminating two separate portions of the same crystal with an X-ray beam of 50 mm diameter allowed overcoming crystal twinning. Crystals diffracted to 3.1 Å resolution with unit cell dimensions of a=b= 61.2 Å, and c= 151.5 Å and space group $P4_3$. The refined structure to 3.1 Å resolution has an R_{factor} of 18.1% and an R_{free} of 26.4% (1). We report on the structure solution, the related methods and a discussion on the instrumentation.

(1) Cianci, M., Groves, M.R., Barford, D. & Schneider, T.R. (2016).Acta Cryst.D72, doi:10.1107/S2059798315010268.

MS01-04

Structural insights into the assembly of spliceosomal U snRNPs

C. Grimm[1], J. P. Pelz[1], A. Chari[1], J. Kuper[2], C. Kisker[2], K. Diederichs[3], H. Stark[4], H. Schindelin[2], U. Fischer[1]
[1]*Biocenter of the University of Würzburg, Biochemistry, Würzburg, Germany*
[2]*Virchow Center, Wuerzburg, Germany*
[3]*University of Konstanz, Konstanz, Germany*
[4]*Max Planck intitute, Göttingen, Germany*

The splicing process excises non-coding introns from the primary transcript (pre-mRNA) and the coding exons are joined. This reaction is catalysed by a multi-megadalton machinery called the spliceosome. It consists of the four small nuclear ribonucleoproteins (snRNPs) U1, U2, U4/6 and U5 as well as a large number of other proteins. Assembly of snRNPs from RNA and protein is an essential pre-requisite for splicesome formation. *In vivo*, this is facilitated in a highly complex biogenesis pathway. Initially, the snRNA is transported from its site of transcription to the cytoplasm. Here, the assembly of common (Sm) and some specific proteins onto U snRNAs takes place. Finally, the mature particle is transported to the nucleus and eventually incorporated into the spliceosome. While the SMN-complex is an RNP assembler and loads Sm proteins onto the snRNA, the PRMT5 complex functions upstream in this pathway. It acts as an assembly chaperone by forcing Sm proteins into a higher order structure (termed 6S complex) required for the subsequent transfer onto the SMN-complex. We solved the crystal structures of two key intermediates of the snRNP assembly machinery. The structure of the early 6S intermediate identifies pICln as an Sm-protein mimic, which enables the topological organization of five Sm proteins in a closed ring. A second structure of the 6S assembly intermediate bound to the SMN-complex components SMN and Gemin2 uncovers the likely mechanism of pICln elimination and Sm protein activation for snRNA binding. Our structural studies provide a solid basis for the understanding of how assembly factors assist the formation of RNA-protein complexes *in vivo*.

Figure: Model of the 8S assembly intermediate
Figure 1

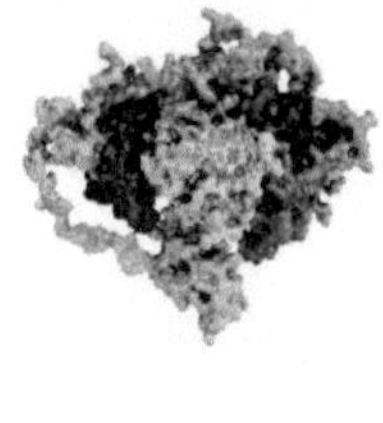

MS02 - Materials: Syntheses and structure I

MS02-01
Structures and thermoelectric properties of compounds in the system $(PbS)_nBi_2Te_2S$

P. Schultz[1], L. Staab[1], O. Oeckler[1]
[1]*IMKM, Chemistry and Mineralogy, Leipzig, Germany*

Bi_2Te_2S, known as the mineral tetradymite, exhibits a layer-like crystal structure known since 1934.[1] Yet, critical reviews suggest a superstructure and deviations from the ideal composition.[2] HRTEM studies show that Bi_2Te_2S slabs may formally incorporate one or more charge neutral PbS layers,[3] resulting in a homologous series $(PbS)_nBi_2Te_2S$. Theoretical calculations include structure optimization and the prediction that the perfectly ordered structures are topological insulators.[4]

The existence of phases $(PbS)_nBi_2Te_2S$ with various values of n was confirmed, and their structures were elucidated by means of single-crystal X-ray diffraction. Cation and anion ordering was analyzed by resonant X-ray scattering (Pb and Bi *K*-edges; ID11, ESRF, Grenoble). Thermoelectric properties were characterized.

Compounds with n = 0, 0.5, 1, 2 were obtained as single-phase samples. Single-phase tetradymite-type (n = 0) samples can only be synthesized with sulfur-rich off-stoichiometric compositions such as $Bi_2Te_{1.68}S_{1.32}$. Superstructure reflections were observed. Both this "tetradymite" and the member with n = 1 ("aleksite") are characterized by only one distinct type slab and crystallize in space group *R*-3*m*. $PbBi_4Te_4S_3$ (n = 0.5, space group *P*-3*m*1) is built up from alternating tetradymite-type (Te-Bi-S-Bi-Te) and aleksite-type (Te-Bi-S-Pb-S-Bi-Te) slabs. $Pb_2Bi_2Te_2S_3$ (n = 2, space group *P*-3*m*1) contains even thicker slabs Te-Bi-S-Pb-S-Pb-S-Bi-Te. Stacking sequences confirm the models used for theoretical calculations;[4] however, small distortions were observed. In accordance with the results from resonant diffraction, bond valence sum calculations confirm a completely ordered cation distribution where all Bi atoms are located near the van der Waals gaps. Such complete ordering is in contrast to many related systems such as $(GeTe)_nSb_2Te_3$. These typically show cation disorder although the trivalent cations tend to concentrate at the van der Waals gaps.

The phases are metals with electrical conductivities s of ~1000 $S\times cm^{-1}$. The thermoelectric properties of all mentioned compounds were determined up to 350° C. Figures of merit $ZT = S^2sT/k$ (S: Seebeck coefficient, k = thermal conductivity) reach values up to 0.4 for n = 0. The thermoelectric properties may be further optimized by doping.

[1] D. Harker, *Z. Kristallogr.* **1934**, *89*, 175-181.
[2] L. Pauling, *Am. Mineral.* **1975**, *60*, 994-997.
[3] N. J.Cook, C. L. Ciobanu, C. J. Stanley, W. H. Paar, K. Sundbald, *Can. Mineral.* **2007**, *45*, 417-435.
[4] I. V.Siokin, T. V. Menshchikova, M. M. Otrokov, S. V. Eremeev, Y. M. Koroteev, M. G. Vergniory, V. M. Kuznetsov, E. V. Chulkov, *JETP Lett.* **2012**, *96*, 322-325.

MS02-02
Crystal structure of hydrated fluorides $MF_2.4H_2O$ (M = Zn, Ni, Co): a combined approach

G. Nénert[1], K. Forsberg[2], C. Martineau[3], T. Tao[4], S. Halasyamani[4]
[1]*PANalytical B. V., Almelo, Netherlands*
[2]*KTH Royal Institute of Technology, STOCKHOLM, Sweden*
[3]*Universite de Versailles, Institut Lavoisier de Versailles , Versailles, France*
[4]*University of Houston, Department of Chemistry, Houston, United States*

Introduction:
The inorganic metal fluorides and oxide-fluorides have significant importance in the development of many new technologies, and are impacting various key points of modern life, that is, energy production and storage, microelectronics and photonics, catalysis, automotive, building, etc. Many research fields and applications benefit from a better knowledge of the relationships occurring between the structure of such compounds and some pertinent physical properties [1].

Based on single crystal X-ray work, $ZnF_2.4H_2O$ has been suggested to crystallize in the polar space group $Pca2_1$ [2]. The possibility for non- linear optical properties of this compound and related phases motivated further investigation of this family.

Among the large possible chemistry of the $MF_2.4H_2O$ family, very little is known about the crystal structure of its members. Compositions such as $MF_2.4H_2O$ family (M = Co, Ni, Fe) have been reported but not characterized structurally [3]. In the absence of single crystals, the structural determination of such materials is particularly complicated as oxygen and fluorine atoms exhibit very similar diffraction cross sections irrespective of the source (X-ray or neutron).

Methods & Results:
Using a combined approach of X-ray diffraction and fluorescence, neutron diffraction, second harmonic generation (SHG) and NMR (^1H, ^{19}F), we show that the crystal structure of these materials is actually centrosymmetric, in contradiction to previous reports. Most importantly, we demonstrate that while $ZnF_2.4H_2O$ is effectively an hydrated fluoride, we give evidence that the real structure of the materials for M = Co, Ni is actually $M(OH,F)_4(H_2O)_2$ with likely unusual oxidation state of Co^{4+} and Ni^{4+}.

Conclusion:
While these materials exhibit a simple chemistry, the absence of single crystal and similar cross section between fluorine and oxygen make these materials a challenge to be investigated from powder diffraction. The key to solve the structure of these materials was to use complementary techniques such as x-ray and neutron diffraction, ^1H & ^{19}F NMR and SHG.

References
[1] Leblanc M., Maisonneuve V., Tressaud A. , "Crystal Chemistry and Selected Physical Properties of Inorganic Fluorides and Oxide-Fluorides" *Chem. Rev.*, 2015,*115*(2), pp 1191-1254
[2] B. V. Bukvetskii, S. A. Polishchuk, V. I. Simonov, "Crystal structure of zinc fluoride tetrahydrate, $ZnF_2.4H_2O$" *Sov. Phys. Crystallogr.* 18(5), 600-602 (1974)
[3] B. A. Lange, H. M. Haendler, "The thermal decomposition of nickel and zinc fluoride tetrahydrates" *J. Inorg. Nucl. Chem.* 1973, vol. 35, 3129-3133; J. Swanepoel, A. M. Heyns, "Bending modes of the water molecules and the M-O stretching modes in the series $MF_2.4H_2O$ (M = Fe, Co, Ni, Zn)" *Spectrochimica Acta*, vol. 47A(2), 243-253 (1991).

MS02-03

Enhanced methods of crystallization: the crossover synthesis from aqueous solution to melt flow

J.- C. Buhl[1]

[1] *Institut für Mineralogie, Leibniz Universität Hannover, Kristallographie, Hannover, Germany*

Modifications of traditional crystallization methods are more and more important to create energy and time efficient optimized procedures [1]. Therefore the crossover synthesis in (aqueous) solution and crystal growth in melt (flux) is demonstrated as a one-pot process. The method based upon two distinctive features, both ruled by controlled heating of a tailored educt mixture. Firstly the nucleation and early growth step proceeds within the hydrate water of the eductsduring heating, comparable with mild hydrothermal conditions. Secondly further crystal growth occurs by a shift of the conditions to a melt flux process at elevated temperature, provided that a suitable flux component is added to the educts.The present work demonstrates this new preparation technique using the sodalite system as a case study.

Zeolite 13-X was selected as Si-Al-source and simultaneously also as "water provider" in the first step of the reaction. Furthermore NaOH granulate is added. NaOH is not only responsible to reach alkaline conditions for sodalites but also for formation of a melt flux at elevated temperature ($T_{max.}$) in the second step of the whole process.

Three examples of sodalite syntheses will demonstrate the effectiveness of the method: 1) preparation of microcrystalline samples of chloride sodalite; 2)Synthesis of sodalite with the taylor-made guest $BO(OH)_2$; 3) Formation of microcristalline nitrite sodalite. In each case a dry powder mixture of zeolite 13-X, NaOH granulate and a certain template salt (exp. 1: NaCl; exp. 2: $NaBO_2 \cdot 4H_2O$; exp. 3: $NaNO_2$) was pressed into a pellet before heated under open conditions. A heating program RT→400°C for chloride and borate sodalite (or RT→300°C for nitrite sodalite) was revealed for syntheses with 90' heating up time and 60' holding time at $T_{max.}$.

SEM images, XRD results as well as FTIR spectra of the synthesis products are summarized in Figures 1 - 2. Microcrystalline chloride sodalite crystallized in remarkable quality after only 2.5 hours (Fig. 1, left image).Borate sodalite with the tailored guest $NaBO(OH)_2$ could be synthesized under these short time conditions too. The enclathrated borate anions can be distinguished by their characteristic B-O and B-(OH) vibrations in the 1500 cm^{-1} - 1200 cm^{-1} region of FTIR spectrum in Fig. 2).

Synthesis of microcrystalline nitrite sodalite (crystals see Fig. 1; proof of nitrite enclathration by ν_{as}= 1261 cm^{-1} in the FTIR spectrum of Fig. 2) as well as formation of a dense membrane of basic nitrite sodalite was also successful. As the latter is a suitable material for the uptake of carbon dioxide [2] membrane production becomes maybe important in future.

Summarizing the results it is shown that the crossover of both reactions (solution + melt) is a suitable method for tailored crystallization and offers many further possibilities. Assynthesis starts without addition of any liquid, hydrolysis sensitive guest anions/templates can be inserted. Direct synthesis with thermal reactive guest species is simply possible by setting the temperature on the specific stability region for simultaneous thermal generation of the desired guest.

References: [1] Buhl, J.-Ch.: On the autothermal synthesis of zeolites. Micropor. Mesopor. Mater. 222 (2016) 73-79. [2] Buhl, J.-Ch.: Basic nitrite sodalite - $Na_8[AlSiO_4]_6(OH \cdot H_2O)(NO_2)$ -a suitable material for the uptake of carbon dioxide. J. Solid State Chem. 94 (1991) 19-26.

Figure 1: SEM images of products of crossover synthesis from aqueous solution to melt flow

Figure 2: X-ray powder patterns and FTIR spectra of the products shown in Fig. 1.

Figure 1

Figure 2

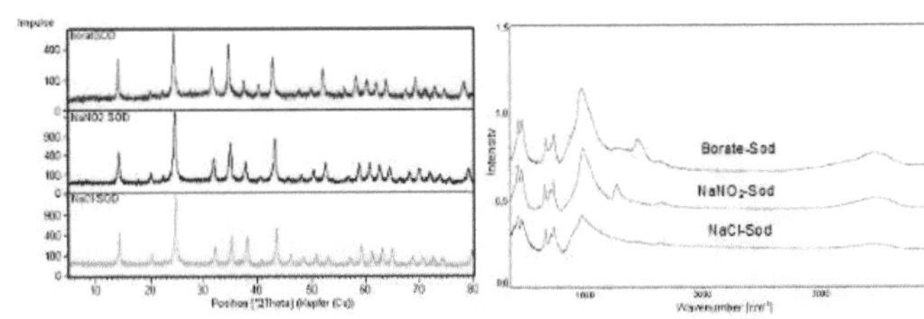

MS02-04

X-ray powder diffraction as a tool to analyze the shape of silver nanoparticles with different morphologies

O. Prymak[1], J. Helmlinger[2], K. Loza[2], M. Heggen[3], M. Epple[2]

[1] *University of Duisburg-Essen / Inorganic Chemistry, Dept. for X-ray Diffraction, Essen, Germany*

[2] *University of Duisburg-Essen / Inorganic Chemistry, Essen, Germany*

[3] *Ernst Ruska-Center and Peter Grünberg Institute / Forschungszentrum Jülich GmbH, Jülich, Germany*

Silver nanoparticles (Ag-NPs) are well known for their interesting optical, electronic, and biological properties. Due to their antibacterial activity, there are many silver-containing consumer products and medical devices. As their properties are not only influenced by their small size but also by their morphology, a shape-controlled synthesis of silver nanoparticles is of special interest. We have synthesized PVP-coated Ag-NPs by a modified polyol process and microwave-assisted reduction, producing well-defined morphologies (spheres, platelets, cubes, and rods). The nanoparticles were characterized by colloidal (DLS) and microscopic methods (scanning and transmission electron microscopy) as well by different powder diffraction methods, including grazing incidence diffraction and pole figure analysis. By Rietveld refinement, the lattice parameters and crystallite sizes were determined. A special focus of the work was the use of different stereographic projections (pole figures) to investigate the crystal growth of Ag-NPs with anisotropic shapes. It was demonstrated that diffraction methods are well suitable to describe not only the size, but also the shape and crystallographic orientation of metallic nanocrystals.

Figure 1: Representative scanning electron micrograph (left) and (200) pole figure (right) of cubic Ag-NPs. The cubes show a distinct preferred orientation on the sample holder.

Figure 1

Figure 2

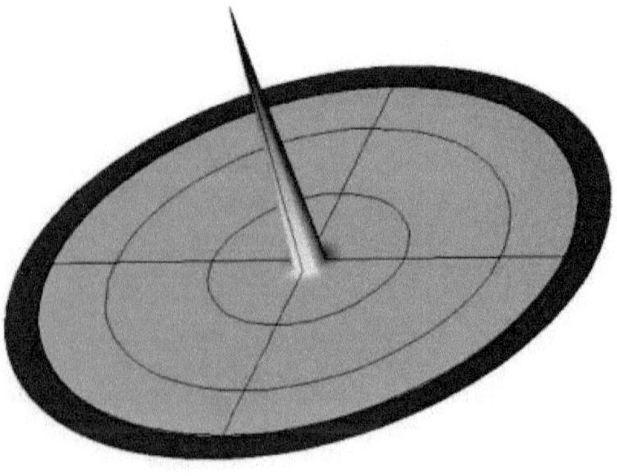

MS03 - Biostructures II: Hot new structures

MS03-01
How does CRM1 interact with FG-repeat proteins of the nuclear pore during passage?

T. Monecke[1], S. A. Port[2], A. Dickmanns[1], R. Hofele[3], H. Urlaub[3], M. S. Weiss[4], R. H. Kehlenbach[2], R. Ficner[1]
[1]*Georg-August-Universität Göttingen, Abteilung für Molekulare Strukturbiologie, Göttingen, Germany*
[2]*Georg-August-Universität Göttingen, Institut für Molekularbiologie, Göttingen, Germany*
[3]*Max-Planck-Institut für biophysikalische Chemie, Bioanalytische Massenspektrometrie, Göttingen, Germany*
[4]*Helmholtz-Zentrum für Materialien und Energie, Makromolekulare Kristallographie (HZB-MX), Berlin, Germany*

Introduction and Objectives
In eukaryotic cells nucleocytoplasmic transport of macromolecules is an active process and proceeds through nuclear pore complexes (NPCs). Main mediators are soluble nuclear transport receptors of the karyopherin-β superfamily termed importins and exportins. CRM1 is the major nuclear export receptor in eukaryotic cells. During translocation through the NPC, nuclear transport receptors transiently interact with phenylalanine-glycine (FG-) repeats of multiple nucleoporins (Nups). On the cytoplasmic side of the NPC, CRM1 tightly interacts with the nucleoporin Nup214. The transient interaction of FG-Nups with transport receptors in general is necessary to overcome the nuclear envelope barrier and thus is crucial for all nuclear transport events.

We sought to characterize the interaction between the major nuclear export receptor CRM1 and multiple FG-repeats of the nucleoporin Nup214.

Methods
Since high conformational flexibility is an intrinsic and indispensable property of nuclear transport receptors, their crystallization and structure determination is particularly challenging. Optimization of constructs, seeding and the development of a sophisticated protocol including successive PEG-mediated crystal dehydration as well as additional post-mounting steps were essential to obtain well diffracting crystals and to solve the crystal structure [1].

Results
Here, we present the crystal structure of a 117 amino acid FG-repeat containing fragment of Nup214, in complex with CRM1, the export cargo Snurportin 1 and the small GTPase Ran at 2.85 Å resolution [2]. The structure reveals eight binding sites for Nup214 FG-motifs on CRM1, with intervening stretches that are loosely attached to the transport receptor. Nup214 binds to N- and C-terminal regions of CRM1, thereby clamping CRM1 in a closed conformation and stabilizing the export complex. The role of conserved hydrophobic pockets for the recognition of FG-motifs was analyzed in biochemical and cell-based assays.

Conclusion
Unraveling the molecular details of nucleoporin-karyopherin interactions, which have to be strong enough to promote transport, but sufficiently weak to avoid stalling of transport complexes within the pore, is key to our understanding of the mechanisms of nucleocytoplasmic transport [2].

With CRM1 as an example, we are beginning to understand the molecular details of transport complexes passing through the permeability barrier of the NPC, a process that involves binding to local FG-regions, but also rapid dissociation from such sites. Based on the principles described above, movement of transport complexes within the pore becomes feasible, without bringing translocation to a standstill due to slow off-rates [2].

References

[1] Monecke, T., Dickmanns, A., Weiss, M. S., Port, S. A., Kehlenbach, R. H., & Ficner, R. (2015). *Acta Crystallographica Section F Structural Biology Communications*. **71**, 1481-1487.
[2] Port, S. A.*, Monecke, T.*, Dickmanns, A., Spillner, C., Hofele, R., Urlaub, H., Ficner, R., & Kehlenbach, R. H. (2015). *Cell Reports*. **13**, 690-702.

MS03-02
Crystal Structure of the Human Cytomegalovirus pUL50-pUL53 Core Nuclear Egress Complex

S. A. Walzer[1], C. Egerer-Sieber[1], K. Hohl[1], M. Sevvana[1], Y. A. Muller[1]
[1]*FAU Erlangen-Nürnberg, Biology, Erlangen, Germany*

Introduction
The Human Cytomegalovirus (HCMV) is a ubiquitous pathogen with global infection rates between 60 and 100% of the adult population. While infection is generally asymptomatic in immunocompetent adults, congenitally infected children as well as immunocompromised adults suffer from various, potentially lethal symptoms [1]. Nuclear replication of HCMV relies on elaborate mechanisms of nucleocytoplasmic egress of viral particles, involving two essential and conserved viral proteins, pUL50 and pUL53: pUL50 and pUL53 heterodimerize and form a core nuclear egress complex (NEC), which is anchored to the inner nuclear membrane and provides a scaffold for the assembly of a multimeric viral-cellular NEC [2].

Objectives
Recently, the viral NEC has attracted the interest of researchers as it represents a regulatory key position of viral replication and, therefore, a putative target for novel antiviral strategies [3]. Here, we present the crystal structure of the pUL50-pUL53 heterodimer as a first step towards knowledge-based drug discovery.

Materials & Methods
pUL50 and pUL53 proteins were coproduced in *E. coli* and copurified via immobilized metal affinity chromatography and size exclusion chromatography. Crystals of both the unlabeled and the Selenomethionine-labeled protein complex were obtained in 0.2 M $CaCl_2$ and 20% w/v PEG 3350 after 10 days at 20 °C.

Data sets were collected at the synchrotron facilities BESSY II (Berlin, Germany) and PETRA III (Hamburg, Germany) and processed with XDS. The structure was solved via single anomalous dispersion phasing using the programs HKL2MAP, SHELX, PHENIX and COOT, and refined in REFMAC.

Results
The crystal structure of the pUL50-pUL53 heterodimer (amino acids 1-175 and 50-292, respectively) was solved at 2.44 Å resolution [4]. Both proteins adopt a globular fold with mixed α and β secondary structure elements (**Fig. 1**).

pUL53-specific features include a zinc-binding site and a hook-like N-terminal extension, the latter representing a hallmark element of the pUL50-pUL53 interaction: The hook-like extension (amino acids 59-87) embraces pUL50 and contributes 80% to the total interface area. pUL53 also shows structural resemblance to the GHKL domain of bacterial sensory histidine kinases despite a low sequence similarity of less than 15%.

The pUL50 structure overall resembles the recently published NMR structure of the murine cytomegalovirus homolog pM50, but reveals a considerable repositioning of the very C-terminal α-helix of pUL50 upon pUL53 binding.

A close examination of the crystal structure indicates partial assembly of pUL50-pUL53 heterodimers to hexameric ring-like structures, possibly providing additional scaffolding opportunities for the NEC (**Fig. 2**).

Conclusion
The structural information on pUL50-pUL53 considerably improves our understanding of the mechanism of HCMV nuclear egress. It may also accelerate the validation of the NEC as a unique target for developing a novel type of antiviral drug and improve options for broad-spectrum antiherpesviral therapies.

References
[1] Griffiths, P., Baraniak, I., and Reeves, M. (2015) *J. Pathol.* **235**, 288-297
[2] Milbradt, J., Auerochs, S., Sticht, H., and Marschall, M. (2009) *J. Gen. Virol.* **90**, 579-590
[3] Marschall, M., and Stamminger, T. (2009) *Future Microbiol.* **4**, 731-742
[4] Walzer, S. A., Egerer-Sieber, C., Sticht, H., Sevvana, M., Hohl, K., Milbradt, J., Muller, Y. A., Marschall, M. (2015) *J Biol Chem.* **290(46)**, 27452-8

Figure 1

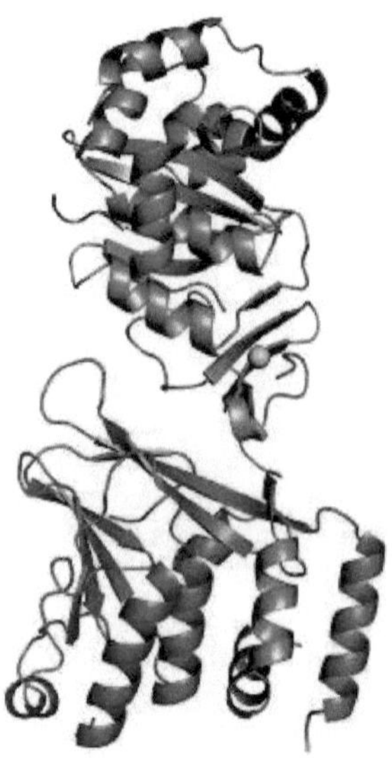

Figure 2

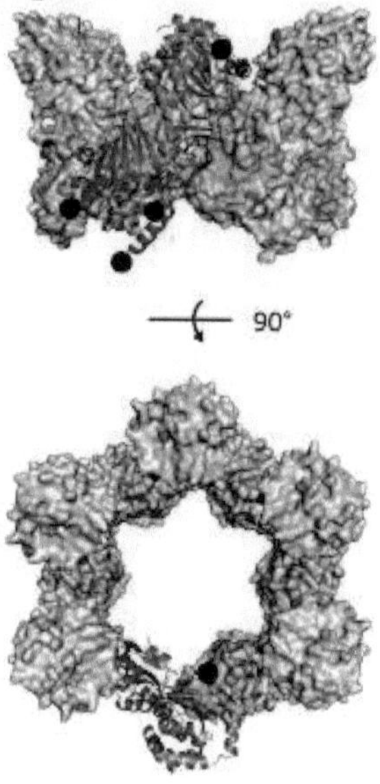

MS03-03
Crystal structure of animal fatty acid synthase KS-MAT domain
K. Paithankar[1], A. Rittner[1], M. Grininger[1]
[1]Goethe University Frankfurt, BMLS Institute of Organic Chemistry and Chemical Biology, Frankfurt am Main, Germany

Fatty acid synthase (FAS) is a multi-enzyme protein that catalyzes the synthesis of fatty acids (Wakil, 1989). These fatty acids are important components of cell membranes, as a storage form of energy and as signalling molecules. Fatty acid metabolism is a primary factor in human obesity, and exhibits pronounced expression levels in types of cancer. Animal fatty synthase (EC 2.3.1.85) is a homodimer with identical polypeptides of 2500 amino acid residues possessing different catalytic domains necessary for biological function. The first steps in fatty acid synthesis are the acetylation of β-ketoaceyl synthase domain (KS) active site by acyl carrier protein (ACP), and the recharging of ACP with a malonyl group by active site of MAT (malonyl/acetyl transferase) domain. Recently, X-ray crystal structures of the FAS and its constituent domains were revealed (Maier,et al., 2008, Pappenberger,et al.,2010). We have recombinantly expressed KS-MAT domain of animal fatty acid synthase and determined X-ray crystal structures of wild-type and mutant KS-MAT domain soaked with different substrates. There are four molecules in the asymmetric unit (Fig. 1) and residues in both KS and MAT active sites are well resolved in the electron density maps. Intra- and inter-domain structural superpositions reveal subtle differences in the active sites and large rigid-body movements in the flexible MAT domain. These new structures are essential to understanding the chemistry of the catalysis. With over 90 % sequence similarity to human fatty acid synthase, these structures may provide a good basis for designing medicinally relevant human FAS inhibitors. Furthermore, high similarity to module 2 in 6-deoxyerythronolide B synthase (DEBS) suggest the structural models can be exploited for the design and synthesis of medicinally important polyketide products.

Wakil, S. J. (1989) Fatty acid synthase, a proficient multifunctional enzyme Biochemistry 28, 4523-4530.
Maier, T., Leibundgut, M, and Ban, N (2008) The Crystal Structure of a Mammalian Fatty Acid Synthase. Science 321, 1315-1322

Pappenberger, G. et al., (397) Structure of the Human Fatty Acid Synthase KS-MAT

Didomain as a Framework for Inhibitor Design J. Mol. Biol. 397, 508-519
Chirala, S. et al. (1997) Animal Fatty Acid Synthase: Functional Mapping and Cloning and Expression of the Domain I Constituent Activities. PNAS 94, 5588-5593

Figure 1

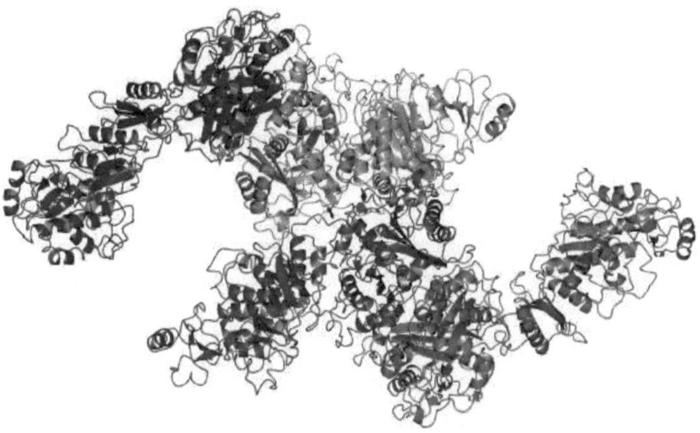

MS03-04

Structural insight into the protein-aided bacterial biofilm formation

Y. Roske[1], A. Diehl[2], P. Schmieder[2], L. Ball[2], î Akbey[2], H. Oschkinat[2], U. Heinemann[1]
[1]*Max-Delbrück-Centrum, Macromoleculare Structures and Interaction, Berlin, Germany*
[2]*Forschungsinstitut für molekulare Pharmakologie, NMR, Berlin, Germany*

Biofilm formation is a social behavior that generates favorable conditions for sustained survival in the natural environment. In the biofilm individual cells are held together by a self-produced extracellular polymeric matrix (ECM) commonly comprised of polysaccharides, proteins and DNA. The ECM protects cells within a biofilm from antibiotics, provides mechanical stability, affects hydrophobicity of biofilm and mediates cell adhesion to surfaces. ECM proteins serve a variety of structural roles and form often amyloid-like fibers. Extensive studies of amyloid fiber formation in humans are known, but less about the assembly of bacterial functional amyloid-like precursors into fibers at molecular level.

Amyloid-like fibers produced by Bacillus subtilis, are composed of two components, the major subunit TasA and the minor subunit TapA (1). Whereas the highly abundant secreted protein TasA can form a fibrillary amyloid-like structure in the biofilm, TapA is important for the proper localization of TasA to the matrix (2).

Here we present the first high-resolution crystal structure of TasA from the Gram-positive B. subtilis in its soluble monomeric form. The crystal structure was determined by anomalous phase determination of Selenomethionine derivative protein crystals and shows a globular central 9 stranded antiparallel β-sandwich with a jelly-roll fold flanked by 6 α-helices and several larger loop regions.

Interestingly, other known major biofilm functional amyloids are synthesized as unstructured monomers which form the amyloid-like structures in biofilms, as CsgA from E.coli or FapC from P. aeruginosa. However, TasA uniquely having a globular monomeric fold as the crystal structure shows, which transformed in the biofilm into a fibrillary structure, suggested being amyloid like. Remarkably, NMR experiments could show that TasA capture indeed three distinct protein conformations as the soluble monomer, oligomeric state (aggregate or gel like) and a fibril form, which could be part of the protein fibrillation pathway in the biofilm.

The structural information obtained on monomeric TasA, combined with the first steps of structural switch towards protein fibrillation constitutes a basis to understand functional amyloids in biofilms, their regulation, biogenesis and possible ways of controlling them.

References

1. Amyloid fibers provide structural integrity to Bacillus subtilis biofilms. Romero D1, Aguilar C, Losick R, Kolter R. Proc Natl Acad Sci U S A. 2010 Feb 2;107(5):2230-4.
2. A major protein component of the Bacillus subtilis biofilm matrix. Branda SS, Chu F, Kearns DB, Losick R, Kolter R. Mol Microbiol. 2006 Feb;59(4):1229-38.

MS04 - Lightning talks of young crystallographers

MS04-01
1.6 Å Structure of PilBac1: Insights into Long-Range Electron Transfer in Bacteria and a Sulfur-SAD Success Story

M. Gorgel[1,2], A. Boeggild[2], J. Ulstrup[2], U. Mueller[3], M. Weiss[3], P. Nissen[2], T. Boesen[2]
[1]*Ludwig-Maximilians University, Gene Center, Munich, Germany*
[2]*Aarhus University, Department for Molecular Biology and Genetics, Aarhus, Denmark*
[3]*Helmholtz Zentrum Berlin für Materialien und Energie, Macromolecular Crystallography (HZB-MX), Berlin, Germany*

Bacterial nanowires are extracellular filaments that can transfer electrons from one cell to another and from a cell to an electron acceptor. This allows the metal reducing bacterium *Shewanella oneidensis* to grow on insoluble and toxic metals such as Fe(III), Mn(IV) and U(VI). These networks have already shown application in environmental sciences and nanotechnology, nevertheless, their optimal implementation requires an understanding of their underlying structure and mechanism as revealed by structural biology and nanoscience.

We therefore determined the 1.6 Å crystal structure of the putative nanowire subunit PilBac1. We phased the structure by sulfur-SAD despite several obstacles such as a low anomalous signal (a theoretical Bijvoet ratio of 0.9% at a wavelength of 1.8 Å), radiation damage-induced reduction of the cysteines and a multiplicity of only 5.5 [1]. The anomalous signal was improved by merging three data sets from different volumes of a single crystal, yielding a multiplicity of 17.5, and a sodium ion was added to the substructure of anomalous scatterers. The overall structure of PilBac1 reveals the common fold of type IV pili, but in a model of the assembled pilus we can clearly define individual clusters of closely spaced aromatic residues (3-5 Å). These short distances might allow for metallic-like electron transfer between the aromatic rings; however, the distances between the clusters will be too high for continuous electron transfer along the pilus and additional factors such as cytochromes will consequently be necessary.

In summary, our structure provides insight into the assembly of (non-conductive) pili in *Shewanella* and at the same time represents an exceptional case for sulfur-SAD phasing despite high radiation damage and a low anomalous signal.

References: 1: Gorgel *et al.* (2015). Acta Cryst. D. 71, 1095-1101.

MS04-02
Temperature-induced phase transition in tetragonal copper sulfide at low temperatures

D. Zimmer[1], L. Bayarjargal[1], J. Zhang[2], E. Alig[3], L. Fink[3], E. Haussühl[1], C. Jin[2], B. Winkler[1]
[1]*Goethe University, Crystallography, Frankfurt am Main, Germany*
[2]*Institute of Physics Chinese Academy of Sciences, Beijing, Germany*
[3]*Goethe University, Chemistry, Frankfurt am Main, Germany*

Cuprous sulfides ($Cu_{2-x}S$; $0 \leq x \leq 0.25$) are nontoxic p-type semiconductors with a suitable band gap for the application as thin film solar cell absorbers [1]. However, their crystal structures and corresponding phase relations are complex due to the high mobility of copper atoms and vacancies within the sulfur sublattice and have therefore been in the focus of many studies for several decades. Since their electrical transport properties strongly depend on the stoichiometry and the crystal structure [2], the experimental and theoretical investigation of this structure-property relation is of great interest.

The aim of this study is to investigate the influence of structural changes in the tetragonal high pressure high temperature polymorph of Cu_2S [3, 4] on its electrical transport properties at low temperatures.

We synthesized polycrystalline samples of tetragonal Cu_2S in a large volume press at 5 GPa and 973 K. The samples were characterized by EDX and powder XRD. The evolution of the crystal lattice as a function of temperature was investigated in situ by powder XRD measurements between 295 and 87 K. Resistance measurements were performed by voltage-driven 4-wire measurements between 300 and 2 K. The heat capacity was determined experimentally by thermal-relaxation calorimetry between 300 and 2 K and theoretically investigated by DFT-based calculations.

The heat capacity and powder XRD measurements confirm previous findings [5], showing a low temperature phase transition. According to our data the transition takes place at ~205 K with a hysteresis of ±15 K (Fig. 1).

The resistance is strongly correlated with the structural transition and changes by several orders of magnitude across the transition, also showing a hysteresis on cooling and heating (Fig. 2).

The authors gratefully acknowledge financial support by the DFG (Projects Wi 1232 & BA 4020/2-1).

Figure 1: Heat capacity of tetragonal Cu_2S

Figure 2: Resistance of a tetragonal Cu_2S sample

[1] R.B. Hall & J.D. Meakin (1979). *Thin Solid Films* **63**, 203.
[2] K. Okamoto & S. Kawai (1972). *Jap. J. Appl. Phy.* **12**, 8, 1130.
[3] A. Janosi (1964). *Acta Cryst.* **17**, 311.
[4] B.J. Skinner (1970). *Econ. Geol.* **65**, 724.
[5] S. Stølen, F. Grønvold & E. F. Westrum Jr. (1990). *J. Chem. Thermodynamics* **22**, 1035.

Figure 1

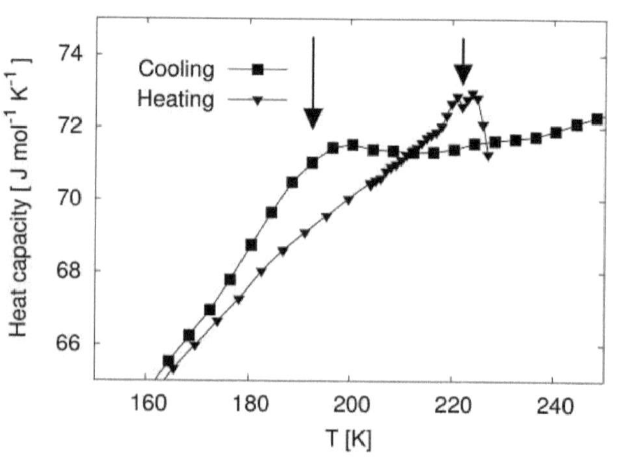

Figure 2

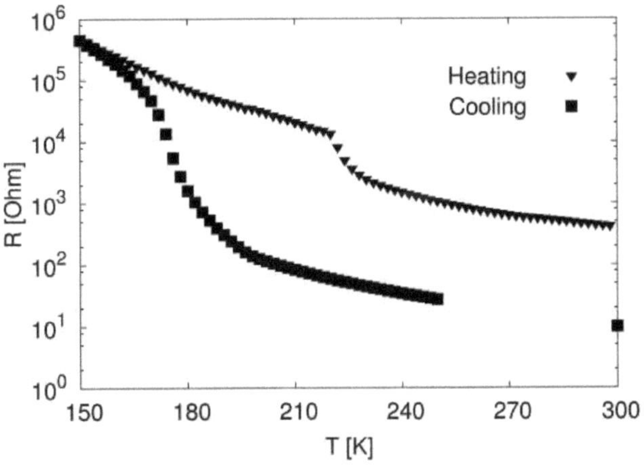

MS04-03

Structural complexity of simple Fe_2O_3 oxide at high pressures and temperatures

E. Bykova[1], M. Bykov[1], C. McCammon[1], S. Ovsyannikov[1], H.- P. Liermann[2], I. Kupenko[3], A. Chumakov[3], R. Rüffer[3], M. Hanfland[3], V. Prakapenka[4], N. Dubrovinskaia[1], L. Dubrovinsky[1]
[1]*University of Bayreuth, Bayerisches Geoinstitut, Bayreuth, Germany*
[2]*Deutsches Elektronen-Synchrotron, P02.2 Extreme Conditions Beamline, Hamburg, Germany*
[3]*European Synchrotron Radiation Facility, Grenoble, France*
[4]*Argonne National Laboratory, GSECARS, Argonne, United States*

The Fe-O binary system is widely studied both in solid state and mineral physics. Oxides of iron are present in nature in the form of FeO (known at normal conditions as wüstite, rock-salt structure), Fe_2O_3 (hematite, corundum structure), and Fe_3O_4 (magnetite, inverse spinel structure). High-pressure behaviour of natural iron oxides was a subject of studies since early 60s but still many controversies, in particular, regarding crystal structures of high-pressure high-temperature (HPHT) polymorphs, remained. For instance, many structures have been proposed for a high-pressure polymorph of Fe_3O_4, among them monoclinic one with octahedrally coordinated iron atoms [1], $CaMn_2O_4$-type [2], and $CaTi_2O_4$-type (space group *Bbmm*, No. 63) structures [3,4]. Orthorhombic $GdFeO_3$ perovskite [5] and Rh_2O_3-II [6] structural types have been proposed as candidates for a high-pressure polymorph of Fe_2O_3 which exists above 45 GPa.

We report here the crystal structures of novel iron oxides, synthesized at HPHT conditions. HPHT polymorphs of Fe_2O_3 and Fe_3O_4, and iron oxides with completely new stoichiometry, as e.g. Fe_5O_7 and $Fe_{25}O_{32}$, were observed in our experiments in laser-heated diamond anvil cells. They were characterized using single-crystal X-ray diffraction and synchrotron Mössbauer source spectroscopy. Our crystallographic data allowed to explain electronic and magnetic changes in Fe_2O_3 and Fe_3O_4 induced by HPHT.

We found similarities in the crystal structures of Fe_5O_7, some high-pressure polymorphs of Fe_2O_3 and Fe_3O_4, and the recently discovered Fe_4O_5 [7] and Fe_5O_6 [8]. Based on the study of Guignard *et al.* [9] we suggest a homological series of iron oxides with a common formula $m\text{FeO}\cdot n\text{Fe}_2\text{O}_3$ ($m \geq 1$, $n \geq 0$). The crystal structures of the iron oxides belonging to the homological series are composed of FeO6 trigonal prisms and octahedra. So far we experimentally observed six representatives of the series. Our results suggest that unusual trigonal prismatic oxygen coordination of iron may be common for iron-oxygen compounds at conditions of the Earth's lower mantle.

References
[1] H.-K. Mao, T. Takahashi, W. A. Bassett, G. L. Kinsland, L. Merrill, *J. Geophys. Res.* **79**, 1165 (1974).
[2] Y. Fei, D. J. Frost, H.-K. Mao, C. T. Prewitt, D. Haeusermann, *Am. Mineral.* **84**, 203 (1999).
[3] C. Haavik, S. Stølen, H. Fjellvåg, *Am. Mineral.* **85**, 514 (2000).
[4] L. S. Dubrovinsky, N. A. Dubrovinskaia, C. McCammon, G. K. Rozenberg, R. Ahuja, J. M. Osorio-Guillen, V. Dmitriev, H.-P. Weber, T. Le Bihan, B. Johansson, *J. Phys. Condens. Matter* **15**, 7697 (2003).
[5] J. S. Olsen, C. S. G. Cousins, L. Gerward, H. Jhans, B. J. Sheldon, *Phys. Scr.* **43**, 327 (1991).
[6] G. Rozenberg, L. Dubrovinsky, M. Pasternak, O. Naaman, T. Le Bihan, R. Ahuja, *Phys. Rev. B* **65**, 064112 (2002).
[7] B. Lavina, Y. Meng, R. T. Downs, P. F. Weck, S. R. Sutton, Y. Zhao, *Proc. Natl. Acad. Sci. USA* **108**, 17281 (2011).
[8] B. Lavina, Y. Meng, *Sci. Adv.* **1**, e1400260 (2015).
[9] J. Guignard, W. A. Crichton, *Mineral. Mag.* **78**, 361 (2014).

MS04-04

High-pressure phase transition and single-crystal elasticity of $SrCO_3$ by Raman spectroscopy and Brillouin spectroscopy

N. Biedermann[1], S. Speziale[2], H. J. Reichmann[2], M. Koch-Müller[2], G. Heide[3]
[1]*European XFEL, HED Group, Hamburg, Germany*
[2]*Deutsches GeoForschungsZentrum GFZ, Potsdam, Germany*
[3]*Technische Universität Bergakademie Freiberg, Institut für Mineralogie, Freiberg, Germany*

Carbonates are key carriers of Carbon into the deep mantle and they were found to be stable at conditions of the lower mantle due to their presence as inclusions in transition-zone diamonds (Brenker et al., 2007). Strontianite ($SrCO_3$), which is a component in all natural carbonates, is isostructural to aragonite, a major high-pressure polymorph of $CaCO_3$, and therefore a model material to investigate the stability and physical properties of carbonates at high pressure and high temperature. Strontianite was synthesized at 4 GPa and 1273 K for 24h in a multi-anvil apparatus. We compressed single-crystals or powder samples in the diamond anvil cell and measured Raman scattering up to 78 GPa. A transition at 26 ± 1 GPa was observed during compression to a new high-pressure structure of $SrCO_3$. This post-aragonite phase was preserved until the highest experimental pressure and in decompression still present down to 18 GPa. An increase in the coordination number for Sr^{2+} produces a clear change in the vibrational spectrum of $SrCO_3$ and is in agreement with previously published high-pressure structures with space group *Pmmn* for aragonite-type carbonates (e.g. Ono et al., 2005; Oganov et al., 2006; Ono, 2007). In addition to our study of the high-pressure polymorphism of $SrCO_3$, we have used Brillouin spectroscopy to determine the elastic properties of synthetic $SrCO_3$ single-crystals at ambient conditions. The present study provides the first experimental data for the individual stiffness tensor c_{ij} of $SrCO_3$. Measuring Brillouin scattering along 69 different crystallographic directions in a set of three platelets we were able to determine c_{11}, c_{12}, c_{13}, c_{22}, c_{23}, c_{33}, c_{44}, c_{55} and c_{66} with 151.3(7), 63.2(9), 30(2), 106.4(7), 41(1), 75(1), 39.8(7), 28.0(3) and 31.5(4) GPa, respectively. Results for the adiabatic bulk modulus are at 59(1) GPa and in comparison with the elastic moduli of other aragonite-type carbonates it confirms a systematic linear relationship between cation radius and bulk modulus. The shear modulus for $SrCO_3$ presents a value of 32.0(5) GPa which is about 10 % smaller in comparison to that of aragonite (Liu et al., 2005). Although pure strontianite is not a relevant phase of the Earth's mantle, measurements of the acoustic velocities as well as studying the high-pressure phase behavior provide necessary reference information for the elasticity and stability of aragonite-type carbonates that might be present in the Earth's interior.

References:

Brenker et al. (2007) Earth and Planetary Science Letter 260, 1.
Ono et al. (2005) Phys Chem Minerals 32, 8.
Oganov et al. (2006) Earth and Planetary Science Letter 241, 95.
Ono (2007) Phys Chem Minerals 34, 215.
Liu et al. (2005) Phys Chem Minerals 32, 97.

MS04-05

On the mechanism and accompanying effects of two-dimensional polymerization in a novel mono-layered monomer crystal

G. Hofer[1], T. Weber[2], A. D. Schlüter[1]
[1]*ETH Zürich, Polymer Chemistry, Zürich, Switzerland*
[2]*ETH Zürich, Laboratory of Crystallography, Zürich, Switzerland*

A breakthrough in the synthesis of sheet-like polymers was recently achieved based on the polymerization of trifunctional monomers [1]. By crystallizing these novel monomers prior to their connection, genuine two-dimensional polymerization can be achieved resulting in a two-dimensional polymer (2DP).

The chiral anthracene-based monomer molecules of the here presented 2DP crystallize in the polar space group *R*3 in a 2-cyanopyridine (*cpy*) solution. The crystal structure allows the anthracene moieties from the nearest neighbouring monomers to face each other close enough to achieve a controlled long-ranged ordered polymerization in the *ab*-plane of the crystal by continuous UV irradiation. This turns the monomer single crystal into a polymer single crystal, a process that can thermally be reversed. During either direction of this phase transition, the space group *R*3 is preserved. Both, polymerization and depolymerization can be frozen at any time by simply removing the crystal from the triggering source thus creating a disordered and meta-stable monomer/polymer single crystal. The polar *cpy* is incorporated into the structure in a partially disordered manner next to the anthracene moieties of monomers within the same layer, Figure 1.

This contribution will address all important issues of this unusual polymerization based on an analysis of the Bragg and diffuse X-ray [2] scattering of various states of the polymerization/depolymerization with the goal to provide a complete understanding of the process and its implications for structure formation and properties. This last aspect concerns the polar *cpy* molecules, which during polymerization undergo gradual positional changes that result in strong effects on the overall dipole moment.

Figure 1: Network structure of the two-dimensional polymer synthesized by [1].

[1] Kory, M. J., Wörle, M., Weber, T., Payamyar, P., van de Poll, S. W., Dshemuchadse, J., Trapp, N., Schlüter, A. D. (2014) *Nat. Chem.*, *6*, 779-784.
[2] Weber, T., Simonov, A. (2012) *Z. Kristallogr.*, *227*, 238-247.

Figure 1

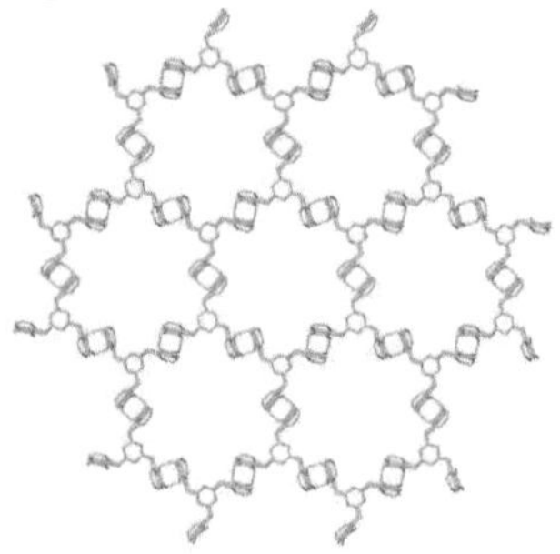

MS04-06

Synthesis, Characterization, and investigation of electronic and optical properties of double perovskite oxide $BaSrMgTeO_6$

M. Ait Haddouch[1], A. abbassi [2], Y. Aharbil[1], H. Labrim[3], Y. Tamraoui[4], F.-E. Mirinioui[4], B. Manoun [4], H. Ez-Zahraouy[2], A. Benyoussef[2,5], L. Laânab[6], S. Benmokhtar[1]
[1]*université hassan II Casablanca, Laboratoire de Chimie Physique des Matériaux LCPM, casablanca, Morocco*
[2]*Université Mohammed V, Laboratoire de Magnétisme et Physique des Hautes Energies L.M.P.H.E. URAC 12, rabat, Morocco*
[3]*Centre National de l'Energie, des Sciences et des Techniques Nucléaires, rabat, Morocco*
[4]*Univ. Hassan 1er, laboratoire des Sciences des Matériaux, des Milieux et de la modélisation (LS3M), Khouribga, Morocco*
[5]*Institute of Nano-materials and Nanotechnologies, MAScIR, rabat, Morocco*
[6]*Université Mohammed V, Laboratoire Conception et Systèmes (LCS), rabat, Morocco*

A double perovskite $BaSrMgTeO_6$ has been synthesized and characterized by physical techniques: X-ray diffraction, Raman diffusion, and diffuse reflectance spectroscopy.The crystal structure was resolved at room temperature by the Rietveld refinement method, and revealed that $BaSrMgTeO_6$ crystallizes in the cubic system, space group Fm-3m with lattice parameter (a=8.018Å), and shows an almost perfect ordering between Mg^{2+} and Te^{6+} cations at the B substructure. The Raman spectra of this crystalline perovskite oxide is interpreted by means of factor group analysis in terms of space group Fm-3m. Assignments of the Mg/Te-O vibrational stretching and bending modes have been made. Based on density functional theory (DFT), and using full potential-linearized augmented plane wave (FP-LAPW) method with the generalized gradient approximation (GGA), implemented in the Wien2k package, we have investigated the electronic and the optical proprieties of such material. The optical band gap obtained with GGA approximation is equal to 2.8 eV, which is in good agreement with our experimental results. We found that the transmittance T is stable and reach the average of 90% in both experiment and theoretical studies.

MS04-07

Controlled Stepwise Synthesis of a Cu MOF From Magnesium Precursor to Mesoporous Material

Q. Guo[1], U. Englert[1]
[1]*Institute of Inorganic Chemistry, Aachen, Germany*

A three-dimensional copper *metal-organic framework* (MOF) has been synthesized from a magnesium precursor and copper sulfate crystalized in space group $I2_13$. The effective free volume calculated by PLATON analysis after removing all the guest molecules amount approximately 54.8 percent of the crystal volume (8773 $Å^3$ of 16007 $Å^3$ unit per cell volume). Topological analysis was carried out by *GTECS3D* to obtain an insight into the structure [*Z. Kristallogr. Suppl.* **2012**, *32*, 117.]. The structure can be viewed as a 4-connected topology with a Vertex symbol

3.3.10₂.10₂.10₃.10₃, which corresponds to a **lcv** net. The 3D intersecting channel system has a cross section of approximately 9 Å. In additional, the oxygen atoms of sulfate ions and methyl groups from the ligand are exposed toward the channel, which facilitate the trapping of polar guests.

Figure 1

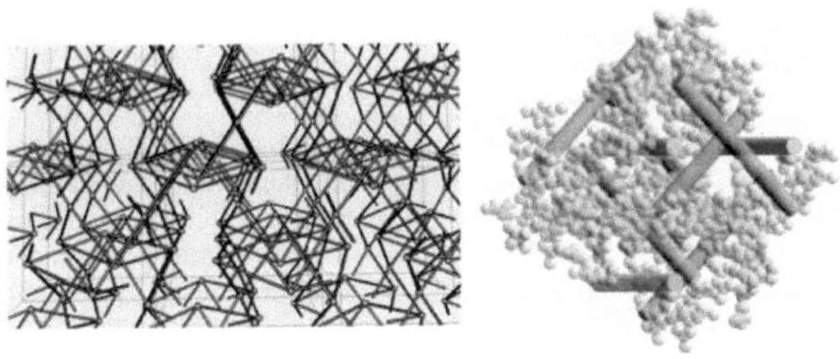

lcv network structure of the Cu MOF (*left*); 3D channel net running along the *a*, *b* and *c* axes (*right*).

MS04-09
Anion recognition by dinuclear Ni(II) complexes with macrocyclic NHC ligands
P. J. Altmann[1], A. Pöthig[1]
[1]*TU München, Catalysis Research Center, Garching, Germany*

The tailored design of ligands to optimise the electronic and steric properties of molecular complexes is a common principle in homogeneous catalysis. In the last two decades, more sophisticated supramolecular concepts have been developed. We recently introduced the family of calix[4]imidazolium[2]pyrazole cyclophanes, which can act as a ligand precursor for bowl-shaped dinuclear transition metal complexes with a defined spatial arrangement of the two centres.[1] A new member of this family is presented, namely a binuclear Ni-NHC complex, which is able to selectively recognise and self-assemble with guests via tennis-ball like encapsulation. NMR titration studies showed, that addition of halides leads to formation capsules with cavities of defined size, in which only chloride and bromide can be incorporated. The solid state structures derived by SC-XRD studies suggest a stabilisation of the capsules through a combination of multiple hydrogen bonding and attractive Coulomb interactions between the guest and the metal centres.

[1] Philipp J. Altmann, Christian Jandl, Alexander Pöthig* *Dalton Trans.*, **2015**, *44*, 11278-11281.

MS04-08
Structure Analysis of Human Prolidase Mutations gives insight into the Prolidase Deficiency disease mechanisms
P. Wilk[1,2], M. Ühlein[1], H. Dobbek[2], M. S. Weiss[1], U. Mueller[1]
[1]*Helmholtz-Zentrum Berlin, BESSY II, Macromolecular Crystallography (HZB-MX), Berlin, Germany*
[2]*Humboldt-Universität zu Berlin, Institute of Biology, Berlin, Germany*

Prolidase is the only enzyme in humans capable of cleaving peptide bonds preceding proline or hydroxyproline residue in dipeptides. It is involved in last steps of collagen catabolism and likely in regulation of peptidic hormones. The diminished or absent prolidase activity was related to a rare autosomal disease, manifested by variety of clinical symptoms, known as Prolidase Deficiency (PD). To date there is no definitive cure to PD, partially due to a limited understanding of the wild-type (wt) enzyme in terms of detailed substrate binding mode and as a consequence catalyzed reaction mechanism, as well as due to lack of any structural information of the mutants. Here we describe the high resolution crystal structures of wt human prolidase in the ligand-free form and in the substrate-bound state providing the missing structural information. We analyze the mode of substrate binding and a mechanism of hydrolysis utilized by prolidase pointing to position of a catalytic bridging molecule. We have also selected and characterized structurally eight clinically relevant mutations introducing single amino acid changes in the prolidase chain. Based on high resolution crystal structures we have identified different mechanisms of enzymes inactivation. Together this information shall ease the understanding of pathologies related to this enzyme.

Figure 1: The overall structure of human prolidase is shown in cartoon representation with subunit A colored according to secondary structure. Mn2+ ions in the active site are shown as magenta spheres and GlyPro substrate as stick representation in orange. The positions of mutated residues in chain A are indicated by cyan spheres. The inlets show two examples of enzymes inactivation mechanism: by losing of Mn2+ ion (Asp276Asn, upper panel) or by structure destabilization (Gly448Arg, lower panel). Structures of mutants are superimposed on wt protein, 2Fo-Fc maps contoured at 2 σ are shown in blue and anomalous difference maps (AnoDe) contoured at 10σ shows positions of manganese ions.

Figure 1

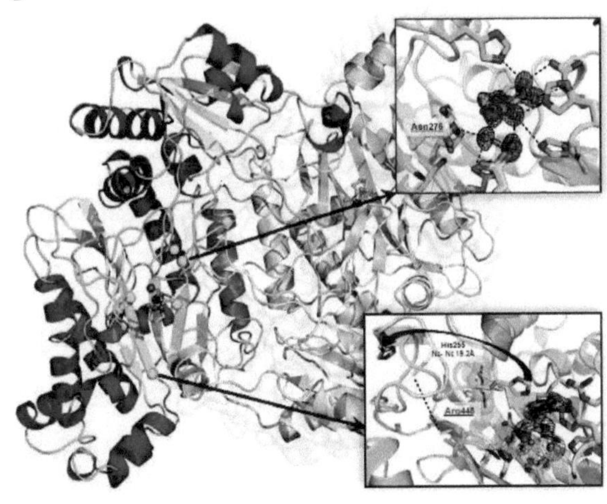

MS04-10
Influence of Modulators for MOF Formation
J. Bergmann[1], H. Krautscheid[1]
[1]*Universität Leipzig, Institut für Anorganische Chemie, Leipzig, Germany*

Introduction
In the last two decades Metal-Organic Frameworks (MOFs) have emerged as a new class of porous materials with great potential for a wide range of applications, including gas storage, separation, transport, heterogeneous catalysis, sensors and drug delivery systems.[1,2]

Objectives
A well-known problem in the synthesis of MOFs is their sensitivity towards extraneous influences such as reaction temperature and time, solvent, pH-value or the presence of modulators. Aim of this work is the systematic investigation of different modulators on the formation of a Cd^{2+} based MOF with a substituted 1,2,4-triazolylisophthalate linker[3] (Fig. 1).

Materials & Methods
To obtain single crystals suitable for X-ray diffraction, solvothermal syntheses were performed in steel autoclaves with water/acetonitrile (1:1; *v*:*v*) as solvent.

Single crystal structure analyses[4] were performed on a STOE IPDS 1 or 2T diffractometer with Mo-K_{α} radiation (λ = 71.073 pm). To verify phase purity, powder X-ray diffractograms were measured on a STOE STADI-P (Cu-$K_{\alpha 1}$; λ = 154.059 pm). Thermal behaviour was investigated by temperature dependent PXRD and TG-DTA (STA F1 Jupiter; NETZSCH).

Results
Depending on the modulator, three closely related modifications of a MOF with the general formula $_\infty^3$[Cd(Me-prz-trz-ia)] {Me-prz-trz-ia$^-$ = 5-(3-methyl-5-(pyrazine-2-yl)-4*H*-1,2,4-triazol-4-yl)isophthalate} could be synthesized and characterized. These three dimensional frameworks with **rtl** topology exhibit microporous channels with calculated pore diameters in the range of 300-550 pm[5] and calculated porosities up to 50 %[6] which make these compounds interesting for gas storage and separation.

Without any modulator a monoclinic phase ($P2_1/n$; $\mathbf{1}^{\mathrm{mP}}$; z = 4) is formed.[7] Using modulators with carboxylic and/or amino groups a tetragonal phase ($P4_2/n$; $\mathbf{1}^{\mathrm{tP1}}$; z = 8) is formed. At 55 °C a phase transition towards the monoclinic modification $\mathbf{1}^{\mathrm{mP}}$ is observed. Synthesis in presence of isophthalic acid or aminoisophthalic acid leads to another tetragonal modification ($P4_2/m$; $\mathbf{1}^{\mathrm{tP2}}$; z = 4).

Conclusion
The formation of MOF crystals is very sensitive regarding the presence of modulators. Especially modulators with one or more donor groups are able to coordinate to the metal ions and pre-organize them after getting replaced by the actual linker, leading to different modifications of related structures.

References:
[1] S. Kitagawa, R. Kitaura, S. Noro, *Angew. Chem.*, **2004**, *43*, 2334.
[2] A. U. Cajka, N. Trukhan, U. Müller, *Chem. Soc. Rev.*, **2009**, *38*, 1284.
[3] D. Lässig, J. Lincke, H. Krautscheid, *Tet. Lett.*, **2010**, *51*, 653-656.
[4] G. M. Sheldrick, *Acta Cryst.* **2008**, *A64*, 112-122.
[5] S. Bhattacharya, K.E. Gubbins, *Langmuir*, **2006**, *22*, 7726.
[6] *PLATON*, A. L. Spek, *J. Appl. Cryst.* **2003**, *36*, 7-13.
[7] J. Bergmann, K. Stein, M. Kobalz, M. Handke, M. Lange, J. Möllmer, F. Heinke, O. Oeckler, R. Gläser, R. Staudt, H. Krautscheid, *Microporous and Mesoporous Materials* **2015**, *216*, 56-63.

Fig. 1: Tested modulators (top) and reaction scheme of MOF synthesis (bottom).

Figure 1

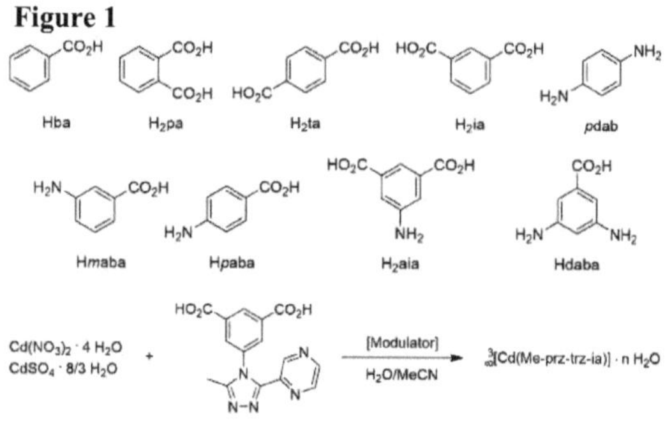

MS04-11
Riddles of a novel Mn-polymer with a tetraazaadamantane-like ligand

D. Premužić[1], M. Hołyńska[1]
[1]*Philipps-Universität Marburg, Fachbereich Chemie, Marburg, Germany*

Symmetry plays an important role in the field of single-molecule magnets (SMMs), where axial symmetry of the system enhances the magnetic anisotropy.[1] Tripodal ligands can be used to impose a three-fold symmetry in their metal complexes. Tris(1-propan-2-onyl oxime)amine (Ox_3H_3) is an example of a tripodal oxime ligand with different known metal complexes, mainly containing Ni.[2] This ligand undergoes an unusual intramolecular cyclization leading to a 4,6,10-trihydroxy-1,4,6,10-tetraazaadamantane product.[3] In combination with Mn, Ox_3H_3 transforms *in situ* and yields different Mn^{IV}-dimers depending on the used additives. In order to improve the magnetic properties and obtain Mn^{III}-complexes with much higher intrinsic magnetic anisotropy, these complexes were reduced with KC_8. In the case of the azide-brigded complex a polymeric compound was synthesized (Fig. **1**). Herein we present the results obtained by various methods including NEXAFS, SQUID, and HF-EPR. The resulting data indicate a successful reduction, but the oxidation state of the Mn is still under discussion.

Fig. **1**: Molecular structure of the title polymeric compound (hydrogen omitted for clarity).

[1] T. Glaser, V. Hoeke, K. Gieb, J. Schnack, C. Schröder, P. Müller, *Coord. Chem. Rev.*, **2015**, *289-290*, 261-278. [2] A. N. Semakin, A. Yu. Sukhorukov, A. V. Lexiv, S. L. Ioffe, K. A. Lyssenko, Y. V. Nelyubina, V. A. Tartakovsky, *Org. Lett.*, **2009**, *11*, 4072-4075. [3] M. J. Goldcamp, S. E. Robison, J. A. Krause Bauer, M. J. Baldwin, *Inorg. Chem.*, **2002**, *41*, 2307-2309.

Figure 1

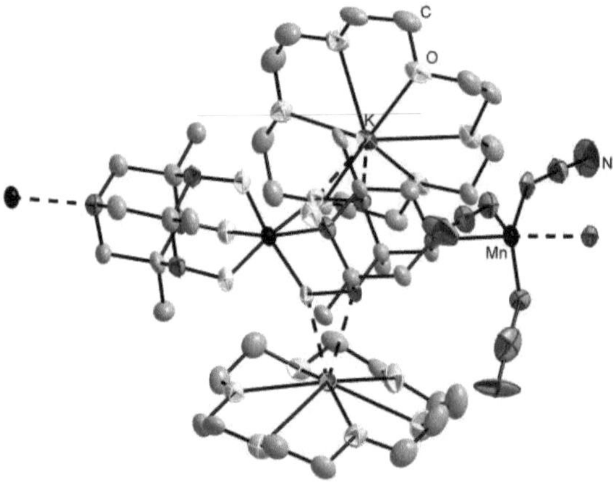

MS04-12
Computational adsorption experiments at the (100)-pyrite-water interface: The influence of surface defects to the H_2O-adsorption model.

S. Meis[1], H. Gies[1], U. Magdans[2]
[1]*Ruhr-University Bochum, Bochum, Germany*
[2]*University of Potsdam, Potsdam, Germany*

Chemical reactions at the (100)-pyrite (FeS_2) surface play an important role in many environmental, geological and chemical processes, e.g. acid mine drainage, bacterial pyrite oxidation, oxidation in aquifers and coal liquefaction and heterogeneous catalysis. To understand these processes it is of essential importance to understand the adsorption properties of pyrite to organic molecules and water. The structures of the dry (100)-pyrite surface and the interface of the (100)-pyrite surface with water were determined with grazing incidence X-ray diffraction experiments (GIXRD) previously[1,2]. An adsorption model for water at the pyrite surface was determined including three adsorption layers, a transitional zone of partial ordered water molecules and water molecules that occupy defect sites of the topmost FeS_2-layer.

In this study the adsorption model of water at the (100)-pyrite surface near iron and sulphur defect sites is analysed. We present computational experiments, especially molecular dynamic simulations using the forcefield COMPASS. The simulations were carried out with the program package Materials Studio 5.0, particularly the program "Forcite", from Accelrys Inc. The surface cell structure determined previously from GIXRD experimental results was used as a start model for the simulations. A surface area of 3x3 surface cells (16,25Å²) and a depth of 2,5 surface cells (13,5Å) of pyrite was selected. A vacuum slab of 30Å was added as adsorption region to permit enough mobility of the H_2O-molecules during the dynamic runs. Different amounts of iron and sulphur defect sites were introduced to the topmost FeS_2-layers of the surface during different simulation runs. Each dynamic run was calculated for 200ps with a step width of 0.2fs.

The introduction of sulphur defects to the surface model has a stronger influence to the H_2O-adsorption model compared to the effect of iron defect sites. By introducing iron defects the adsorption model shows no differences compared to a surface free of defects. Only a small decrease in the distance between the topmost FeS_2-layer and the first H_2O-adsorption layer about 0.2-0.4Å was determined. The sulphur vacancies show a stronger influence, as H_2O-molecules could be observed in the topmost FeS_2-layers occupying the defect sites of sulphur dimers. Additional adsorbed H_2O-molecules occur between the topmost FeS_2-layer and the first H_2O-adsorption layer that is located 2.0(3)Å above the topmost FeS_2-layer. By adding iron and sulfur defects to the three topmost FeS_2-layers it was possible to obtain results in perfect agreement with the previously determined experimental model from GIXRD-data, with molecular dynamic simulations after Philpott et al.[3] and UHV-experiments carried out by Guevremont et al.[4].

References:
[1] Meis S, Magdans U, Torrelles X, Gies H. *Mineralogical Magazine* (2013), 77(5), 1733
[2] Meis S, Magdans U, Torrelles X. *Acta Cryst.* (2011) A67, C338-C339
[3] Philpott M R, Goliney I Y, Lin T T. *J. Chem. Phys.* (2004), 120(4), 1943-1950
[4] Guevremont J M, Strongin, Schoonen M. A. *Am. Mineral.* (1998), 83, 1246-1255

MS04-13

A Racemic Solution Crystallizing as a Disordered Crystal Structure of Opposite Enantiomers: Pseudoracemates in a Sohncke and a Centrosymmetric Space Group

R. Herrmann[1], B. Braun[1], T. Braun[1]
[1]*Humboldt-Universität zu Berlin, Berlin, Germany*

The crystallization of a racemic solution results mainly in *racemic crystals* (heterochiral) or *conglomerates* (homochiral). The unit cell of a *racemic crystal* contains the (*S*) and the (*R*) enantiomer in an equal amount in a well-ordered manner. Furthermore, *conglomerates* can be obtained as (*S*) and (*R*) handed crystals as a result of spontaneous resolution. Additionally, a rare outcome of the crystallization of a racemic solution is the formation of *pseudoracemic crystals* (heterochiral), in which both the (*S*) and the (*R*) enantiomer are present in the crystal structure in a disordered manner.[1] A *pseudoracemate* is a solid solution of opposite enantiomers which can crystallize in any space group. Their existence was first formulated by *Kipping and Pope*.[2]

Figure 1: Various types of the crystallization of a racemic solution.

All three types of crystalline racemates differ in their melting point diagrams and properties. That is why their accurate characterization is crucial for the pharma industry in the context of chiral drug design. Due to the fact that in general only one of the enantiomers has biological activity, the other is inactive, inhibits the desired effect or is even toxic.

We were fortunate to come across the structure of (*R*)- and (*S*)-*exo*-2-[Pt(C_7H_{11}){Ge(2-$C_6H_4PPh_2$)$_3$}] which was investigated by X-ray diffraction (see Scheme 1).[3] The compound under discussion crystallizes as a *pseudoracemic* THF solvate in the monoclinic chiral space group $P2_1$ with a small excess of the (*R*)-*exo*-2 enantiomer. The asymmetric unit contains one independent molecule on a general position, whereas no other particularly interesting intra- or intermolecular interactions are present.

Scheme 1: (*R*)- and (*S*)-*exo*-2-[Pt(C_7H_{11}){Ge(2-$C_6H_4PPh_2$)$_3$}].

Particularly interesting and exceptionally rare is the fact that the norbornyl complex crystallizes as a *pseudoracemate* in a Sohnke space group, therefore giving a heterochiral crystal in a chiral space group.

1. H. D. Flack, *Helvetica Chimica Acta* **2003**, *86*, 905.
2. F. S. Kipping, W. J. Pope, *J. Chem. Soc. Trans.* **1897**, *71*, 989.
3. R. Herrmann, B. Braun, T. Braun, *manuscript in preparation.*

Figure 1

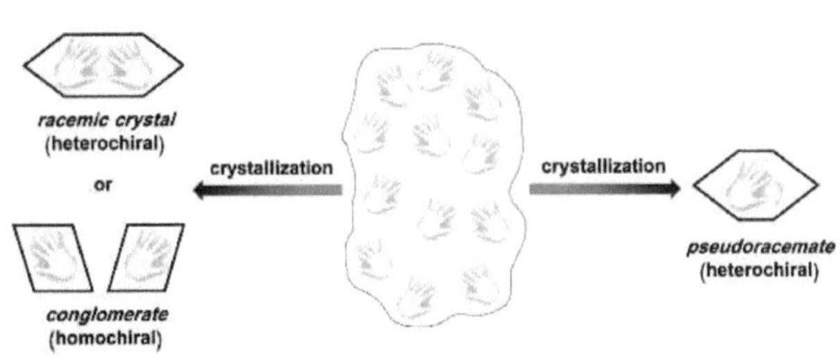

Figure 2

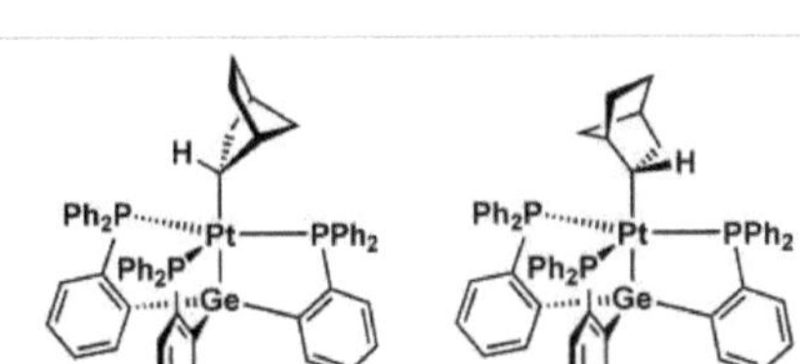

MS04-14

Crystallographic and energetic evaluation tools for the prediction of ionic conduction of Al in oxides

F. Meutzner[1], T. Nestler[1], A. A. Kabanov[2], M. Zschornak[1], W. Münchgesang[1], T. Leisegang[1,2], V. A. Blatov[2], D. C. Meyer[1]
[1]*TU Bergakademie Freiberg, Institut für Experimentelle Physik, Freiberg, Germany*
[2]*Samara State Aerospace University, Samara, Russia, Russian Federation*

An ionic conduction in crystalline oxides is characterisable as jumps of mobile ions between interconnected crystallographic sites in oxygen-coordinated voids. The environment needs to allow passage between jump sites. Considering this theory of ionic conduction, crystallographically motivated evaluation tools can be utilised and combined into a high-throughput identification method for suitable structures for ionic conduction. An energetic consideration in the end allows an in-depth analysis of the ionic conduction process in order to verify the most-promising compounds.

Three approaches can be synergetically used to allow a time-efficient and consistent methodology for the identification of an arbitrarily chosen ion. Voronoi-Dirichlet partitioning (VDP) combined with data mining offers a fast high-throughput screening of crystallographic databases on the basis of a geometrical division of the crystal structure into domains assigned to each constituent of the structure, pointing towards possible void networks.

Bond-valence methodology (BVM) as a chemically motivated evaluation tool assists in the evaluation process as the second step, determining the most probable migration ways by calculating bond-valence sums for each point in the crystal structure. Bond-valence energy landscape calculations then allow for an energetic estimation of jump energies.

The most promising structures are fully electronically modelled by density functional theory (DFT) calculations *ab initio*. Calculating migration step energies subsequently enables a full assessment of jump energies from equilibrated structures as well as electronic conductivities and diffusion coefficients. As it is a very time-consuming method, DFT is applied as the final step.

This work presents the first results of the identification of promising oxidic Al-ion conductors involving all three approaches and their comparison.

This work is funded by the German BMBF (CryPhysConcept: 03EK3029A) and BMUB (BaSta: 0325563B).

MS04-15

Disordered Intermetallic Phases visited by Solid-State NMR and Quantum Mechanical Calculations

B. Mausolf[1], F. Haarmann[1], V. Peters[1], A. Korthaus[1]
[1]*Institute of Inorganic Chemistry at RWTH Aachen University, Arbeitskreis für moderne Strukturanalytik, Aachen, Germany*

Structure-property relations became more and more important for materials and solid state scientists. This is particularly true for disordered materials and materials whose applicability is controlled by incorporated defects. Especially for intermetallic compounds the bonding situation and the atomic structure are in the focus.[1] A combination of XRD, NMR and DFT calculations turned out to be suitable for this purpose during a systematic study of binary intermetallics.[2] Recently, this approach was extended on disordered ternary phases.[3]

The various information about an intermetallic system which are available from quantum mechanical (QM) calculations are in the focus of this contribution. In the presence of a defect, the atomic structure in the vicinity reacts with a change of the atomic arrangement. The atomic shifts due to a structural optimization and along with this the changes of interatomic distances are valuable information which is neither accessible by diffraction methods nor by NMR alone.

Referring to NMR as a local probe, the electric field gradient (EFG) serves as sensitive experimental measure. QM calculations give the EFG resolved for each atom in the structure model and therefore are essential for the interpretation of the experimental results. In addition, the tensor orientation and the sign of the EFG are available by theoretical methods. This offers the possibility to interpret orientation dependent NMR experiments on single crystals or aligned powder samples. Furthermore, the calculated orbital contributions of the EFG give a hint for the change of the bonding situation due to incorporated defects.

Studying the details of the electronic structure leads to an deeper understanding of changes in the bonding situation. For this purpose we analyze the changes of the total, atomic and integrated density of states (DOS). In addition, we focus on theoretical derived formation energies and the efficient use of density functional theory (DFT) codes in this contribution.

We focus on Zintl-like intermetallic materials: These contain metals lying close to the Zintl boarder which form an anionic framework coordinated by alkali- and alkaline-earth metal cations.[4] The applied DFT codes VASP[5] and Wien2k[6] offer the necessary tools and implementations for the above mentioned investigations. For the generation of suitable structure models and super lattice structures we follow the Bärnighausen formalism based on group subgroup relations summarized in the International Tables for Crystallography.[7,8,9]

The system $Cu_{1-x}Al_2$, containing Cu vacancies[10] is currently investigated with our strategy. Details of the atomic and electronic structure in combination with Results from XRD and NMR experiments are presented. Using the high sensitivity of NMR in combination with DFT calculations, we present a systematic way to investigate disordered intermetallic materials.

Literature

[1]Yu. Grin, *Wiss. Z. Techn. Univers. Dresden*, **49**(1), 16, 2000.
[2]F. Haarmann, K. Koch, P. Jeglic, O. Pecher, H. Rosner, Y. Grin, *Chem. Eur. J.*, **17**, 7560 - 7568, 2011.
[3]O. Pecher, B. Mausolf, K. Lamberts, D. Oligschläger, C. Niewieszol, U. Englert, F. Haarmann, *Chem. Eur. J.*, **21**, 1-13, 2015.
[4]F. Laves, *Z. Nat.*, *17*, 244-255, 1941.
[5]G. Kresse, J. Furthmüller, *Phys. Rev. B*, **54**, 11169-11186, 1996.
[6]P. Blaha, K. Schwarz, G. K. H. Madsen, D. Kvasnicka, and J. Luitz, WIEN2k, Vienna *Univ. of Techn.*, 2001.
[7]H. Bärnighausen, MATCH, **9**, 1980.
[8]Int. Tables for Cryst. A, IUCr, 2006.
[9]Int. Tables for Cryst. A1, IUCr, 2006.
[10]F. Haarmann, M. Armbrüster, Yu. Grin, *Chem. Mater.*, **19**, 1147-1153, 2007.

MS04-16

In situ high pressure single crystal X-ray diffraction study of clinoferrosilite

A. Pakhomova[1], L. Ismailova[1], M. Bykov[1], E. Bykova[1], T. Boffa Ballaran[1], L. Dubrovinsky[1]
[1]*Bayerisches Geoinstitut, Bayreuth, Germany*

Pyroxenes, $M1M2Si_2O_6$, are one of the major constituent minerals of the Earth's upper mantle and therefore they have an important influence on its mineralogy, thermodynamics and the geological structure. A well-constrained correlation between chemical compositions, crystal structures and elastic properties of pyroxenes is hence necessary for the modeling their behavior in the Earth's interior. In the present study we report high-pressure single-crystal diffraction (SC XRD) data of the pyroxene Fe end-member: clinoferrosilite (CFs). CFs is known to transform from $P2_1/c$ into $C2/c$ phase between 1.48 and 1.75 GPa (Hugh-Jones *et al.*, 1996), however little is known about its structural behavior at Earth's mantle pressures.

Single crystals of CFs synthesized at 9.5 GPa and 1100 °C were loaded together with a ruby sphere or an Au foil for pressure estimation into diamond anvil cells (DAC). Neon was used as pressure-transmitting medium. *In situ* high-pressure SC XRD experiments were performed at the synchrotron stations P02.2 at Petra III and ID09 of European Synchrotron Radiation Facility.

The $C2/c$ crystal structure of CFs consists of three crystallographically distinct polyhedra: two octahedra $Fe1O_6$, $Fe2O_6$ and a tetrahedron SiO_4. The tetrahedra are connected via a bridging O3 atom forming chains running along the *c* axis. The octahedra form edge-sharing chains parallel to the silicate chains. Tetrahedral and octahedral "layers" alternate along the *a* axis forming a three-dimensional crystal structure by sharing the O1 and O2 atoms of the SiO_4-..-SiO_4 chain. With increasing pressure the crystal structure accommodate the compression by kinking of the tetrahedral chains as reflected in the decrease of the O3-O3-O3 angle from 137.54 to 133.5° up to 30 GPa. Between 30 and 35.8 GPa CFs undergoes a first order phase transformation into a $P2_1/c$ phase associated with volume change of 6%. The $C2/c \rightarrow P2_1/c$ transition is induced by the rearrangement of half of the layers of corner-sharing SiO_4 tetrahedra: in these the Si atoms become octahedrally coordinated and a new type of layer is formed by polymerization of $Si2O_6$ polyhedra via common edges. The remaining chain of corner-sharing tetrahedra $Si1O_4$ is significantly more kinked in the $P2_1/c$ phase as indicated by the small value of the O3-O3-O3 angle of 128.7°.

The isothermal bulk moduli K_0 at zero pressure and its first pressure-derivative K_0' for the $C2/c$ phase were calculated using a third-order Birch-Murnaghan equation of state (BM EoS) and were found to be 113(3) and 6.3(3) GPa, respectively. The axis compressibilities reveal a strong anisotropy: the *a* axis is the stiffest while the *b*-axis is the softest.

Comparison between the structural compression of CFs and those of diopside and hedenbergite will be discussed in order to shed light on the influence of Ca and Mg substitutions on the elastic behavior of pyroxenes.

MS04-17

Millisecond time resolved diffraction study of $SrCO_3$ at high pressures and temperatures

M. Stekiel[1], L. Bayarjargal[1], W. Morgenroth[1], R. Luchitskaia[1], B. Winkler[1]
[1]*Goethe University, Institute of Geoscience, Frankfurt am Main, Germany*

Strontium carbonate, $SrCO_3$, is isostructural to aragonite. At ambient pressure, $SrCO_3$ transforms at 1185 K to a disordered rhombohedral phase [1,2] (calcite-V structure). Upon further heating, $SrCO_3$ transforms into a cubic phase [3] (NaCl structure) at 1690 K. At ambient temperature, $SrCO_3$ undergoes a pressure induced phase transition, but the exact transition pressure is still a matter of discussion, since reported results are in the range of 10 to 35 GPa [4,5,6,7]. Rapoport & Pistorius [8] studied the aragonite-type to calcite-V type transition from ambient pressure up to 4 GPa, and found that the transition temperature increases with pressure up to 1500 K at 4 GPa.

In this study a previously unexplored p-T range of the $SrCO_3$ phase diagram was investigated. Synthetic $SrCO_3$ sample was grown by hydrothermal method.Powder diffraction experiments were performed at the Extreme Conditions Beamline P02.2 at PETRA III,Hamburg, inthe double-sided laser-heated diamond anvil cell (DAC). The sample was heated up to 1600 K at 8 GPa. Then, diffraction images were collected every 2.5 ms using the LAMBDA detector [9], while the two lasers were switched off. This allowed us to observe the time-dependence of the cooling process.

Figure 1 shows representative, time resolved diffractograms. At 8 GPa and 1600 K rhombohedral $SrCO_3$ was observed (t=0 ms), the reflections are labeled with "R" in the figure. Additionally, at these conditions, at least seven reflections (labeled "h") from a previously unknown phase were observed. Indexation gave an orthorhombic unit cell, and we tentatively assign these reflections to a new high (pressure, temperature) phase of $SrCO_3$. Other reflections are from NaCl (labeled "N") which was used as a pressure medium, and Pt and W metals ("M") which were employed as a heat absorber and gasket, respectively. The horizontal L1 and L2 lines indicate the time when the upstream and downstream laser of the double-sided laser heating system were turned off, respectively. The cooling process lasts a few milliseconds only. Surprisingly, upon rapid cooling at 8 GPa $SrCO_3$ does not transform back tothe aragonite structure, stable at ambient conditions, but to another, yet unknown phase with poor crystallinity ("a").

The study is supported byprojects DFG-ANR WI1232/41-1, and BMBF 05K13RF1.

[1] J. J. Lander, *J. Chem. Phys.*, vol. 17, pp. 892-901, 1949.
[2] K. Strømme, *Acta Chem. Scand. A*, vol. 29, p. 105, 1975.
[3] E. H. Baker, *J. Chem. Soc.*, pp. 2525-2526, 1962.
[4] S. Ono, M. Shirasaka, T. Kikegawa and Y. Ohishi, *Phys. Chem. Miner.*, vol. 32, pp. 8-12, 2005.
[5] S. Arapan and R. Ahuja, *Phys. Rev. B*, vol. 82, p. 184115, 2010.
[6] M. Wang, Q. Liu, S. Nie, B. Li, Y. Wu, J. Gao, X. Wei and X. Wu, *Phys. Chem. Minerals*, vol. 42, pp. 517-527, 2015.
[7] C.-C. Lin and C.-g. Liu, *Eur. J. Mineral.*, vol. 9, pp. 785-792, 1997.
[8] E. Rapoport and C. Pistorius, *J. Geophys. Res.*, vol. 72, pp. 6353-6357, 1967.
[9] T. Hatsui and H. Graafsma, *IUCrJ*, vol. 2, pp. 371-383, 2015.

Figure 1

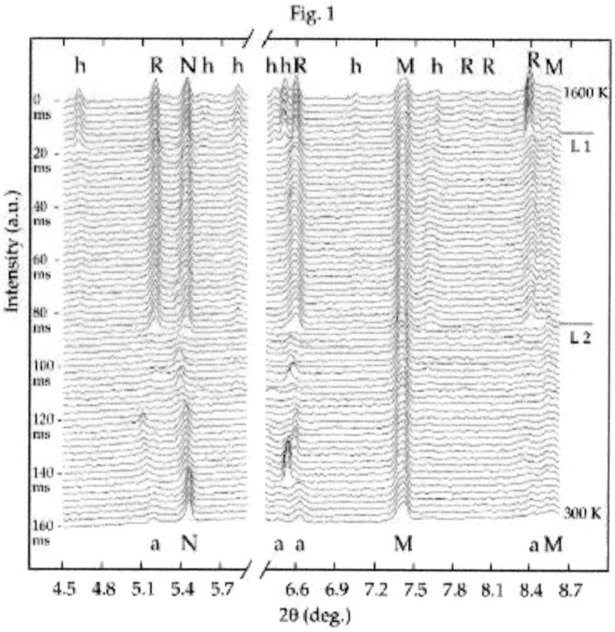
Fig. 1
h
R
N
h
h
h
h
R
h
M
h
R
R
R
M
1600 K
L1
L2
300 K
a
N
a
a
M
a
M
Intensity (a.u.)
0 ms
20 ms
40 ms
60 ms
80 ms
100 ms
120 ms
140 ms
160 ms
4.5
4.8
5.1
5.4
5.7
6.6
6.9
7.2
7.5
7.8
8.1
8.4
8.7
2θ (deg.)

MS05 - Electron crystallography

MS05-01

Transmission electron microscopy on materials of the deep Earth's interior to understand the working of our dynamic planet

K. Marquardt[1], M. Palot[2,3], J. Rudloff[4], G. Pearson[3], F. Brenker[4], F. Nestola[5], S. Jacobsen[6], N. Miyajima[1], C. A. McCamon[1], J. Harris[7], F. Kaminsky[8]
[1]*Universität Bayreuth, Bayerisches Geoinstitut, Bayreuth, Germany*
[2]*Université Jean Monnet, Saint-Etienne, France*
[3]*University of Alberta, Edmonton, Canada*
[4]*Goethe University, Frankfurt am Main, Germany*
[5]*Università di Padova, Padua, Italy*
[6]*Northwestern University, Evanston, United States*
[7]*University of Glasgow, Glasgow, Europ鋆che Union*
[8]*KM Diamond Exploration Ltd., West Vancouver, Canada*

Diamonds directly probe material of the Earth's lower mantle (> 660 km) and transport these materials to the Earth's surface, encapsulated in the diamond's chemical inert structure. Such material inclusions provide insight into processes taking place in the inaccessible parts of our planet. The general objective of our studies is to gain insight into the global dynamics of our planet including deep Earth material cycles, for example the N-, H-, and C-cycles.

Here we present findings from different natural diamonds with inclusions ranging from (i) olivine - the dominant upper mantle phase, over (ii) nitrogen - a main impurity in diamond used for their classification and indicative of the depth of their formation and (iii) (Mg,Fe)O ferropericlase (fp) - the second most abundant phase of the Earth's lower mantle.

From these inclusions we extracted electron transparent lamella using FIB machining. Subsequent characterization was done using a Titan™ G² 80-200 with ChemiSTEM™ technology operated at 200 kV acceleration voltage at the Bayerisches Geoinstitut. Four SDDs are symmetrically mounted in the microscope column around the optical axis. Each XFlash 5030 detector has an active area of 30 mm^2, adding up to an overall detector area of 120 mm^2 covering 0.7 srad solid angle.

(i) The milky appearance of our second diamond sample results from several thousands of 30-200 nm-sized N-bearing inclusions. Such defect structures were previously investigated and found to be either empty or filled with nitrogen. Our EDS nitrogen maps yield statistical information on nitrogen abundance. We will use the data collected from the milky diamond to discuss nitrogen abundance and its suitability to infer the depth of diamond formation. (ii) Lower-mantle fp inclusion show exsolved magnesioferrite (mf - spinel structure, fig.1) evenly dispersed in the fp as well as nucleated at dislocations formed at the surface of the fp. The microstructure and composition of the mf places the formation of mf at temperatures of well above 1600°C, which is roughly the temperature at the 660km discontinuity that marks the boundary between upper and lower mantle. We interpret the microstructure and composition of mf exsolutions as indices for incorporation of fp into the diamond before the formation of exsolutions. This indicates an origin from the upper lower mantle or deeper.

In summary TEM allows enhancing our knowledge of processes and properties in the deep interior of our planet by exploring nm-scale features in samples brought to the Earth's surface within diamonds.

Fig.1: Spatial relation between matrix precipitates of spinel and elongated precipitates of mf in fp. a,b) Scanning transmission electron (STEM) image tuned to give Z-contrast. b) Enlargement of the boxed region in a. The broken lines indicate principal crystallographic planes of the mf, {111} and {202}. c) Selected area electron diffraction highlighting the intergrowth relation between fp and mf.

Figure 1

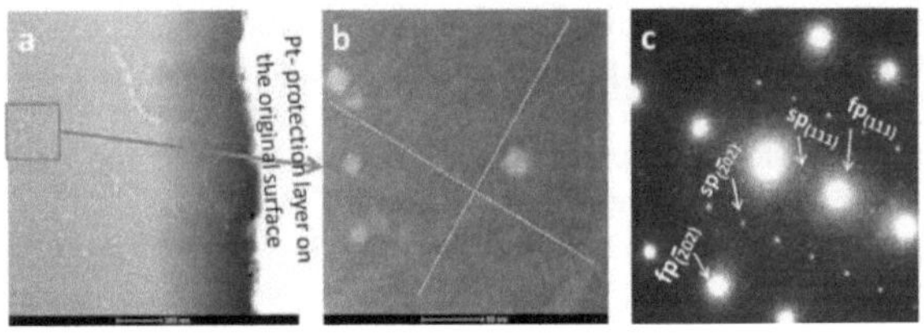

MS05-02

A hybrid TEM approach to study non-periodic 3D structures in nano materials

B. Barton[1], Y. Krysiak[1], H. Zhao[1], U. Kolb[1]
[1]*Gutenberg University, Physical Chemistry, Mainz, Germany*

Introduction: Non-periodic features such as surfaces, vacancies or dislocations have significant influence on the properties of crystalline (nano) materials, especially in the case of nanoparticles and thin films with a large portion of surface atoms. Other examples are promoter atoms at the edge of industrial MoS_2 nano catalysts [1] and self-interstitials and out-of-plane distortions in semiconductors [2,3]. Analyzing these non-periodic structures by diffraction is difficult since translational symmetry is imposed. On the other hand, state-of-the-art (scanning) transmission electron microscopy ((S)TEM) enables 2D imaging of nano materials with deep sub-Å resolution [4] without the principal need for translational symmetry. However, 3D atomic resolution electron tomography could only be successfully applied in a few cases [5,6], since 1st — recording tilt series from an area of some tens of nm requires outstanding optical and mechanical stability and 2nd — the required dose of 5,000 - 10,000 $e^-/Å^2$ per image leads to significant beam damage, such as sputtering from the surface.

Objectives: A technique is desirable that yields both periodic and aperiodic 3D structural information for nano materials at the atomic scale, limits radiation dose, and can be applied using widely accessible instrumentation. Periodic 3D structures of single nanoparticles have been successfully described ab-initio by TEM using automated electron diffraction tomography (ADT) [7] while applying minimal doses of ~20 $e^-/Å^2$. At the same time, it was shown that 2D atomic resolution in-line holograms can be reconstructed from TEM focal series using dose rates as low as 12 $e^-/Å^2s$ [8]. Our experimental approach combines these techniques, while aiming to describe non-periodic atom sites as pertubations of the translational symmetry.

Materials & Methods: First, we employ ADT to obtain an ab-initio 3D crystal structure of the object. Next, we reconstruct the complex exit electron wave along relevant zone axes in an area containing non-periodic features, such as the particle surface. Using automated a-posteriori aberration correction, the projected Coulomb potential can be approximated with minimal artifacts and delocalization. By superposition of exit waves and projected ADT structures, we then characterize the crystal perturbations to approximate the non-periodic atomic structure.

Results: We tested the technique using a mid-range TEM operating at 300 kV with a field emission source, while no optical aberration corrector was needed. Depending on the number of holograms required, the accumulated electron dose of ~ 20,000 $e^-/Å^2$ was significantly lower than in the case of real space electron tomography. We present results of the ab-initio non-periodic structural analysis for aluminum borate and gadolinite nanoparticles, as well as of calcite growth on Au seeds.

Conclusion: After the described TEM method has been successfully tested at room temperature in vacuum, it can be extended in-situ using e.g. heating or biasing of the sample, cryo TEM, or environmental TEM. In the future, we aim to study a wide range of functional materials, including nano catalysts for water splitting in solar fuel generators.

References

[1] Y. Zhu et al., Angewandte Chemie International Edition 53.40 (2014): 10723-10727.
[2] D. Alloyeau et al., Physical Review B 80.1 (2009): 014114.
[3] N. Alem et al., Physical review letters 106.12 (2011): 126102.
[4] C. Kisielowski et al., Microscopy and Microanalysis 14.05 (2008): 469-477.
[5] S. Van Aert et al., Nature 470.7334 (2011): 374-377.
[6] M.C. Scott et al., Nature 483.7390 (2012): 444-447.
[7] E. Mugnaioli et al., Ultramicroscopy 109.6 (2009): 758-765.
[8] B. Barton et al., Microscopy and Microanalysis 18.05 (2012): 982-994.

Figure 1: Combining 2D electron holography and 3D diffraction tomography to study non-periodic features of nanocrystal. (a) Phase of the exit wave near the surface of a GdO-NCN nanoparticle reconstructed from a focal series recorded on a FEI Tecnai F30 at 300 kV. (b) Boxed area in (a) overlaid with a EuO-NCN homologue model viewed along [100]. The inset plot shows a line profile of the phase as indicated by the dashed arrow. (c) Power spectrum of the phase image (a). (d) Simulated electron diffraction pattern from the EuO-NCN single crystal at 300 kV.

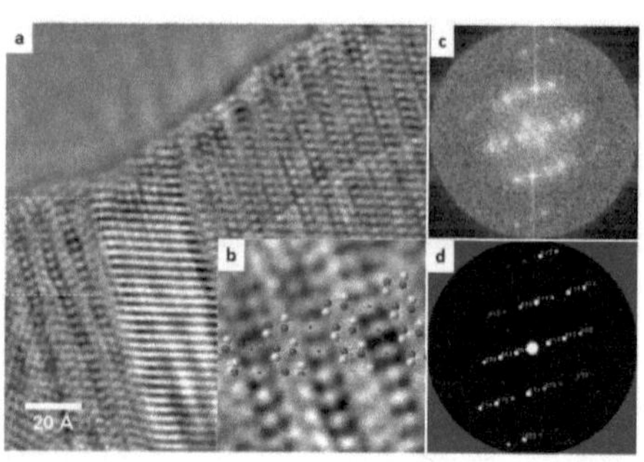

MS05-03

Structural characterisation of $Al_4B_2O_9$ by automated electron diffraction tomography

H. Zhao[1], K. Hoffmann[2,3], Y. Krysiak[1], B. Barton[1], H. Schneider[2], R. X. Fischer[2], U. Kolb[1]
[1]*Institute of Physical Chemistry, Johannes Gutenberg-University Mainz, Mainz, Germany*
[2]*Crystallography, Department of Geosciences, University of Bremen, Bremen, Germany*
[3]*Institute of Inorganic Chemistry, University of Bremen, Bremen, Germany*

Introduction

Structural characterization of materials is a key step to understand their physical properties in order to optimize industrial applications. Transmission electron microscopy provides structural and elemental information over several orders of magnitude. Electron nano diffraction, as opposed to X-ray powder diffraction, can provide atomic structure information from a single nanocrystal. Electron diffraction has advantages over X-ray powder diffraction in the case of structures with large unit cells, superstructures, low crystallinity as well as for samples with impurities or multiple phases. Automated electron diffraction tomography (ADT) combines nano diffraction, tilt series acquisition and beam precession to reconstruct the three-dimensional atomic structure with minimal dynamic scattering artifacts.[1,2]

Objectives

The structure of aluminum borate ($Al_4B_2O_9$), an important mullite-type ceramic material used for industry, was derived based on a model from the known boralsilite structure and refined with Rietveld method from X-ray powder diffraction data.[3] In the main structure the edge-connected AlO_6 octahedra expand along *c*-axis and are cross-linked by AlO_4, AlO_5, BO_3 and BO_4 groups. However, the occupancies of two oxygens (O5 and O10) stayed unclear in such a complex structure. In addition, single crystal X-ray diffraction was not possible.

Materials & Methods

Synthetic $Al_4B_2O_9$ samples contain needle like crystals in the range of few tens of nanometers up to some hundred nm, which are suitable for electron microscopy. ADT measurements were performed on single crystals of two samples prepared with different initial boron contents.

Results and Conclusion

3D data reconstruction delivered monoclinic unit cells in space group *C* 2/m (a = 14.8 Å, b = 5.5 Å, c = 13.1 Å, and β = 90.7°), respectively. *Ab-initio* structure solution of $Al_4B_2O_9$, using a direct method, delivered for both samples the same structure (Fig. 1). The reconstruction clearly shows one fully occupied position of O10 (circles in Fig. 1) but no significant signal for the second postulated oxygen O5 in the channel. Furthermore, only for the sample prepared with an excess of boron, electron diffraction data from some crystals revealed modulated streaking in ***b*** direction (Fig. 2), which has been assigned to a superstructure with a threefold ***b***-axis and an additional disorder.

Figure 1. Electron diffraction potential map of aluminum borate and the correspondent atomic model along the (010) direction. The circles show the fully occupied oxygen O10 position.

Figure 2. Three-dimensional reconstructed reciprocal space of disordered $Al_4B_2O_9$ viewed in projection along the (001) direction.

References

[1] U. Kolb, T. Gorelik, C. Kübel, M. T. Otten, D. Hubert, *Ultramicroscopy* **2007**, *107*, 507-513.
[2] U. Kolb, T. Gorelik, M. T. Otten, *Ultramicroscopy* **2008**, *108*, 763-772.
[3] R. X. Fischer, V. Kahlenberg, D. Voll, K. J. MacKenzie, M. E. Smith, B. Schnetger, H. J. Brumsack, H. Schneider, *Am. Mineral.* **2008**, *93*, 918-927.

Figure 1

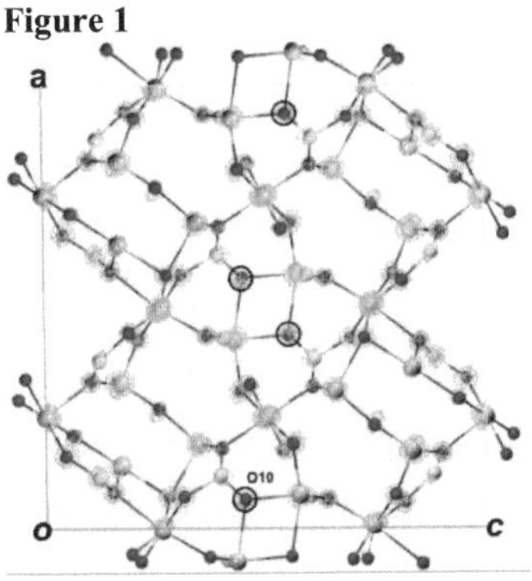

Figure 2

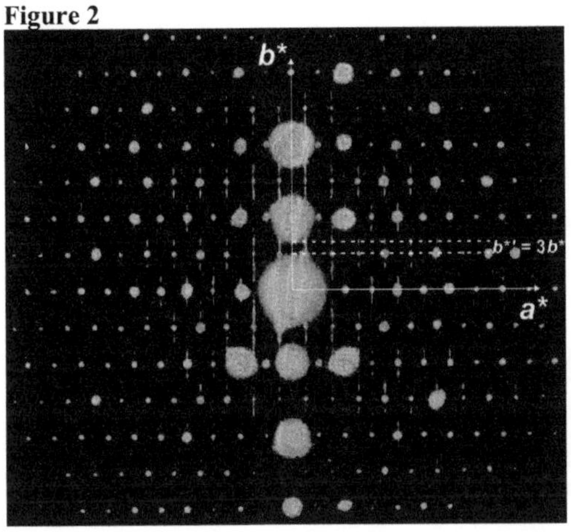

MS05-04

Investigation of layered and porous nanomaterials by diffraction tomography, simulations and HRTEM

Y. Krysiak[1], B. Barton[1], H. Zhao[1], R. Neder[2], U. Kolb[1]
[1]*Johannes Gutenberg University, Institute of Physical Chemistry, Mainz, Germany*
[2]*Friedrich-Alexander-Universität Erlangen-Nürnberg, Chair for Crystallography and Structural Physics, Erlangen, Germany*

Periodicity is often broken with decreasing particle size, in which case the most common diffraction techniques no longer deliver interpretable structure information. In particular, layered and porous nanomaterials show disorder, pseudo symmetry and stress/strain effects, which often give rise to unique properties but make structural analysis difficult. For such materials, automated electron diffraction tomography[1] (ADT) can deliver 3D structure information of nanoparticles at atomic resolution. For example, a super structure in $CaCO_3$ polymorph vaterite, twinned domains for boron oxynitride or intergrown phases inside nm scaled crystal of a new layered pentasil borosilicate MZ-35 could be solved with ADT.[2-4]

For an all-encompassing structural access the qualitative description from intergrown phases, twinned or disordered layered structures will comprise a more quantitative analysis. The combination of transmission electron microscopy (TEM) imaging and diffraction together with simulations allows a complete structural description including disorder.

ADT samples the reciprocal space by recording off-zone diffraction patterns, while the crystal rotates around a tilt axis. First, a nm-sized crystallite is chosen through scanning TEM (STEM) imaging and diffraction patterns are acquired using a 30-100 nm probe. The reciprocal lattice points of the reconstructed 3D volume are indexed and their intensities extracted for ab-initio structure solution. Disorder and distortion or intergrown phases are simulated using the DISCUS software package[5] through accurate analysis of the structure's variance. The results are confirmed by Rietveld Methods[6] and comparison to TEM holograms[7] reconstructed from focal series.

Zeolite Beta[8] was chosen as a test sample. The structures of a polymorphic mixture of Zeolite Beta A and B was solved ab-initio from a single ADT dataset. The disorder type was identified by exit wave reconstruction and confirmed by DISCUS simulations. Based on these results it was possible to solve the aperiodic structure of an unknown zeolite ab-initio. In addition, we present strong layer shifts in a rare earth cyanamid[9].

The 3D atomic structure of different compounds forming highly distorted and disordered nanoparticles could be successfully described by combining ADT, layer shift simulations and TEM imaging. In the future, we aim to fit the simulations automatically on diffraction data and apply the technique on a wide field of compounds such as alloys, organic pigments, and photoactive organics.

[1] E. Mugnaioli, T. Gorelik, U. Kolb, *Ultramicroscopy* **2009**, *109*, 758-765.
[2] E. Mugnaioli, I. Andrusenko, T. Schüler, N. Loges, R. E. Dinnebier, M. Panthöfer, W. Tremel, U. Kolb, *Angew. Chem.* **2012**, *124*, 7148-7152.
[3] R. Arletti, E. Mugnaioli, U. Kolb, F. Di Renzo, *Microporous Mesoporous Mater.* **2014**, *189*, 64-70.
[4] S. Bhat, L. Wiehl, L. Molina-Luna, E. Mugnaioli, S. Lauterbach, S. Sicolo, P. Kroll, M. Duerrschnabel, N. Nishiyama, U. Kolb, et al., *Chem. Mater.* **2015**, *27*, 5907-5914.
[5] T. Proffen, R. B. Neder, *J. Appl. Crystallogr.* **1997**, *30*, 171-175.
[6] H. M. Rietveld, *J. Appl. Crystallogr.* **1969**, *2*, 65-71.
[7] B. Barton, B. Jiang, C. Song, P. Specht, H. Calderon, C. Kisielowski, *Microsc. Microanal.* **2012**, *18*, 982-994.
[8] J. M. Newsam, M. M. J. Treacy, W. T. Koetsier, C. B. D. Gruyter, *Proc. R. Soc. Lond. Math. Phys. Eng. Sci.* **1988**, *420*, 375-405.
[9] Y. Hashimoto, M. Takahashi, S. Kikkawa, F. Kanamaru, *J. Solid State Chem.* **1996**, *125*, 37-42.

Figure 1: Intergrown polymorphic mixture of Zeolite Beta A and B: HRTEM hologram combined with structure solution models.

Figure 1

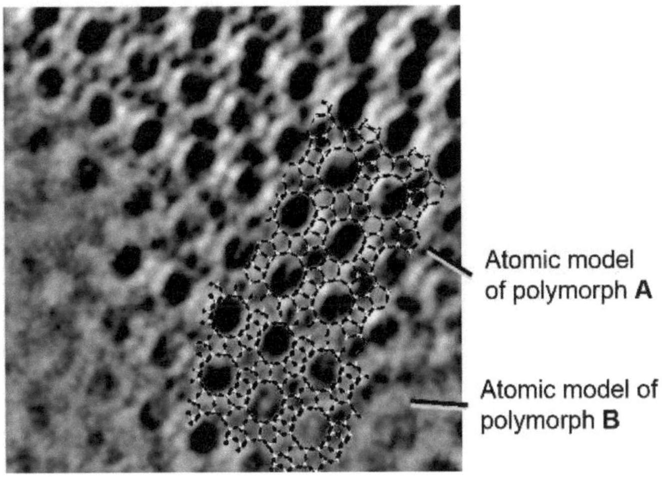

MS05-05

Orientation Analysis of Non-Centrosymmetric Crystal Structures using Electron Backscatter Diffraction

A. Winkelmann[1], G. Nolze[2]
[1]Bruker Nano GmbH, Berlin, Germany
[2]BAM, Federal Institute for Materials Research and Testing, Berlin, Germany

Crystal structures that show a broken inversion symmetry in their cubic and hexagonal modifcations are of great relevance for optoelectronic and photonic applications. This is why methods which are sensitive to non-centrosymmetric structures are important for analysis of these materials in technological applications.

We demonstrate polarity-sensitive orientation mapping of non-centrosymmetric phases by Electron Backscatter Diffraction (EBSD) in the scanning electron microscope (SEM). The method overcomes the restrictions of kinematic orientation determination by EBSD, which is limited to the centrosymmetric Laue-groups according to Friedel's rule.

Using examples of zinc-blende-type and wurtzite-type compounds, we apply a quantitative pattern matching approach based on simulations using the dynamical theory of electron diffraction implemented in the EBSD software "ESPRIT DynamicS" (Bruker Nano).This data analysis procedure results in a distinct assignment of the local orientation according to the non-centrosymmetric point group of the crystal structure under investigation. Extending the point group sensitivity to space group discrimination, we also demonstrate the chirality dependence of EBSD patterns from quartz.

[1] Aimo Winkelmann, Gert Nolze, Point-group sensitive orientation mapping of non-centrosymmetric crystals, Applied Physics Letters 106, 072101 (2015)
[2] G. Nolze, C. Grosse, A. Winkelmann, Kikuchi pattern analysis on non-centrosymmetric crystals, Journal of Applied Crystallography 48, (2015) 1405-1419
[3] Aimo Winkelmann, Gert Nolze, Chirality determination of quartz crystals using Electron Backscatter Diffraction, Ultramicroscopy 149, 58 (2015)

Figure 1: Crystal structure of zinc-blende compounds

Figure 2: Electron Backscatter Diffraction Pattern of GaP (b) and comparison to two simulations corresponding to two equivalent Laue-group orientations connected by a 90 degree rotation around [001] (a,c). Due to the non-centrosymmetric crystal structure and the sensitivity of EBSD to point groups, one can distinguish orientation (c) as the correct one by the higher value of the normalized cross correlation coefficient with respect to the experiment. The relative changes between simulations (a) and (c) is shown in panel (d).

Figure 1

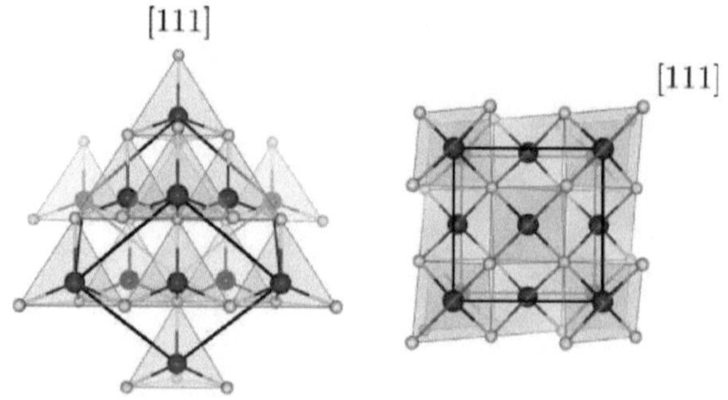

Figure 2

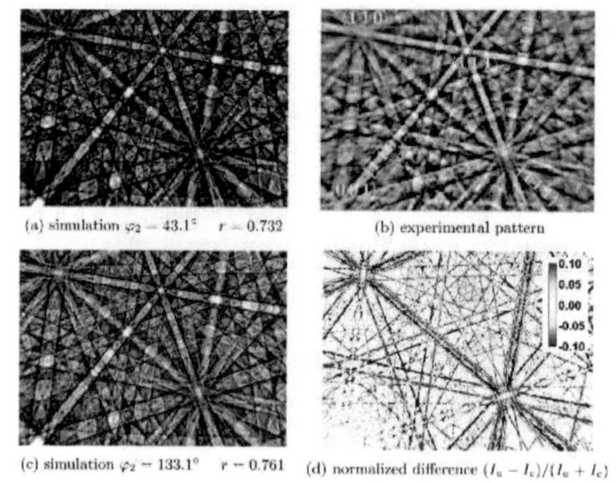

MS06 - Computational and theoretical crystallography

MS06-01
Being able to detect disorder is a question of resolution (also in small-molecule crystallography)

B. Dittrich[1]
[1]*Heinrich Heine Universtitaet Duesseldorf, Anorganische Chemie und Strukturchemie, Düsseldorf, Germany*

A brief overview into disorder phenomena in small molecule crystallography is given. Disorder in the solid state is not just a nuisance in modelling crystal structures. Dynamics of combined atomic or molecular motion and phonon modes is also important for solid-state behavior, and this plays an underlying role in materials properties (e.g. in Nb3X superconductors). For charge density studies disorder is especially problematic, since parameters describing the occupancy of a disordered atom correlate with multipole parameters (or refined wavefunctions). De-convolution of electron density and atomic displacement parameters therefore becomes impossible, leading to unreliable electron-density distributions. We describe methodology to model disorder in molecular compounds based on the invariom aspherical scattering-factor database [1]. Model electron densities provided help distinguishing between dynamic and static disorder. Procedures have already been tested in selected cases [2], mostly to model disordered solvent, and are now freely available in the program molecoolQt [3]. The example structure of cefaclor, where the whole molecule is disordered, is discussed. A second example structure highlights our ability to detect disorder in high-resolution single-crystal X-ray diffraction: the antibiotic spectinomycin crystallized as the dihydrochloride pentahydrate [4], for which data to unusually high resolution of 1.5 in sin theta / lambda have been collected at the Daresbury synchtroton at a temperature of 30 K illustrates that disorder can be overlooked even in a charge density study at a resolution in sin theta / lambda of 1.0 Å-1. That being able to resolve disorder depends on the experimental resolution is well known in macromolecular crystallography; we report analogous findings for small-molecule crystallography at resultions higher by one order of magnitude. Our conclusion is that disorder can go undetected even at high resolution. To get reliable electron densties by experiment it needs to be taken into account in the modelling process.

[1] B. Dittrich, C. B. Hübschle, K. Proepper, F. Dietrich, T. Stolper and J. J. Holstein, Acta Cryst. B69 (2013) 91.
[2] B. Dittrich, T. Koritsanszky, A. Volkov, S. Mebs, P. Luger, Angew. Chem. Int. Ed. 46 (2007) 2935; D. Kratzert, D. Leusser, J. J. Holstein, B. Dittrich, K. Abersfelder, D. Scheschkewitz, D. Stalke, Angew. Chem. Int. Ed. 52 (2013) 4478; K. Pröpper, J. J. Holstein, C. B. Hübschle, C. S. Bond, B. Dittrich, Acta Cryst. D69 (2013) 1530; P. Luger, B. Dittrich, R. Tacke, Org. Biomol. Chem. 13 (2015) 9093.
[3] C. B. Hübschle, B. Dittrich J. Appl. Cryst. 44 (2011) 238.

MS06-02
Benchmarking plane-wave DFT calculations for the structure prediction of porous SiO_2 and $AlPO_4$ polymorphs

F. Evers[1], M. Fischer[1,2]
[1]*University of Bremen, Crystallography, Department of Geosciences, Bremen, Germany*
[2]*MAPEX Center for Materials and Processes, Bremen, Germany*

Density-functional theory (DFT) has become a widely used method for the computational study of various properties of solids. In order to apply DFT to advanced problems, such as the investigation of materials under pressure, or the structure prediction of novel materials, it has to be ensured that a given approach reproduces the ground-state structures of known materials with sufficient accuracy. The equilibrium structures predicted by DFT are particularly sensitive to the choice of the exchange-correlation functional, and to the inclusion (or omission) of a dispersion correction. Previous benchmarking studies for periodic solids have addressed, for example, the structures of Al_2SiO_5 polymorphs and H-bonded aluminum hydroxides,[1] sheet silicates,[2] and cation-exchanged zeolites containing water.[3] However, no rigorous comparison of DFT-optimised structures to experimental data has yet been attempted for microporous all-silica zeolites and aluminophosphates, a fact that is somewhat surprising given the high potential of these materials for applications in catalysis and separation.

Available experimental data for calcined structures of all-silica zeolites and aluminophosphates were first extracted from the ZeoBase database.[4] DFT structure optimisations were then performed for a subset of systems, ruling out structures with very large unit cells or signs of structural disorder. Non-porous α-quartz and berlinite ($AlPO_4$) were included for comparison. The calculations made use of the plane-wave DFT code CASTEP.[5] Three GGA-type functionals without dispersion correction (PBE, PBEsol, WC) and two dispersion-corrected variants of the PBE functional (PBE-D, PBE-TS) were compared.

As expected on the basis of previous studies, the PBE functional performs poorly, overestimating the lattice parameters by several per cent for many systems. While PBEsol and WC functionals improve upon this behaviour, they are outperformed by the dispersion-corrected versions of PBE in the majority of cases. As the identification of general trends is not straightforward, systems in which particularly large deviations are found are discussed on a case-by-case basis. Furthermore, it is assessed to what extent the optimisation of the scaling factor in the dispersion correction could improve the performance.

References

(1) Demichelis, R.; Civalleri, B.; D'Arco, P.; Dovesi, R. *Int. J. Quantum Chem.* **2010**, *110*, 2260-2273.
(2) Tunega, D.; Bučko, T.; Zaoui, A. *J. Chem. Phys.* **2012**, *137*, 114105.
(3) Fischer, M. *Z. Kristallogr.* **2015**, *230*, 325-336.
(4) Baur, W. H.; Fischer, R. X. *ZeoBase*, **2013**.
(5) Clark, S. J. et al. *Z. Kristallogr.* **2005**, *220*, 567-570.

MS06-03
Prediction of Metal Coordination by Data Mining of experimental Crystal Structures

D. Hofmann[1], L. Kuleshova[2]
[1]*CRS4, Pula, Italy*
[2]*FlexCryst, Uttenreuth, Germany*

While the crystal structure prediction of small organic molecules is possible, the prediction of large biomolecules (more as 20 degrees of freedom) and the prediction of organometallics is still an open question. For small organics the stereochemistry is in general known, however, this is for coordination complexes very often unknown. Therefore, a succesful prediction of this complexes must include the prediction of the coordination of the metal. From the theoretical point of few we face the problem that not only intermolecular interactions must be predicted correctly, but bonds as well.

In our approach the required potentials are described as a Taylor-Series of (1/r^2). The inter- and intra-molecular interactions are merged in one effective potential as it was already elaborated for the reactive force field for water (1). For the parametrization of the potentials we used a multistep procedure of data mining on the experimental crystal structures of the Cambridge Structure Data Base. In a first step we analyzed the radial distribution function of the interatomic distances to eliminate unusual and susceptive crystal structures. The retained crystal structures have have been used to derive an approximation of the force field by using the gradients and solving the obtained equation system by singular value decomposition. The result is a good guess for a further refining of the potentials by classification (2). The parametrization by classification is based on the idea that observed atomic distances must be preferable to unobserved and the parameters are obtained by training the potential till the scoring function recognizes correctly, if a given crystal structure is invented or observed. We will demonstrate the method for the prediction of nickel-complexes as example. We can show that a the present state of development the data mining force field is able to predict the correct coordination among the proposed candidates. A future project might allow us to predict the right coordination on the first rank.

1) Hofmann, Detlef WM, Liudmila Kuleshova, and Bruno D'Aguanno. "A new reactive potential for the molecular dynamics simulation of liquid water." *Chemical Physics Letters* 448.1 (2007): 138-143.
2) Hofmann, Detlef WM. "Data Mining in Organic Crystallography." *Data Mining in Crystallography*. Springer Berlin Heidelberg, 2010. 37-58.

Figure 1) The crystal structure of the solvate of this nickel-complex (reference code DUNHEB) shows unusal Cl..Cl.contacts and is foundais unlikely by the analysis of the radila distribution function in this structure. The calculated voids suggest a shift of the solvates.

Figure 2) The predicted coordination versus the experimental coordination of a nickel complex.

Figure 1

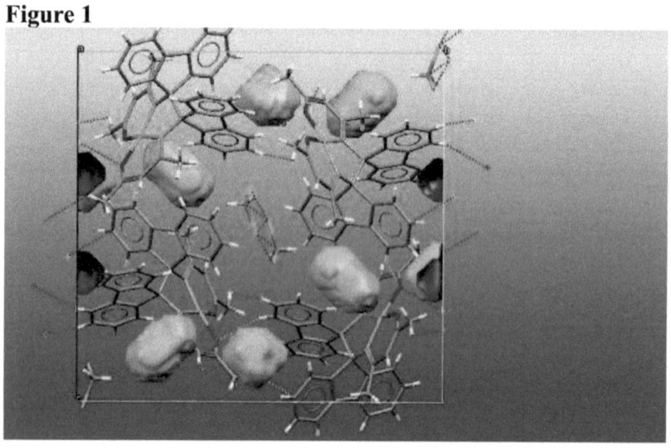

Figure 2

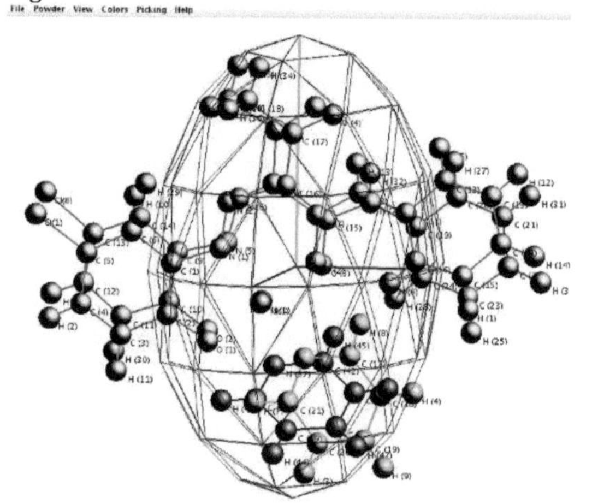

MS06-04

Refining angular- and wavelength-dispersive neutron time-of-flight powder-diffraction data

P. Jacobs[1], A. Houben[1], W. Schweika[2], A. Tchougréeff[1,3], R. Dronskowksi[1]
[1]*RWTH Aachen University, Institute of Inorganic Chemistry, Aachen, Germany*
[2]*Forschungszentrum Jülich, Jülich Center for Neutron Science and Peter Grünberg Institute PGI, Jülich, Germany*
[3]*Lomonosov Moscow State University, Department of Chemistry, and Poncelet Laboratory of Mathematics in Interaction with Physics and Informatics, Independent University of Moscow, Moscow Center for Continuous Mathematical Education, Moscow, Germany*

Introduction

We present an extension of the well-established Rietveld refinement method for modeling two-dimensional neutron time-of-flight powder-diffraction data, i.e., intensity as function of diffraction angle 2θ *and* wavelength λ. Hereby, the novel approach also takes into account the varying resolution function with respect to the angular- and wavelength-dispersive data. By doing so, common data-reduction steps and the loss of high-resolution information introduced by integration may be avoided.

Objectives

For the future instrument POWTEX at FRM II and, also, for the future powder diffractometer DREAM at the ESS, the novel approach will fully exploit the best resolution properties which would be lost by averaging in today's standard refinement procedures. The flexibility of this approach in including a varying resolution function in two dimensions has also an important benefit for the instrument design, simply because the detector shape is much more compact and, at the same time, more economical in terms of construction costs, i.e., by a cylindrical shape and covering of a large solid angle.

Materials & Methods

We used data obtained from the POWGEN instrument at SNS, Oak Ridge, that were recorded in event mode. In contrast to the *d*-binning generated by the standard data-reduction procedure as implemented in MantidPlot (logarithmic binning), the events were now binned in 2θ as well as in λ using bin sizes of 0.1° and 0.001 Å. Subsequently, the sample pattern was corrected for background and finally calibrated by the vanadium pattern to account for detector efficiency and the wavelength-dependent intensity distribution. The final diffraction pattern is depicted in Figure 1. The next step involved finding a suitable description of the analytical profile function in 2D. A common trait of all current profile functions is that they depend on one variable only (in most cases either 2θ, TOF, or *d*). Obviously, analyzing two-dimensional data sets will require the reflection profile to be a function of two variables. Thus, by directly implementing the derived 2θ *and* λ-dependent analytical function into the common Rietveld method, one may straightforwardly fit the multi-dimensional measured data.

Results

We have demonstrated a Rietveld refinement of angular- and wavelength-dispersive 2D data sets. To do so and similar to the common procedures on current instruments, "standard" samples served as input to determine the instrumental profile function as a function of 2θ and λ. Once established on the basis of real data, this instrumental profile function will be provided to the user and can be used to refine additional sample effects.

Conclusion

With the fundamentals of 2D data profiling laid out, future developments will aim at incorporating additional sample effects (e.g., preferred orientation, absorption and suchlike) into the multi-dimensional, analytical parametrization. The given proof-of-concept will hopefully motivate the incorporation of this novel approach into existing programs such as FullProf, GSAS, MAUD, TOPAS, or JANA, which will definitely be needed for future user applications.

Figure 1: Diffraction pattern I(2θ,λ) of a standard diamond sample using a two-dimensional (left) and quasi-two-dimensional (right) representation.

Figure 1

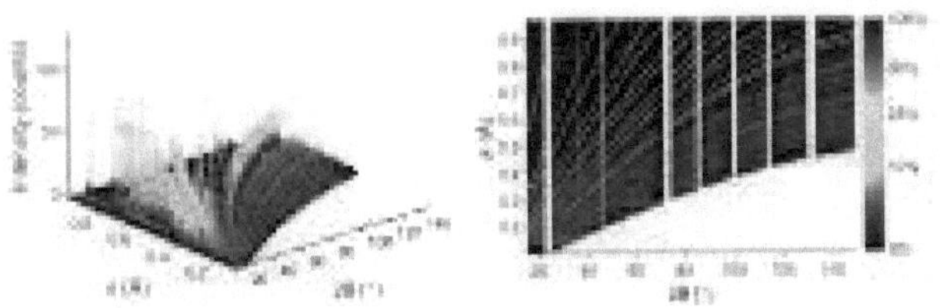

MS07 - Crystallography, crystal chemistry, and application of layered materials

MS07-01
Structure solution and refinement of stacking faulted NiCl(OH)

S. Bette[1], R. E. Dinnebier[2]
[1]*TU Bergakademie Freiberg, Institute of Inorganic Chemistry, Freiberg, Germany*
[2]*Max Planck Institute for Solid State Research, Stuttgart, Germany*

Introduction
NiCl(OH) is a layered double hydroxide phase (LDH), that belongs to a row of basic metal chlorides *M*ClOH, with M = Mg, Ni, Cu, Zn, Fe, Mn, Cd, Ca, Sn. Because of the slow crystallization and the occurrence of stacking fault disorder the crystal structure of NiCl(OH) has not been solved yet.

Objectives
The present work focuses on the structural characterization of NiCl(OH) and the detection and modelling of stacking faults occurring in the relating crystal structure.

Materials & Methods
Two samples of pure NiCl(OH) were produced by hydrothermal synthesis and characterized by chemical analysis, IR-spectroscopy, high resolution laboratory X-ray powder diffraction and scanning electron microscopy. *DIFFaX* simulations were employed to investigate the effects of stacking faults, on the diffraction patterns. A recursion routine for creating distinct sequences of rigid body like layers in real space with faults (global optimization) and a Rietveld compatible approach (local optimization), was realized and implemented into a *macro* for the program TOPAS.

Results
The crystal structure consists of layers of edge sharing, distorted $NiCl_{6x}(OH)_{6-6x}$ octahedra stacked in an (AγB)(BαC)(CβA) type pattern. Crystal intergrowth with (AγB)(AγB) stacked sections (C6-like, β-$Ni(OH)_2$ related), besides an inversion of the stacking order (C19-like, $NiCl_2$ related), was determined as the main features of the microstructure of NiCl(OH) (Fig. 1). The obtained microstructure models were supported by high resolution SEM images and *DIFFaX* simulations.

Conclusions
A routine for creating stacking sequences in a supercell, containing shifts among the diverse stacking patterns with distinct probabilities was developed for the program TOPAS. Including this into a global optimization method (*simulated annealing*) an appropriate structure model is obtained by *recursion*. Hence, a method is provided for refining powder patters of stacking faulted samples, which is *directly* related to the microstructure. The routine was successfully applied to two samples of stacking faulted NiCl(OH) [1].

Figure 1 Illustration of the stacking patterns apparent in the microstructure of NiCl(OH).

References

[1] S. Bette, R. E. Dinnebier, D. Freyer *J. Appl. Cryst.* **2015**, *48, in press.*

Figure 1

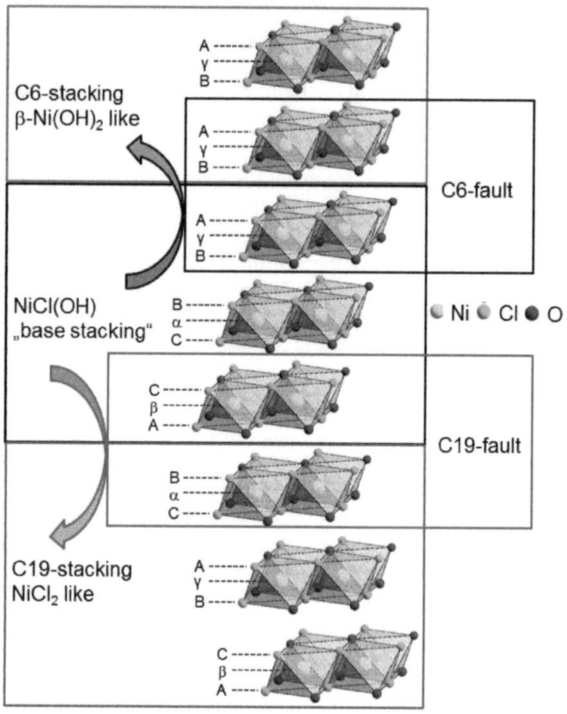

MS07-02
Intercalated Layered Superconductors: Iron Selenide as an Unusual Example

K. Yusenko[1], S. Margadonna[1]
[1]*Swansea University, College of Engineering, Swansea, Great Britain*

Layered *quasi*-2D-superconductors play an important role in our understanding of general trends in structure-properties relations in superconducting materials. Since superconductivity with critical temperature up to 50 K was found in iron-pnictide systems, evident progress in understanding the phenomenon has been achieved [1]. Tetragonal β-FeSe with layered structure and transition temperature T_c = 8 K has been found as one of the most promising model system for a deeper understanding of the unconventional pairing mechanism in iron-based superconductors. It has been shown that its superconducting transition is extremely sensitive to applied pressure, reaching the value of 39 K at 7 GPa [2]. Recently, an interlayer intercalation of alkali/alkali-earth metals (mainly, Li, Na, K, Cs, Rb, Ca, Ba, Sr, Eu, and Yb) between layers at low temperature using a solution of in liquid ammonia has been employed for the synthesis of phases with critical temperatures between 30 and 46 K (*e.g.* [3]). Due to particular significance of the β-FeSe reaction in liquid ammonia solution for the preparation of single-phase vacancy-free intercalates, investigations of composition/properties relations and dependence on the preparatory conditions is of timely importance. Herein, we report our recent investigation of Na and Ba intercalates prepared using reaction in liquid ammonia with polycrystalline β-FeSe [4-5]. Both reactions were investigated using time *in situ* time resolved powder X-ray diffraction. Figure 1 shows a formation of two intermadiate phases upon Ba intercalation between FeSe layers. *In situ* magnetic susceptibility measurements were used as a complementary technique to understand structure-properties relations. Figure 2 summerizes magnetic susceptibility measurements on the intermediate phases forming before a stabilization of ambient phase $Ba_{0.3}FeSe$.

Alkali metal's intercalates show high variations in interlayer distances and stability depending on preparatory conditions, which plays a fundamental role in their superconducting properties. Na intercalated phases show the most complex phase diagram where metal content, reaction temperature and time play important role in a formation of superconducting and magnetic phases with various amount of intercalated ammonia. All phases can be characterised by high level of planar defects such as stacking faults and intergrowth.

Phase relations in Na and Ba intercalates and presence of unstable intermediates clearly show the importance of considering the size and chemical nature of the metal employed during the ammonothermal synthesis of superconducting b-FeSe intercalates. Exact metal content (and doping level) appears to be playing an important role in determining the value of the T_c, which can then be further tuned by modulation of the interlayer spacing.

References: [1] Malavasi, Margadonna (2012): Chem. Soc. Rev. 41, 3897; [2] Margadonna *et al.* (2009): Phys. Rev. B 80, 064506; [3] Burrard-Lucas *et al.* (2013): Nature Mater. 12, 15; [4] Yusenko, Margadonna *et al.* (2015): ChemCommun 51, 7112; [5] Yusenko, Margadonna *et al.* (2015): Inorg. Chem, *submitted.*

Figure 1

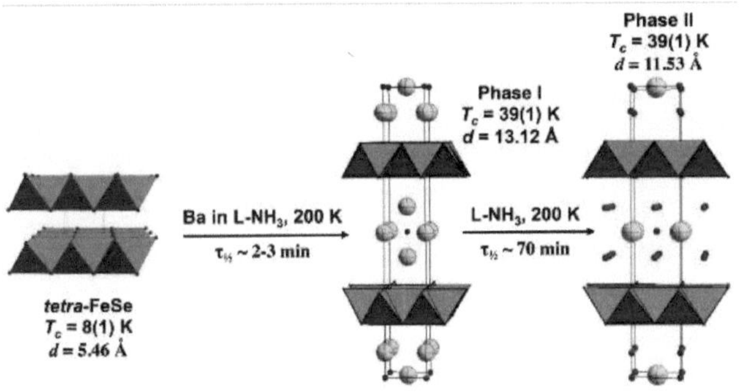

Figure 2

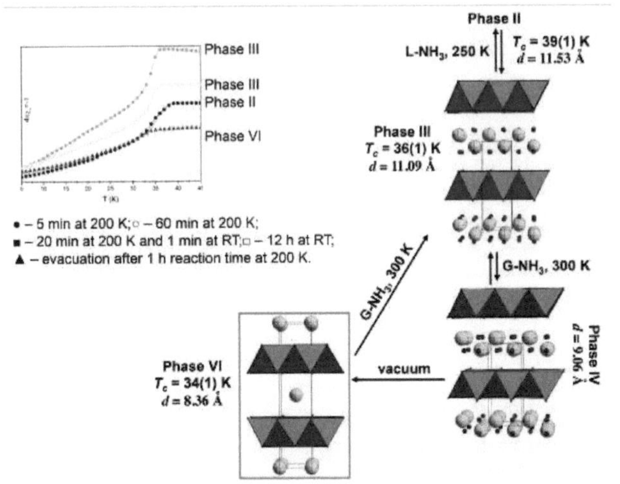

MS07-03
Layer groups databases and tools on the Bilbao Crystallographic Server

G. de la Flor[1], L. García[2], G. Madariaga[2], M. Aroyo[2], L. Elcoro[2], J. M. Perez-Mato[2]
[1]*Universidad del País Vasco UPV/EHU, Física Aplicada I, Donostia-San Sebastian, Spain*
[2]*Universidad del País Vasco UPV/EHU, Física de la Materia Condensada, Leioa-Bilbao, Spain*

Introduction

Materials with diperiodic symmetry (layer and multilayer) have attracted great interest due to their interesting properties and possible technological applications. Also bidimensional sections of crystal structures are fundamental for the study of bicrystals [1] in which two crystals, usually with the same structure but with different orientations, share a common boundary or interface. Twinned crystals can be considered special cases of bicrystals where the orientations of twin boundaries are restricted by symmetry. The *scanning tables* [2] provide the layer symmetry (*layer groups*) of the sections of the 230 space groups and allow the structural analysis of layer and multilayer materials and the analysis of their correlated properties. On the *Bilbao Crystallographic Server* (www.cryst.ehu.es) [3-4], a free web server that grants access to crystallographic databases and tools to study different kinds of problems related to crystallography and solid state physics, there is a section dedicated to *Subperiodic groups: Layer, Rod and Frieze Groups* which can be very helpful for this kind of studies.

Objectives

The aim of this contribution is to announce the current state of the programs and databases available on the *Bilbao Crystallographic Server* for the study of materials with layer group symmetry.

Materials and Methods

The *Bilbao Crystallographic Server* includes at present several tools which access to layer groups databases containing basic crystallographic information of the 80 layer groups [2], the Brillouin-zone, the tables of **k**-vectors for all the layer groups that form the background and a classification of the irreducible representations of layer groups. A new tool has also been developed to identify the layer symmetry of space-groups sections.

Results

Several programs in the server are dedicated to *layer groups.* GENPOS, WPOS, MAXSUB and KVEC provides access to the general positions, Wyckoff positions, maximal subgroups and Brillouin-zone databases, respectively. SECTIONS identifies, for a given space group, the full set of possible layer symmetries of periodic sections defined by their common normal vector and located within the unit cell. The program accepts as input a space group given in any arbitrary *setting.* The results are expressed in both the given basis of the input space group and the *standard setting* of the sectional groups as given in [2].

Conclusion

The *Bilbao Crystallographic Server* is in constant improvement and development, offering free of charge tools to study an increasingly number of crystallographic systems which now also include the ones with layer symmetry.

[1] R.C. Pond & W. Bollmann (1979), *Philos. Trans. R. Soc. London Ser. A,* **292,** 449-472.
[2] *International Tables for Crystallography Vol. E: Subperiodic-Groups* (2002), edited by V. Kopský and D. B. Litvin. Dordrecht: Kluwer Academic Publishers.
[3] M. I. Aroyo et al. (2006) *Z. Kristallog.* , **221**, 15-27

[4] M. I. Aroyo et al. (2011) *Bulgarian Chemical Communications*, **43**(2), 183-197.

MS07-04

Characteristics of $Cu_2ZnSn(S,Se)_4$ based thin film solar cells grown by thermal evaporation and thermal treatment

R. Gunder[1], K. Neldner[1], L. E. Valle-Rios[1], R. Caballero[2], M. León[2], J. Merino[2], E. Garcia-Llamas[2], Y. Sánchez[3], E. Saucedo[3], S. Schorr[1,4]
[1]*Helmholtz Zentrum Berlin, Crystallography, Berlin, Germany*
[2]*Universidad Autónoma de Madrid, Madrid, Spain*
[3]*Catalonia Institute for Energy Research, Barcelona, Spain*
[4]*Freie Universität Berlin, Institut für Geologische Wissenschaften, Berlin, Germany*

$Cu_2ZnSn(S,Se)_4$ kesterite type polycrystalline thin films were deposited by thermal evaporation onto Mo coated glass substrates, which subsequently underwent a thermal treatment using various conditions. Upon evaporation the substrate temperature was set at 250 °C. For comparison purposes substrate temperatures at 150 °C and 300 °C, respectively, have been applied as well in case of one thin film each. As starting material well pre-characterized kesterite powder of different off-stoichiometry types has been employed [1, 2] which, in turn, was synthesized by solid state reaction [e.g. 2]. The thermal treatment was done both in graphite box under Ar + S atmosphere as well as in evacuated silica ampoules ($p \approx 10^{-13}$ Pa), at which different temperatures, annealing times and heating rates have been applied. Prior to the fabrication of solar cell devices, including deposition of CdS buffer layer by chemical bath and sputtering of either ZnO/ITO or i-ZnO/AZO, the superficial segregation of Sn(S,Se) secondary phase detected by grazing incidence X-ray diffraction (GIXRD) has been removed by etching with KCN. The effect both of compositionally different starting material as well as different annealing procedures on the properties of thin films was carried out by means of GIXRD, SEM, XRF and GDOES in order to obtain information on microstructure, phase content and cross-sectional elemental distribution, respectively. By this method solar cell efficiencies up to 2.7 % were achieved so far, which will be discussed in consideration of the results provided by the structural and compositional characterization. And yet, the impact of the various parameters on the solar cell performance still has to be unscrambled more comprehensively.

[1] Lafond, A., et al., *Crystal Structures of Photovoltaic Chalcogenides, an Intricate Puzzle to Solve: the Cases of CIGSe and CZTS Materials*. Zeitschrift für Anorganische und Allgemeine Chemie, 2012. **638**(15): p. 2571-2577.
[2] Neldner, K., Valle-Rios, L.E., et al., *Existence of off-stoichiometric single phase kesterite*. Journal of Alloys and Compounds, 2015. **657**: p. 408-413.

MS08 - Storage and battery materials

MS08-01
Mechanochemical synthesis of a microporous MOF-74 metal-organic framework studied by *in situ* synchrotron X-ray diffraction

K. Uzarevic[1,2], G. Ayoub[2], P. Julien[2], T. Friscic[2]
[1]*Ruder Boskovic Institute, Division of Physical Chemistry, Zagreb, Croatia*
[2]*McGill University, Department of Chemistry, Montreal, Canada*

Mechanochemical synthesis[1] of microporous metal-organic frameworks (MOFs) recently emerged as a rapid, green and scalable alternative to lengthy and often energy-consuming solvothermal procedures. The field is in its infancy, with just a handful of model MOFs reported so far. All these reports, however, show significant benefits of mechanochemistry in terms of invested energy, reaction yields, applicability of poorly soluble, inexpensive oxide or carbonate precursors, and the use of only small amount of liquids to gain MOFs of excellent porosity.[2,3] Mechanistic studies of these solvent-free reactions are, however, still limited largely to zinc-based zeolitic imidazolate frameworks (ZIFs).[4]

Here we report the mechanistic study of mechanochemical synthesis of microporous MOF-74 from ZnO. The MOF-74 frameworks (also known as a CPO-27)[5] are a particularly important family of MOFs made from bivalent metal nodes and 2,5-dihydroxyterephthalic acid anions. Structures of these MOFs are characterized by honeycomb architecture and free metal sites, with reports showing high catalytic acitivty and potential for gas separation. *In situ* XRPD studies, performed at the European Synchrotron Radiation Facility (ESRF) and Deutsches Elektronen Synchrotron (DESY) revealed that the mechanosynthesis of Zn-MOF-74 proceeds over two steps, initially yielding a non-porous coordination polymer which reacts with remaining ZnO to yield Zn-MOF-74 whose porosity exceeds previously reported materials. Such unprecedented close-packed-to-open reaction pathway can be ascribed to difference in acidity of hydroxyl and carboxylate functionalities on the ligand, and we believe can lead to completely new MOF materials with unprecedented properties.

[1] James et al, Chem. Soc. Rev. **2012**, 41, 413.
[2] Beldon et al, *Angew. Chem. Int. Ed.* **2010**, *49*, 9640.
[3] Crawford et al, *Chem. Sci.*, **2015**, *6*, 1645.
[4] a) Friščić et al, *Nature. Chem.* **2013**, *5*, 66; b) Užarević et al, *J. Phys. Chem. Lett.*, **2015**, *6*, 4129.
[5] a) Rosi et al, *J. Am. Chem. Soc.* **2005**, *127*, 1504; b) Dietzel et al, *Chem. Commun.*, **2006**, 959.

Figure 1

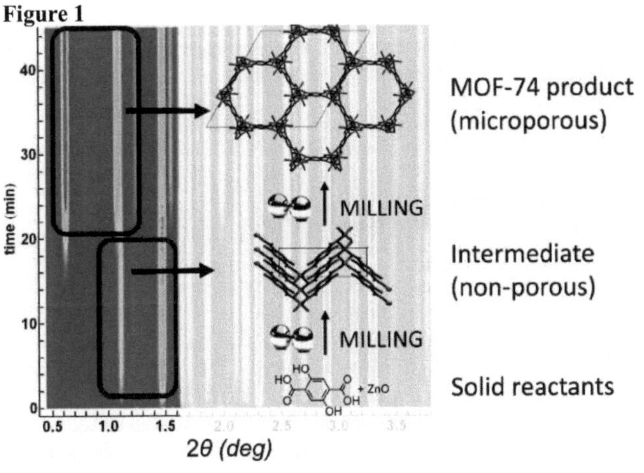

MS08-02
In situ solid-state NMR on Li- and Na-ion battery materials

O. Pecher[1], C. P. Grey[1]
[1]*University of Cambridge, Department of Chemistry, Cambridge, Great Britain*

Due to the high abundance, low costs and very suitable redox potential, Na-ion batteries (NIBs) should open new avenues of research and engineering as complementary alternatives to Li-ion batteries (LIBs).[1] This shift has to be accompanied with a deeper understanding of the chemical reactions involving the multiple cell components. The application of a non-invasive analysis tool that can follow the reactions *in operando* is therefore highly desired. A fundamental technique for this approach is *in situ* solid-state nuclear magnetic resonance (NMR) due to its high chemical specificity and sensitivity to crystalline, amorphous as well as metastable/short-lived phases.[2] However, *in situ* NMR on LIBs/NIBs does not come without its challenges, e.g., significantly different shifts of the multi-component samples, changing sample conditions during cycling, signal broadening due to paramagnetism as well as interferences between the NMR and external cycler circuit that might impair the real-time monitoring of the processes.[2-4]

In this matter, we are developing and exploring the use of a new Automatic Tuning Matching Cycler (ATMC) *in situ* NMR probe system that addresses many of these issues (Figure).[4] We applied the new *in situ* NMR methodology to paramagnetic $LiFePO_4$ (LFP) and $Na_3V_2(PO_4)_2F_3$ cathodes as well as Na—Na cells and Sn anodes. ^{7}Li *in situ* NMR reveals significant changes of the strongly paramagnetic broadened LFP line shape in agreement with the structural changes due to delithiation[5]. Additionally, ^{31}P *in situ* NMR shows a full separation of the electrolyte and cathode NMR signals (Figure). ^{31}P *in situ* NMR on $Na_3V_2(PO_4)_2F_3$ as a cathode in a NIB shows a significant shift and changes in the number of ^{31}P signals during (de)sodiation, revealing changes of local P environments that in part have not been seen in *ex situ* NMR investigations.[6] ^{23}Na *in situ* NMR was adjusted on symmetrical Na—Na cells during galvanostatic plating. An automatic adjustment of the NMR carrier frequency on the Na metal and electrolyte peak during the *in situ* experiment ensured on-resonance conditions and enabled an interleaved investigation of both species including the application of slightly different NMR set-ups. The measurements allow to monitor the formation of different Na metal species as well as a quantification of Na electrolyte consumption during the electrochemical experiment.

[1] K. Kubota, S. Komaba, *J. Electrochem. Soc.* **2015**, *162*, A2538. [2] F. Blanc et al., *Acc. Chem. Res.* **2013**, *46*, 1952. [3] N. M. Trease et al., *Solid State Nucl. Magn. Reson.* **2012**, *42*, 62. [4] O. Pecher et al., *J. Magn. Reson.* **2015**, *submitted.* [5] H. Liu et al., *Science* **2014**, *344*, 1252817. [6] Z. Liu et al., *Chem. Mater.* **2014**, *26*, 2513-2521.

Figure 1

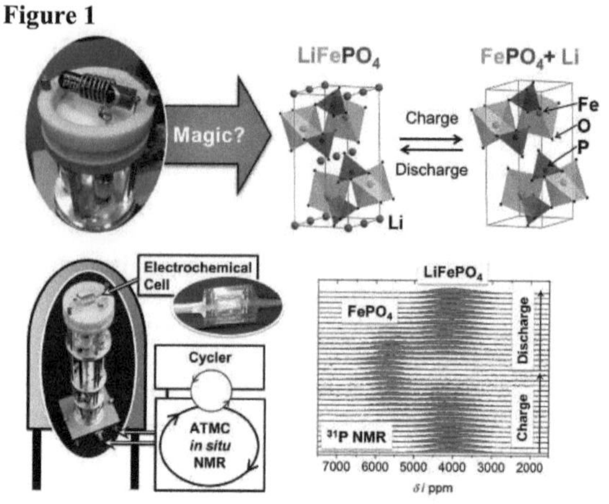

Figure. **(Top)** Is it all about "magic" when it comes to an application of *in situ* NMR to reveal structural changes during charge/discharge, e.g. on $LiFePO_4$ cathodes? **(Bottom)** Scheme of the ATMC *in situ* NMR approach and ^{31}P *in situ* NMR spectra of LFP during C/10 charge/discharge.

MS08-03

A new cubic modification of the fast Li-ion conductor $Li_7La_3Zr_2O_{12}$

R. Wagner[1], G. Redhammer[1], D. Rettenwander[1], A. Senyshyn[2], W. Schmidt[3], M. Wilkening[3], G. Amthauer[1]
[1]*Universität Salzburg, Chemistry and Physics of Materials, Salzburg, Austria*
[2]*Technische Universität München, Heinz Maier-Leibnitz Zentrum, Garching b. München, Germany*
[3]*Technische Universität Graz, Institute for Chemistry and Technology of Materials, Graz, Austria*

Garnet-structured Li-ion conductors such as $Li_7La_3Zr_2O_{12}$ (LLZO) are promising candidates for solid-state electrolytes to be used in next generation Li-ion battery concepts [1, 2]. LLZO does not only show a high Li-ion conductivity of up to 10^{-3} S cm^{-1} at room temperature (RT), it also provides a superior mechanical and chemical stability. In particular, its electrochemical inertness in a wide potential window of up to 6V and its stability against Li-metal makes LLZO exceptionally well suited to be used as a protecting layer for Li-metal based batteries.

For pure LLZO, two structural polymorphs have been described in literature: a tetragonal low-temperature modification, space group (SG) $I4_1/acd$ (no. 142), with a fully ordered arrangement of Li^+ ions; and a garnet-structured cubic high-temperature modification with SG *Ia*-3*d* (no. 230) that exhibits a disorder in the Li^+ distribution [3, 4]. The cubic modification can be stabilized at RT by the introduction of supervalent cations. These cations cause a reduction of the Li^+ content that leads to the introduction of additional vacancies at the Li^+ sites.

Among these cations, Ga^{3+} turned out to be a promising candidate [5, 6]. The electrochemical properties of Ga-stabilized LLZO are superior compared to LLZO stabilized with similar cations. The reason for this behavior is, however, still not fully understood. Investigations on Ga-stabilized LLZO single crystals may help to shed light on this question.

Coarse-grained Ga-stabilized LLZO samples with different Ga contents were prepared by solid state reaction. Single crystals were extracted and characterized by single crystal X-ray diffraction (SC-XRD). Li-ion dynamics were investigated by ^7Li NMR relaxometry.

SC-XRD shows that Ga-stabilized LLZO crystallizes in the acentric cubic SG *I*-43*d*, no. 220 (see Figure 1). This is the first definite record of this cubic modification for LLZO materials. The phase transition might be caused by the site occupation behavior of Ga^{3+}. Interestingly, ^7Li NMR relaxometry indicates an additional Li-ion diffusion process for Ga-stabilized LLZO with SG *I*-43*d* compared to Al-stabilized LLZO with SG *Ia*-3*d* prepared under identical conditions.

The new cubic modification has to be taken into consideration to explain the superior electrochemical performance of Ga-stabilized LLZO compared to its Al-stabilized counterpart.

Figure 1. Simulated precession images of the 0kl layer for LLZO showing SG *Ia*-3*d* (A) and LLZO showing SG *I*-43*d* (B). The presence of Bragg peaks with k = odd and l = odd in (B) is evident. These peaks are forbidden in SG *Ia*-3*d* but allowed in SG *I*-43*d*. Two sections are marked with rectangles to serve as an example.

[1] V. Thangadurai; D. Pinzaru; S. Narayanan; A. K. Baral, *J Phys Chem Lett* **2015,** *6,* 292.
[2] R. Murugan; V. Thangadurai; W. Weppner, *Angew Chem Int Edit* **2007,** *46,* 7778.
[3] J. Awaka; N. Kijima; H. Hayakawa; J. Akimoto, *J Solid State Chem* **2009,** *182,* 2046.
[4] J. Awaka; A. Takashima; K. Kataoka; N. Kijima; Y. Idemoto; J. Akimoto, *Chem Lett* **2011,** *40,* 60.
[5] D. Rettenwander; C. A. Geiger; M. Tribus; P. Tropper; G. Amthauer, *Inorg Chem* **2014,** *53,* 6264.
[6] D. Rettenwander; J. Langer; W. Schmidt; C. Arrer; K. J. Harris; V. Terskikh; G. R. Goward; M. Wilkening; G. Amthauer, *Chem Mater* **2015,** *27,* 3135.

Figure 1

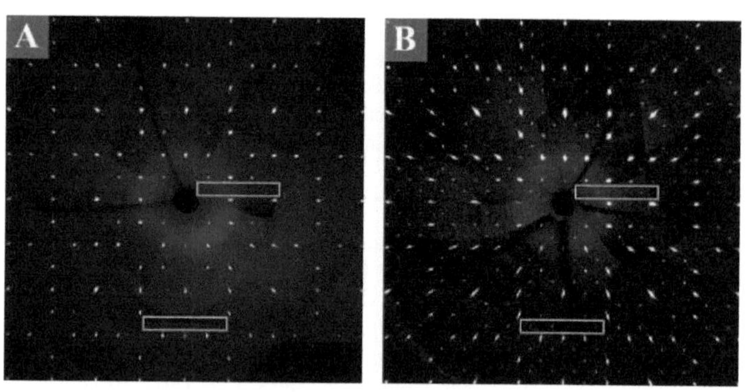

MS08-04

Crystalline solid electrolytes for all-solid-state batteries - complexity, conductivity, crystallography

T. Leisegang[1,2], F. Meutzner[1], T. Nestler[1], M. Zschornak[1], A. A. Kabanov[2], W. Münchgesang[1], R. Schmid[1], V. A. Blatov[2], D. C. Meyer[1]
[1]*TU Bergakademie Freiberg, Institut für Experimentelle Physik, Freiberg, Germany*
[2]*Samara State Aerospace University, Samara Center for Theoretical Materials Science, Samara, Russian Federation*

The already enormous economic relevance of electrochemical energy storage is proven by market studies, which assume annual growth rates of more than 5 %, based on sales revenue of approximately 60 billion Euros in the year 2015. Even today, over 130 years after the invention of the Pb-acid battery, this type of energy storage has with more than 93 % the largest share of the worldwide electrochemically stored energy. The technology is mature and the recycling rate of its components is now almost 100 %. Nevertheless, Pb is not very suitable to meet the anticipated demands for the expansion of particularly stationary energy storage, since the achievable volumetric energy densities are comparably small and the Pb reserves - with a projected range of less than 20 years - make usage on a larger scale difficult. In addition, due to its toxicity and the REACH regulations, the use of Pb is increasingly avoided.

A large variety of other electrochemical energy storage concepts have been tailored for specific applications: On the one hand Li-ion batteries for example have been optimized for mobile applications ranging from mobile phones to electric vehicles. But the Li-ion technology's learning curve now is flat and the Li resources are scarce. On the other hand, the resource efficient Na-S accumulators, which have been developed for stationary applications to account for the fluctuating nature of renewable energies, are promising but the cost per stored energy is still at a high level. A technology jump, which is probably only possible with other mobile ions of oxidation states higher than +1 and an all-solid state concept, would meet the requirements for today's industry and society. One of the most challenging battery components is the solid electrolyte.

Here, the status quo in the field of crystalline solid electrolyte materials for all-solid state batteries is examined. In this context, the availability of chemical elements is rated using a dedicated algorithm and the ionic conductivities of known ion conductors are correlated with common crystallographic features and structure complexity, determined with the program *ToposPro*. Furthermore, in order to derive promising solid electrolyte materials for an arbitrarily chosen ion, a reasonable and time-efficient combination of data mining and crystal chemical as well as crystal physical approaches is presented: Voronoi-Dirichlet partitioning (VDP), bond-valence methodology (BVM) and density functional theory (DFT). While VDP allows for a comparably fast high-throughput screening of crystallographic databases on the basis of a geometrical division of the crystal space detecting possible void networks, the BVM as a chemically motivated tool assists in the evaluation process by determining the most probable migration paths by calculating bond-valence energy landscapes. With DFT then a given crystal structure is modeled *ab initio* in order to determine ionic migration barriers, electronic conductivities and diffusion coefficients from relaxed structures. The methodical approach is demonstrated for Na^+ mobile ions and final perspectives based on high-valent mobile ions such as Al^{3+} are presented.

This work is funded by the German Federal Ministry of Education and Research within the project CryPhysConcept (03EK3029A) and the German Federal Ministry for the Environment, Nature Conservation, Building and Nuclear Safety within the project BaSta (0325563B).

MS09 - Biostructures III: Protein function and regulation

MS09-01

Specificity determinants in a transiently interacting system: Mechanism of Lys11-linked ubiquitin chain formation

S. Lorenz[1]
Rudolf-Virchow-Zentrum für Experimentelle Biomedizin
Universität Würzburg

The extraordinary versatility of ubiquitin in cellular signaling is associated with its ability to form polymeric chains of different topologies that trigger specific functional outcomes. Specificity in ubiquitin chain formation is thus critical for homeostasis. Lys11-linked ubiquitin chains target cell cycle regulators for proteasomal degradation during mitosis. The ubiquitin-conjugating enzyme Ube2S catalyzes the formation of these chains with high specificity, despite its extremely weak affinity for ubiquitin. We reveal the structural underpinnings of this mechanism by combining X-ray crystallographic, NMR spectroscopic, computational and functional studies.

MS09-02

Regulating proteolytic activity of DegQ from *Legionella*

G. Hansen[1], A. Schubert[1], R. Wrase[1], R. Hilgenfeld[1]
[1]*University of Lübeck, Institute of Biochemistry, Lübeck, Germany*

HtrA proteins are important protein-quality factors in the bacterial periplasm [1]. A common feature of all HtrAs is their modular organisation comprising a serine protease domain and at least one PDZ domain. All characterized HtrA proteins assemble into complex oligomers consisting of typically 3 to 24 monomers, which allow a tight regulation of proteolytic activity [2,3]. Here, we provide evidence that the assembly of proteolytically active, higher-order complexes of DegQ from *Legionella* is triggered by the binding of substrate-derived peptides. Crystal structures of inactive 3-mers and active 12-mers of DegQ reveal molecular details of elements of a conserved allosteric activation cascade that defines distinct protease ON and OFF states. Results from DegQ variants harbouring structure-based amino-acid substitutions indicate that peptide binding to the PDZ1 domain is critical for proteolytic activity but not for the formation of higher-order oligomers.

[1] Page, M. J. & Di Cera, E. (2008). Evolution of peptidase diversity. *J Biol Chem* **283**, 30010-4.
[2] Clausen, T., Kaiser, M., Huber, R. & Ehrmann, M. (2011). HTRA proteases: regulated proteolysis in protein quality control. *Nat Rev Mol Cell Biol* **12**, 152-62.
[3] Hansen, G. & Hilgenfeld, R. (2013). Architecture and regulation of HtrA-family proteins involved in protein quality control and stress response. *Cell Mol Life Sci* **70**, 761-75.

MS09-03

Allostery and catalysis of UDP-sugar pyrophosphorylases suggest a new approach to anti-parasitic treatments

J. Cramer[1], J. Führing[2], F. Routier[2], A.- C. Lamerz[2], J. Schneider[3], P. Baruch[3], R. Gerardy-Schahn[2], R. Fedorov[3]
[1]*Hannover Medical School, Institute for Cellular Chemistry / Institute for Biophysical Chemistry, Hannover, Germany*
[2]*Hannover Medical School, Institute for Cellular Chemistry, Hannover, Germany*
[3]*Hannover Medical School, Institute for Biophysical Chemistry / Research Division for Structural Analysis, Hannover, Germany*

UDP-sugar pyrophosphorylases play a central role in cellular carbohydrate biosynthesis. In mammals, UDP-glucose pyrophosphorylase (UGP) is the only enzyme capable of activating glucose-1-phosphate (Glc-1-P) to UDP-glucose (UDP-Glc), a metabolite located at the intersection of virtually all metabolic pathways in the mammalian cell. As metabolic switches UGPs are subject to elaborate regulatory mechanisms. Elucidating these mechanisms is essential for the fundamental understanding of cellular carbohydrate biosynthesis and for the development of new concepts to specifically inhibit individual enzymes.

Here we report comprehensive studies of the catalytic mechanism and allosteric regulation of UGPs from the human pathogen *Leishmania major*[1] and from the host[2,3] using a combination of kinetic X-ray crystallography, biochemical, biophysical and theoretical chemistry methods. The new data on UGP from *Leishmania major* (LmUGP) filled critical gaps in the knowledge of the UGP mechanism and allowed reconstruction of the complete enzymatic cycle. Results were integrated into a model of UGP activity describing structural changes throughout the cycle, mechanisms of substrate binding, catalysis, and product release. Our study revealed mechanisms of allosteric regulation common in nucleotidyltransferases and, in particular, the mechanical control of the chemical reaction in the active site.

In contrast to monomeric LmUGP, human enzyme (hUGP) has an octameric quaternary structure that we show to be essential for its activity[2]. We determined the structure of hUGP in complex with its product, which is the first structure of the active state of octameric UGP[3]. This structure provided first insights into the active site architecture and revealed the substrate binding mode and functionally important interactions in the octameric enzyme. We demonstrate that oligomerization extends the functionality of hUGP in several ways: it (i) facilitates an intermolecular stabilization of the sugar moiety in the active site, (ii) enhances protein stability, (iii) enables positive cooperativity in the reverse reaction, and (iv) allows regulation of the hUGP octamer by modification of a single subunit.

Due to their central position in carbohydrate biosynthesis, UGPs are attractive targets for drug development against protozoan parasites. Comparative analysis of allosteric activation mechanisms in LmUGP and hUGP helped us identify an allosteric site on the surface of LmUGP that can be used as a target for developing specific inhibitors. The potential of this site for LmUGP inhibition was confirmed by mutagenesis experiments and the pattern for a lead compound has been generated. Additionally, the allosteric mechanisms of UGPs discovered here helped in investigating the activation mechanisms in the OAS/cGAS triggers of innate immunity in another study by Lohöfener J, Steinke N, Kay-Fedorov P, Baruch P, Nikulin A, Tishchenko S, Manstein DJ, and Fedorov R.[4]

[1] Führing J, Cramer JT, Routier FH, Lamerz A-C, Baruch P, Gerardy-Schahn R and Fedorov R. (2013) *ACS Catalysis*. **3**:2976-2985.

[2] Führing J, Damerow S, Fedorov R, Schneider J, Münster-Kühnel AK, Gerardy-Schahn R. (2013) *Glycobiology*. **23**(4):426-37.

[3] Führing JI, Cramer JT, Schneider J, Baruch P, Gerardy-Schahn R, and Fedorov R. (2015) *Sci. Rep*. **5**:9618.

[4] Lohöfener J, Steinke N, Kay-Fedorov P, Baruch P, Nikulin A, Tishchenko S, Manstein DJ, and Fedorov R. (2015) *Structure*. **23**(5):851-862.

MS09-04

Designed Armadillo repeat proteins serve as scaffolds for the rational assembly of peptide binders with picomolar affinities.

P. Mittl[1]
[1]*University Zürich, Department of Biochemistry, /ürich, Switzerland*

X-ray crystallography is a powerful tool for the design of proteins with novel functionalities, ranging from enzymes that catalyze novel chemical reactions[1] to artificial protein recognition modules with tailored selectivities. Since the groundbreaking experiments by Köhler and Milstein almost four decades ago, monoclonal antibodies are the prevailing class of protein recognition modules in biomedical research, and diagnostics. Currently the development of new binders depends on the immunization of lab animals or combinatorial biochemistry techniques. Both approaches are elaborate and require sufficient amount of target protein, which renders the specific recognition of peptides by far more complex than the recognition of nucleic acids, where a suitable detection probe can be predicted straight from the nucleotide sequence.

We are developing a system where a binder for an unstructured peptide can be designed straight from the sequence of the target peptide. This system is based on designed Armadillo repeat proteins (dArmRP). More than 20 individual crystal structures of dArmRPs have been determined to optimize the design of the scaffold, prevent swapping of caps[2,3], optimize binding pockets[4] and ultimately reveal the modularity of the system[5]. Crystal structures confirm the logarithmic relationship between dissociation constant and target peptide length, which suggests a linear relationship between free energy of binding and number of dipeptide-binding modules. The modularity of the system offers the unique opportunity to generate very tight binders with low picomolar and perhaps even better affinities by adding more internal repeats to the dArmRP. Crystal structures of $(KR)_n$ peptides fused to the termini of globular domains reveal that dArmRPs recognize target peptides in a position independent manner. Because complexes of dArmRPs and $(KR)_n$-tagged globular domains are amendable to crystallization dArmRPs may serve as crystallization chaperones. X-ray crystallography also confirmed that dArmRPs can be fused to other protein domains using rigid-helix or hybrid-repeat design strategies, thereby increasing the molecular surface of the chaperone to improve the formation of stable crystal contacts.

1. Blomberg R, Kries H, Pinkas DM, Mittl PR, Grutter MG, Privett HK, Mayo SL, Hilvert D (2013) Precision is essential for efficient catalysis in an evolved Kemp eliminase. Nature **503**:418-421.

2. Madhurantakam C, Varadamsetty G, Grütter MG, Plückthun A, Mittl PR (2012) Structure-based optimization of designed Armadillo-repeat proteins. Protein Sci **21**:1015-1028.

3. Reichen C, Madhurantakam C, Plückthun A, Mittl PRE (2014) Crystal structures of designed armadillo repeat proteins: Implications of construct design and crystallization conditions on overall structure. Prot Sci **23**:1572-1583.

4. Reichen C, Madhurantakama C, Hansen S, Grütter MG, Plückthun A, Mittl PRE (2015) Crystal structures of designed Armadillo Repeat Proteins show propagation of inter-repeat interface effects. Acta Cryst. D, under review.

5. Hansen S, Tremmel D, Madhurantakam C, Reichen C, Mittl PRE, Plückthun A (2015) Structure and energetic contributions of a designed modular peptide-binding protein with picomolar affinity. Proc. Nat. Acad. Sci. USA, under review.

MS10 - Structural chemistry at non-ambient conditions

MS10-01

Exploring the phase, crystallisation and composition diagrams of ionic liquids

F. Fabbiani[1], S. Saouane[1]
[1]*Georg-August-Universität Göttingen, GZG, Abt. Kristallographie, Göttingen, Germany*

The number of high-pressure studies devoted to the crystallisation of ionic liquids has been steadily increasing in the past five years or so. The majority of studies are spectroscopic and only a handful of compounds have also been characterised by diffraction methods. At sufficiently high pressures, crystalline phases have also been reported to amorphise. For instance, 1-butyl-3-methylimidazolium hexafluorophosphate, one of the most widely studied ionic liquid at ambient-pressure conditions, shows pressure-induced amorphisation above 5.8 GPa.[1]

By virtue of their physico-chemical properties, *e.g.* melting point and inertness, ionic liquids are particularly suited for *in situ* crystallisation studies both at low-temperature and at high-pressure conditions. In this contribution the results of our own investigations on the polymorphism of a series of imidazolium-based ionic liquids at both low-temperature and high-pressure conditions, up to 1.0 GPa, will be presented.[2, 3] We have obtained single-crystal X-ray structures for all compounds studied and have complemented our diffraction work by Raman spectroscopy. We have found that in many cases the sealed and confined environment of the diamond-anvil cell is ideally suited for *in situ* crystallisation, offering in general better control over other *in situ* techniques used by us, such as low-temperature crystallisation with a heating/freezing stage or in a capillary. The results of our own investigations point to very rich phase, crystallisation and composition diagrams.

[1] H. Abe, T. Takekiyo, N. Hatano, M. Shigemi, N. Hamaya and Y. Yoshimura, J. Phys. Chem. B 2014, 118, 1138–1145.
[2] S. Saouane and F. P. A. Fabbiani, Cryst. Growth Des. 2015, 15, 3875–3884.
[3] S. Saouane, S. E. Norman, C. Hardacre, and F. P. A. Fabbiani, Chem. Sci. 2013, 4 1270–1280.

MS10-02

High quality structures at high pressure? Insights from diamonds and diamond-anvil cells

R. Angel[1], S. Milani[1], D. Pasqual[1], P. Nimis[1], F. Nestola[1], R. Miletich-Pawliczek[2]
[1]*University of Padova, Geosciences, Padova, Italy*
[2]*University of Vienna, Vienna, Austria*

The entrapment of single crystals of gem-quality minerals within natural diamonds provides the opportunity to clearly identify the factors that limit the quality of structure refinements from single crystals held at high pressures in diamond-anvil pressure cells.

Full intensity datasets were collected from many ~100um-sized single-crystal olivines trapped deep within natural diamonds of up to 5mm in size with a Rigaku-Oxford-Diffraction SuperNova diffractometer equipped with a Mo-target microsource, and a Pilatus 200K area detector. Intensity datasets were obtained by integration of the raw detector images with the Crysalis software, which implements a number of procedures inherited from point detector methods, including use of 'learnt profiles' and local background evaluation.

After correction for the combined absorption effects of the irregularly-shaped diamond and other inclusions, and elimination of outliers due multiple diffraction events involving the diamond, the internal agreement factors on the datasets are similar to those from crystals in air. The estimated standard deviations on refined structural parameters and bond lengths are the same as from refinements to datasets collected from olivine crystals in air. In addition, the refined site populations give the correct chemical composition, and the refined displacement parameters of the atoms are consistent with the expected rigid-body behaviour of SiO4 groups in the structure. These results demonstrate that very accurate and very precise intensity data can be collected with laboratory instruments from single crystals held within environmental containers.

Intensity datasets collected from single crystals held inside diamond-anvil pressure cells with the same data collection procedures and the same data integration and reduction procedures yield data of the same quality as measured by Rint. However, they exhibit much larger esd's on bond lengths and angles, and the refined atomic displacement parameters are often unphysical. Comparison with the refinements of olivines trapped in diamonds therefore indicates that it is not failures in the data collection or data reduction methods that are responsible for the lower quality of high-pressure structures. With modern diffractometers, detectors and data reduction software, the reduction in quality of high-pressure structures is solely due to lack of data resolution arising from the physical obstruction by the body of the diamond-anvil cell. The keys to obtaining high-quality structures at high pressure are thus the correct choice of crystal orientation to obtain high resolution data, and the collection of highly redundant data so as to accurately correct the data for absorption and diffraction events.

This work was supported by ERC starting grant 307322, INDIMEDEA, to F. Nestola.

MS10-03

Structural solution of the high-pressure phase of $LaPO_4$

J. Ruiz-Fuertes[1], A. Hirsch[2], B. Winkler[3], A. Friedrich[4], L. Bayarjargal[3], W. Morgenroth[3], L. Peters[2], G. Roth[2]
[1]*Universitat de Valencia, Fisica Aplicada, Burjassot, Spain*
[2]*RWTH Aachen University, Institut für Kristallographie, Aachen, Germany*
[3]*Goethe-Universität Frankfurt, Institut für Geowissenshaften, Frankfurt am Main, Germany*
[4]*Julius-Maximilians-Universität Würzburg, Institut für Anorganische Chemie, Würzburg, Germany*

$LaPO_4$ monazite has a great potential as a host material for solid state lasers [1] and is a promising candidate for oxidation-resistant ceramics [2, 3]. In addition, due to the crystal-chemical similarity between lanthanide and actinide elements, it is being intensively investigated for its use in the conditioning of radioactive waste [4]. One of the fundamental properties of lanthanide orthophosphates that is still not completely understood is their stability at extreme conditions. While a significant effort has been invested in the high-pressure study of orthophosphates with zircon-type structure ($I4_1/amd$), in monazite-type ($P2_1/n$) orthophosphates their lower symmetry and higher stability with respect to increasing pressure has limited the number of high-pressure structural studies to just one powder x-ray diffraction study [5]. Lacomba-Perales et al. [5] have shown that $LaPO_4$ undergoes a structural phase transition to an orthorhombic phase (proposed to be a barite-type structure with space group *Pnma*) at around 26 GPa and also that within the lanthanide orthophosphates $LaPO_4$ is the one with the lowest phase transition pressure. Solving the high-pressure phase of $LaPO_4$ would therefore provide insight into the pressure behavior of the family of monazite orthophosphates and is the goal of the present study.

Figure 1. Projections along the [010] direction of the monazite-type structure (left) and of the post-barite-type structure (right) of $LaPO_4$ as derived from our experiments at 27 GPa. Comparable sections of the crystal structures are highlighted in red.

Single-crystal *x*-ray diffraction experiments of $LaPO_4$ have been performed at different pressures up to 31 GPa at the P02.2 Extreme Conditions Beamline at PETRA III (DESY, Hamburg). We have found that $LaPO_4$ transforms at 26 GPa from the monazite-type structure to an orthorhombic structure in good agreement with Ref. 5. However, our structural solution has shown that instead of a centrosymmetric barite-type structure, the high-pressure phase of $LaPO_4$ is an acentric structure similar to the post-barite structure proposed by Santamaría-Pérez et al. for $BaSO_4$ [6]. The acentricity of the high-pressure phase of $LaPO_4$ has been proven by second harmonic generation (SHG) experiments, which show the emergence of SHG above 25 GPa. *Ab initio* calculations also confirm the structural phase transition from the monazite-type to the post-barite-type structure at 26 GPa in $LaPO_4$, but predict only very small enthalpy differences between $LaPO_4$ polymorphs with the barite- and post-barite-type structure above 10 GPa.

J.R.-F thanks the Alexander von Humboldt Foundation for a postdoctoral fellowship. DESY Photon Science is gratefully acknowledged. A.H. thanks the BMBF (02NUK021E) for funding. PETRA III at DESY is a member of the Helmholtz Association (HGF).

[1] E. Nakawaza and F. Shiga, Jap. J. Appl. Phys. **42**, 1642 (2003).
[2] R. S. Hay, Ceram. Eng. Sci. Proc. **21**, 203 (2001).
[3] J. Wang, Y. Zhou, and Z. Lin, Appl. Phys. Lett. 87, 051902 (2005).
[4] H. Li, S. Zhang, S. Zhou, and X. Cao, Inorg. Chem. **48**, 4542 (2009).
[5] R. Lacomba-Perales, D. Errandonea, Y. Meng, and M. Bettinelli, Phys. Rev. B **81**, 064113 (2010).
[6] D. Santamaría-Pérez, L. Gracia, G. Garbarino, A. Beltrán, R. Chuliá-Jordán, O. Gomis, D. Errandonea, Ch. Ferrer-Roca, D. Martínez-García, and A. Segura, Phys. Rev. B 84, 054102 (2011).

Figure 1

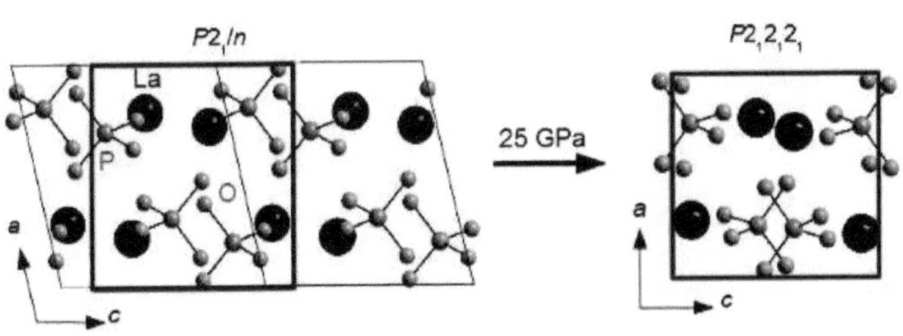

MS10-04
Structure and phase stability of Fe,Al bearing bridgmanite at high pressures and temperatures

L. Dubrovinsky[1], L. Ismailova[1], E. Bykova[1], M. Bykov[1], N. Dubrovinskaia[1]
[1]*Bayreuth University, BGI, Bayreuth, Germany*

The only so far unambiguously confirmed phase transition in $(Mg,Fe)(Si,Al)O_3$ bridgmanite is that into a $CaIrO_3$-type structured phase (called often post-perovskite, PPv) above ~125 GPa and high temperatures. Recently, in laser-heated diamond anvil cell experiments, Zhang et al.[1] found that at pressures of 85-101 GPa and temperatures of 2200~2400 K iron-bearing silicate perovskite is unstable; it loses its iron-rich component and disintegrates into nearly Fe-free $MgSiO_3$ perovskite and an Fe-rich $(Mg,Fe)SiO_3$ phase ("H-phase") with a previously unknown hexagonal structure. These findingscan be of a primary importance forgeophysical, geochemical and geodynamic models of the lower mantle below the depth of 2000 kilometer.

Using single-crystal X-ray diffraction in LHDACs we have investigated the behavior of synthetic bridgmanite samples with different compositions (particularly, $(Mg_{0.83}Fe_{0.17})(Si_{0.94}Al_{0.06})O_3$, $(Mg_{0.88}Fe_{0.12})(Si_{0.96}Al_{0.04})O_3$, and $(Mg_{0.6}Fe_{0.4})(Si_{0.63}Al_{0.37})O_3$) at pressures up to 125 GPa and temperatures over 3000 K. We demonstrate that an accurate structural refinement is possible at conditions corresponding to the lower parts of the Earth's lower mantle. Despite a long duration of the samples' heating at pressures above 100 GPa, the quality of single crystals of bridgmanite does not worsen, but any sign of the "H-phase" formation is not observed up to 125 GPa.

References

[1] L. Zhang, Y. Meng, W. Yang, et al., *Science*, 2014, **344**, 877-882

MS11 - Molecular structural chemistry: Syntheses, structure and applications

MS11-01

Strukturelle, Thermodynamische und kinetische Aspekte der Trimorphie von Betamethason-Valerat

C. Näther[1], I. Jess[1], L. Seyfarth[2], K. Bärwinkel[2], J. Senker[2]
[1]*Universität Kiel, Institut für Anorganische Chemie, Kiel, Germany*
[2]*Universität Bayreuth, Bayreuth, Germany*

Bei Betamethason-Valerat handelt es sich um ein Glucocorticoid, welches auf Grund seiner antiallergischen und immun-repressiven Eigenschaften in der medizinischen Therapie eingesetzt wird. Die Einkristallstruktur der kommerziell erhältlichen Form **I** war unbekannt und es gab auch keinerlei Hinweise auf die Existenz zusätzlicher Modifikationen.

Lösungsvermittelte Umwandlungsreaktionen über mehrere Wochen in unterschiedlichen Lösungsmitteln führten nicht zur Bildung einer neuen Modifikation, jedoch konnte durch Rühren in Methanol ein Solvat dargestellt werden. Beide Verbindungen kristallisieren in der orthorhombischen Raumgruppe $P2_12_12_1$ mit Z = 4 Molekülen in der Elementarzelle.

Beim Aufheizen des Solvates kommt es unter Abspaltung der Lösungsmittelmoleküle zur Bildung einer neuen Modifikation (Form **III**), welche bei weiterem Aufheizen in eine weitere Modifikation (Form **II**) umwandelt. Untersuchungen mit der Dynamischen Differenz-Kalorimetrie (DDK) sowie der Thermomikroskopie belegen, dass zunächst Form **III** schmilzt und dann aus dieser Schmelze Form **II** auskristallisiert,

Die Kristallstrukturen von Form **II** und **III** wurden durch Kombination von ^{13}C-NMR-Spektoskopie, theoretischen Methoden und der Röntgen-Pulverbeugung bestimmt. Form **II** kristallisiert ebenfalls orthorhombisch (RG = $P2_12_12_1$; Z = 4), wohingegen Form **III** in der monoklinen Raumgruppe $P2_1$ mit zwei kristallographisch unabhängigen Molekülen kristallisiert. In allen Modifikationen werden sowohl Unterschiede in den Konformationen als auch in den Packungen der Moleküle beobachtet.

Lösungsvermittelte Umwandlungsexperimente ausgehend von Gemengen aller Modifikationen belegen, dass Form **II** und Form **III** in Form **I** umwandeln und diese daher die bei Raumtemperatur thermodynamisch stabile Modifikation darstellt. Heizratenabhängige DDK-Messungen ergeben darüber hinaus keinerlei Hinweise auf eine polymorphe Fest-Fest-Umwandlung. Weitere DSC-Messungen ergeben, dass Form **I** den höchsten Schmelzpunkt und die höchste Schmelzenthalpie aufweist und sich daher monotrop zu Form **II** und **III** verhält. Die kommerziell erhältliche Form **I** stellt somit die über den gesamten Temperaturbereich thermodynamisch stabile Modifikation dar. Von allen drei Modifikationen besitzt Form **III** den niedrigsten Schmelzpunkt und auch die niedrigste Schmelzenthalpie, weswegen sich diese auch monotrop zu Form **II** verhält.

MS11-02

Porous Supramolecules

A. Virovets[1], E. Peresypkina[1], M. Scheer [1]
[1]*University of Regensburg, Lst. Prof. M. Scheer, Regensburg, Germany*

An inorganic analogue of ferrocene, pentaphosphaferrocene, $[Cp^RFe(\eta^5\text{-}P_5)]$ (Cp^R = η^5-C_5R_5, R = Me (Cp*), CH_2Ph (Cp^{Bn}), PhC_4H_9 (Cp^{BIG})), is able to coordinate Cu^+ and Ag^+ cations resulting in either coordination polymers or in giant supramolecules [1-7]. The self-assembled spherical supramolecules of 2.1 - 4.6 nm in diameter can be isolated in astonishingly high yields. Special crystallization technique allows obtaining single crystals suitable for the X-ray diffraction study with laboratory equipment. A tetrahedral coordination of Cu^+ together with the predetermined five-fold symmetry of the *cyclo*-P_5 ligand favors the formation of giant cages with fullerene [1,2] or fullerene-like [3, 4] topologies. Furthermore, the ability of copper halides for aggregation opens the gate to more sophisticated supramolecules with multi-layered inorganic core built up with hundreds of metal, halogen and P atoms [5, 6].

The pentaphosphaferrocene-based supramolecules display two kinds of porosity. Similarly to the MOFs they are *porous* being able to encapsulate and store various guests. The hosts' shells are large enough to isolate small, unstable P_4 and As_4 molecules (cf. Fig., [6]), as well as the much larger fullerene C_{60}.

The other type of porosity is related to the statistical vacancies in some metal and halide positions. Since the self-assembly process of such complicated molecules is not entirely specific, the structural variations in the replication become possible (cf. Fig.). For this reason, the crystals of the supramolecules consist of similar co-crystallized forms, in which isomerism and different degree of the *porosity* can occur. They represent unique examples of *vacant molecular structures* [7] similar to well-known covalent compounds with statistical distribution of framework atoms.

This work was supported by ERC grant AdG339072-SELFPHOS.

[1] F. Dielmann C. Heindl, F. Hastreiter, E.V. Peresypkina et al (2015) *Chem.-A Eur. J.*, **21**, 6208.
[2] S. Heinl E. V. Peresypkina, J. Sutter, M. Scheer (2015) *Angew. Chem. Int. Ed.*, **54**, 13431. [5] E. V.
[3] J. Bai, A. V. Virovets, M. Scheer (2003), *Science*, **300**, 781.
[4] M. Scheer, J. Bai, B. P. Johnson et al (2010), *Chem. Eur. J.* **18**, 2092.
[5] C. Heindl, E. V. Peresypkina, A.V. Virovets et al (2015) *J. Am. Chem. Soc.*, *137*, 10938.
[6] C. Schwarzmaier A. Schindler, C. Heindl et al (2013)*Angew. Chem. Int. Ed.*, **52**, 10896.
[7] E. Peresypkina, C. Heindl, A. Schindler et al (2014) *Z. Kristallogr.*, **229**, 735, and references therein.

Figure 1

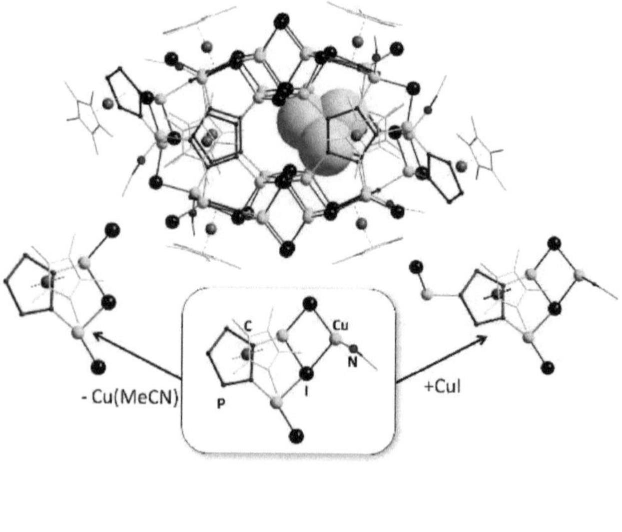

MS11-03

A kryptoracemate and three of its alternative racemic polymorphs

B. Braun Cula[1]
[1]*Humboldt Universität zu Berlin, Berlin, Germany*

So-called *kryptoracemates*[1,2] or *false conglomerates*[3] are racemic crystals crystallizing in chiral space groups.

Four polymorphs of an acetonitrile have been crystallized and structurally analyzed: In addition to three racemic crystals, a rare kryptoracemate has been obtained. The kryptoracemate contains two independent molecules of opposite chirality in the asymmetric unit which differ significantly in conformation.

The kryptoracemate as well as two of its polymorphs contain several independent molecules. The flexible hinge allows for different conformations, but this conformational flexibility is no prerequisite for crystallization.

This is the first example of a kryptoracemic structure co-existing with four of its conformational polymorphs, well-ordered racemic structures.[4]

References

1 R. Bishop and M. L. Scudder, Crystal Growth & Design, 2009, **9(6)**, 2890.

2 L. Fábián and C. Pratt Brock, Acta Crystallographica Section B, 2010, **66**, 94 and references therein.

3 I.Bernal, J. Cai, , S. S. Massoud, S. Watkins and F. Fronczek, J. Coord. Chem. 1996, **38**, 165; I. Bernal, J. Cai and J. Myrczek, Acta Chim. Hung., 1995, **132**, 451. I. Bernal, F. Somoza and V. Bahn, J. Coord. Chem., 1997, **42**, 1; J. Cai, J. Myrczek, H. Chun and I. Bernal, J. Chem. Soc. Dalton Trans., 1998, 4155; I. Bernal and S. Watkins, Acta Crystallographica Section C, 2015, **71**, 216.

4. R. Laubenstein, M.-D. Şerb, U. Englert, G. Raabe, T. Braun and B. Braun, Chem. Comm. 2016, **DOI:** 10.1039/C5CC08992A

Figure 1

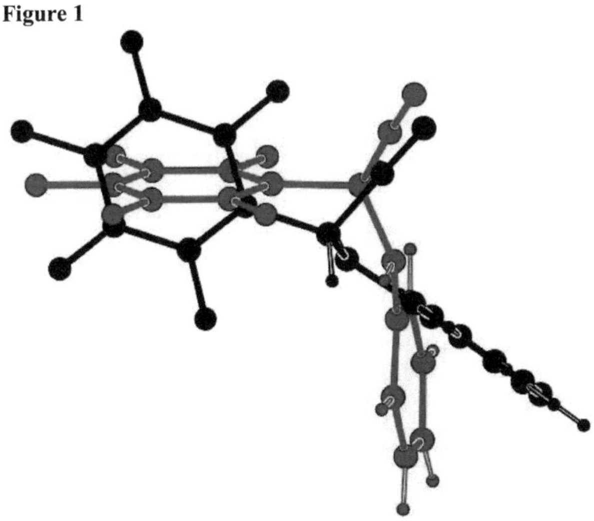

MS11-04

Molecular Structures of Dimeric and Trimeric Aminoalanes

T. Bernert[1], M. Brix Ley[1], J. Ruiz-Fuertes[2], M. Fischer[3], M. Felderhoff[1], C. Weidenthaler[1]
[1]*Max-Planck-Institut fuer Kohlenforschung, Department of Heterogeneous Catalysis, Muelheim an der Ruhr, Germany*
[2]*Goethe-Universität, Institut für Geowissenschaften, Frankfurt am Main, Spain*
[3]*University of Bremen, Fachgebiet Kristallographie, Bremen, Germany*

Interactions of $H^{\delta+}$ and $H^{\delta-}$ are a promising strategy for hydrogen storage. The combination of both species in a single substance can be achieved in Lewis-acid/base adducts, which show a reversible dehydrogenation, as discussed by Stephan and Erker [1]. Lewis-acid/base adducts of nitrogen and aluminium or boron could be economically and environmentally beneficial candidates for hydrogen storage, as already discussed for NH_3BH_3 [2] or polymers based on aryl bisboranes [3]. Within the group 13 elements, Lewis-acid/base adducts with different substituents often form molecular structures build up by dimers or trimers [Figure 1 (a)]. Since the reactivity of these compounds towards small molecules like H_2 [4] or CO_2 [5] depends strongly on the bond lengths between the corresponding Lewis acid and the base, a knowledge of their molecular structures is crucial for their applications.

Recently, we determined the crystal structure of diethylaminoalane by a combination of powder X-ray diffraction, Raman spectroscopy and DFT calculations. This approach has led to a reliable molecular structure determination of diethylaminoalane in the solid state and to further information about the environment and behavior of the Al-H bonds. The structure of diethylaminoalane was compared with other amino-adducts of group 13 elements as well as aminoalanes with different substituents [6]. Diethylaminoalane crystallises in $P2_1/c$ with a = 7.4020(2) Å, b = 12.9663(3) Å, c = 7.2878(2) Å and β = 90.660(2)° at 293 K. The molecular structure consists of dimers [Figure 1 (b)], formed by a four membered Al_2N_2 ring. While the molecular structure is very similar to the amine adducts of other group 13 elements, the crystal structure differs significantly.

Figure 1: (a) Typical molecular structures of amino adducts of group 13 elements. (b) Molecular structure of diethylaminoalane with assignment of the asymmetric unit.

References:
[1] D. W. Stephan, G. Erker (2010), *Angew. Chem., Int. Ed.*, 49, 46 - 76.
[2] A. Staubitz, A. P. M. Robertson, I. Manners (2010), *Chem. Rev.*, 110, 4079 - 4124.
[3] A. Ledoux, P. Larini, C. Boisson, V. Monteil, J. Raynaud, E. Lacote (2015), *Angew. Chem., Int. Ed.*, 54, 15744 - 15749.
[4] P. P. Power (2010), Nature, 463, 171-177.
[5] J. A. B. Abdalla, I. M. Riddlestone, R. Tirfoin, S. Aldridge (2015), *Angew. Chem., Int. Ed.*, 54, 5098 - 5102.
[6] T. Bernert, M. B. Ley, J. Ruiz-Fuertes, M. Fischer, M. Felderhoff, C. Weidenthaler, *submitted to Acta Crystallogr. B.*

Figure 1

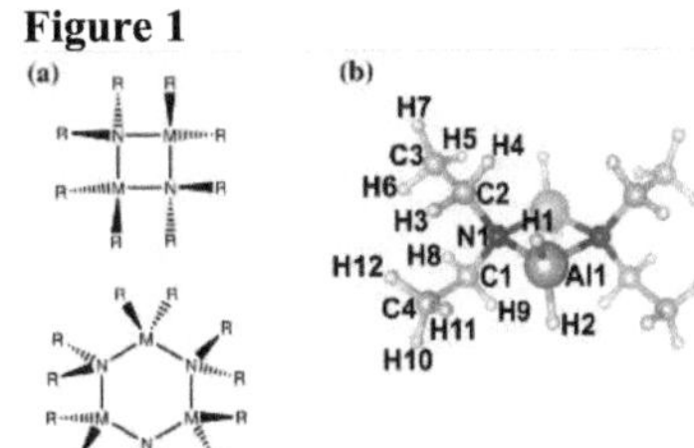

MS12 - Biostructures IV: Protein-nucleic acid interactions

MS12-02
Structural and functional analysis of the spliceosomal RNA helicase Prp43

M. J. Tauchert[1], J.- B. Fourmann[2], R. Lührmann[2], R. Ficner[1]
[1]*Georg-August University Göttingen, Molecular Structural Biology, Göttingen, Germany*
[2]*Max Planck Institute for Biophysical Chemistry, Cellular Biochemistry, Göttingen, Germany*

In higher eukaryotes, the vast number of introns is removed from pre-mRNAs via the spliceosome, a highly dynamic multi-megadalton single-turnover ribonucleoprotein complex, which is composed of five uridine-rich small nuclear RNAs and numerous proteins [1]. Apart from conformational changes, also the proteinaceous composition of the spliceosome changes dramatically during one cycle of splicing and thereby the accurate orchestration of various compositional and conformational rearrangements is essential. Key players of the major rearrangement steps are RNA helicases, which belong to the helicase superfamily SF2.

The DEAH-box helicase Prp43 (pre-mRNA processing factor 43) is involved in spliceosomal disassembly and is required for proper dismantling of the yeast intron-lariat spliceosome [2]. During this process, Prp43's ATPase and helicase activity is stimulated by Ntr1 (nineteen complex-related protein 1), which interacts with Prp43 via its G-patch motif [3]. This G-patch contains several conserved glycine residues and is intrinsically disordered, but adopts a defined structure after the binding to its interaction partner.

Here, we present the crystal structure of Prp43 in an ADP-bound state from the thermophilic ascomycete *Chaetomium thermophilum* refined at 2.9 Å resolution. We demonstrate via spliceosomal disassembly assays that ctPrp43 is the authentic orthologue of Prp43 from *Saccharomyces cerevisiae*, which can be explained by a high level of structural similarity and surface conservation. In addition, we provide a biochemical characterization of ctPrp43 in order to understand the activation of ctPrp43 by Ntr1 and RNA.

Figure 1

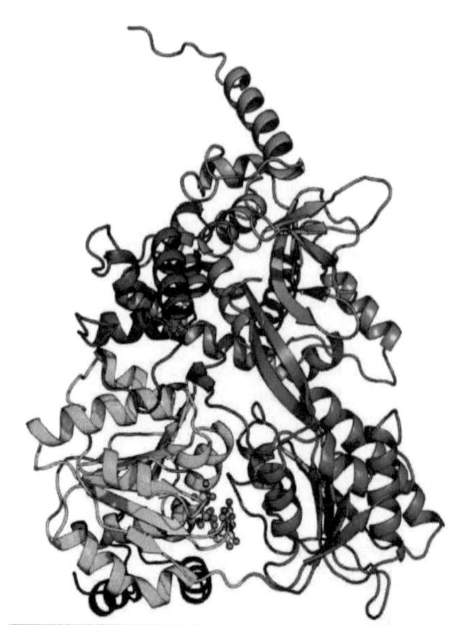

MS12-03
Structural basis of transcription antitermination

N. Said[1], E. Anedchenko[2], K. F. Santos[1], O. Dybkov[3], H. Urlaub[3,4], G. Weber[1], M. C. Wahl[1]
[1]*Freie Universität Berlin, Strukturbiochemie, Berlin, Germany*
[2]*Humboldt-Universität zu Berlin, Molecular Cell Biology, Berlin, Germany*
[3]*Max Planck Institut für biophysikalische Chemie, Cellular Biochemistry, Göttingen, Germany*
[4]*Georg-August-Universität, Institute for Clinical Chemistry, Göttingen, Germany*

Transcription is a highly regulated process that in bacteria is catalyzed by a DNA-dependent RNA polymerase (RNAP) comprising five subunits (α2, β, β', and ω). The main stages of transcription, i.e. initiation, elongation and termination, and transitions between these stages are controlled by specific transcription factors. Several bacteriophages reprogram the host transcriptional apparatus to support their own life cycles [1]. E.g., phage λ encodes the N protein that hijacks the host transcriptional machinery to switch the phage gene expression program from immediate-early to delayed-early genes. Upon transcription of nut RNA signal sequences (composed of boxA and boxB elements) on the λ genome, λN together with host transcription factors NusA, NusB, NusE and NusG assembles an elaborate ribonucleoprotein (RNP) complex on the surface of RNAP [2]. The λN-RNP remains associated with RNAP during further transcription by an RNA looping mechanism, rendering the enzyme resistant to rho-dependent and intrinsic terminators that can be located thousands of base pairs downstream of a nut site [3]. While structures of RNAP, individual Nus factors and some sub-complexes involving Nus factors, parts of λN or part of nut RNA have been elucidated [4], the overall architecture of a λN-based transcription antitermination complex and how antitermination is elicited are still enigmatic. Here, we have applied an integrative structural biochemical strategy to reveal the molecular basis of λN-mediated transcriptional antitermination. We elucidated the crystal structural of a complex containing λN, NusA, NusB, NusE and nut site RNA and probed the observed interactions by structure-guided mutagenesis. The results revealed how NusA and nut RNA provide the major scaffolds for the assembly, how the intrinsically unstructured λN protein folds upon integration into the complex and how a multitude of weak, binary interactions gives rise to a stable, multi-subunit RNP. We then reconstituted a complex comprising RNAP, template DNA, a transcript containing a nut site, λN, NusA, NusB, NusE and NusG, and delineated the ensuing molecular neighborhoods by cross-linking/mass spectrometry. Together with the λN-NusA-NusB-NusE-nut RNA crystal structure as well as structures of RNAP and NusG, these data enabled us to construct a hybrid model of a complete, functional λN-based transcription antitermination complex. Testing the resulting model by targeted mutagenesis in combination with in vitro transcription assays revealed key molecular principles underlying λN-dependent transcription antitermination.

References

1. Nudler, E & Gottesman, M. E. (2002). Transcription termination and anti-termination in E. coli. Genes to Cells 9, 2831-2845
2. Greenblatt, J., Nodwell, J. R., and Mason, S. W. (1993). Transcriptional antitermination. Nature 364, 401-406
3. Bubunenko B, Court DL, Al Refaii A, Saxena S, Korepanov A, et al. (2013). Nus transcription elongation factors and RNase III modulates small ribosome subunit biogenisis in Escherichia coli. Mol. Microbiol. 87, 382-93
4. Björn M. Burmann & Paul Rösch (2011) The role of E. coli Nus-Factors in transcription regulation and transcription: translation coupling. Transcription, 2:3, 130-134.

MS12-04

Structural and Biochemical Insights into Transcriptional Regulation Mediated by Grainyhead-like 1

Q. Ming[1], Y. Roske[1], U. Heinemann[1]
[1]*Max-Delbrück Center for Molecular Medicine, Macromolecular Structure & Interaction, Berlin, Germany*

Grainyhead is one of the best studied genes encoding transcription factors of the CP2 family (TFCP2) that contribute to regulation of gene expression from early embryonic development to terminal cell differentiation. In the human genome there are three *Grainyhead* homologs, named as Grainyhead-like 1, 2, and 3 (Grhl1, Grhl2, and Grhl3). They are expressed in the surface ectoderm and in other epithelial tissues[1], regulating genes involved in cell-junction formation and proliferation, controlling the development and differentiation of multicellular epithelia[2]. Here we direct the focus on structural studies of Grhl1, which plays key roles in the functioning of the epidermis. Published research reveals that mice lacking *Grhl-1* exhibit an abnormal hair coat, defective hair anchoring, and palmoplantar keratoderma with Spina bifida and severe skin-barrier defects[3]. The Grhl1-3 proteins are composed of three domains, an N-terminal transactivation domain (TAD), a central DNA-binding domain (DBD) and a C-terminal dimerization domain (DD).

It is the main objective of this project to determine the mechanism how Grhl1 recognizes and binds to the target DNA with structural biology approaches. As of yet there are no structural reports on any member of the CP2 protein family, which stimulates our effort to elucidate the Grhl1 protein folding and protein-DNA interaction. We also try to address the disease-related point mutations in Grhl1 to provide useful information for further functional studies of Grhl protein.

We used X-ray crystallography to analyze the Grhl1 structure and biochemical methods like electrophoretic mobility shift assays (EMSA) and isothermal titration calorimetry (ITC) to characterize its DNA binding. In cooperation with others we are validating our conclusions based on these experiments using cellular experiments.

So far, we determined the structures of Grhl1 DBD and of DNA-bound Grhl1 DBD at resolutions of 2.5 Å and 2.9 Å, respectively. The DBD structure was phased using SeMet single-wavelength anomalous diffraction (SAD). Grhl1 DBD adopts an immunoglobulin-like fold, with two β-sheets packing in the hydrophobic core. In the protein-DNA complex structure each duplex DNA is bound to two Grhl1 DBD molecules in a symmetric arrangement. Based on the structures we generated Grhl1 mutants with amino-acid replacements in the DNA interface and studied their DNA binding. We could map the Grhl1 mutation R427Q, linked to lung and stomach cancer, to the DNA-binding interface.

We wish to share the results of our structural and biochemical studies on Grhl1 in the conference. We also hope to present some on-going studies related to the Grhl1-DNA interaction, for instance the protein oligomerization behavior in the presence of the C-terminal dimerization domain.

References

[1] Auden, A., Caddy, J., Wilanowski, T., Ting, S. B., Cunningham, J. M., & Jane, S. M. (**2006**). Spatial and temporal expression of the Grainyhead-like transcription factor family during murine development. Gene Expression Patterns, 6, 964-970.
[2] Rifat, Y., Parekh, V., Wilanowski, T., Hislop, N. R., Auden, A., Ting, S. B. Jane, S. M. (**2010**). Regional neural tube closure defined by the Grainyhead-like transcription factors. Developmental Biology, 345, 237-245.

[3] Wilanowski, T., Caddy, J., Ting, S. B., Hislop, N. R., Cerruti, L., Auden, A., Zhao, L. L., Asquith, S., Ellis, S., Sinclair, R., Cunningham, J. M., and Jane, S. M. (**2008**). Perturbed desmosomal cadherin expression in grainy head-like 1-null mice. EMBO Journal, 27, 886-897.

MS13 - Time resolved crystallography

MS13-02
Time resolved diffraction study of the Zintl phase SrGe under 50 bar hydrogen pressure and elevated temperatures

H. Auer[1], H. Kohlmann[1]
[1]*Leipzig University, Leipzig, Germany*

Zintl phases are polar intermetallic compounds of an group one or two metal and a more electronegative element that exhibits a polyanionic partial structure. Hydrogen acts as an oxidizing agent. For group 14 Zintl phases it facilitates the formation of additional element-element or element-hydrogen bonds. The Zintl phases AeTt, Ae = (Ca),Sr,Ba; Tt = Si,Ge form hydrogen rich β-hydride phases $AeTtH_x$ with hydrogen content 1.0 <= x <= 2.0 [1,2,3]. The existence of an hydrogen poor α-hydride phase was suggested for the BaGe-H(D) system [3]. This scope is extended to the SrGe-H(D) system using time resolved diffraction under non-ambient conditions.

Time resolved diffraction under 50 bar hydrogen pressure and elevated temperatures was applied to the SrGe - H(D) system. *In situ* neutron diffraction was done in (leuco-)sapphire single crystal cells up to a temperature of 620 K. Data were collected at the D20 high- intensity diffractometer at Institute Laue-Langevin (ILL), Grenoble, France. Additionally *in situ* synchrotron diffraction was done in quartz capillaries up to 775 K at KMC2 beamline at BESSY II, Berlin, Germany.

The formation of the hydrogen rich β-hydride phase ($SrGeD_{1.214(4)}$) starting at 420 K could be shown to be a one step reaction. Further heating results in a subsequent decomposition in two hydrogen poorer α'- (520 K, $SrGeD_{0.524(12)}$) and α-hydride phases (570 K, $SrGeD_{0.232(4)}$). Upon cooling, the system takes up hydrogen again forming the β-hydride phase. The beginning of an irreversible decomposition into $SrGe_2$ and SrH_2 can be monitored above 670 K.

All phases show a strong structural relationship. They crystallize in the orthorhombic crystal system. The main structural feature of the SrGe-H(D) system are parallel in plane zigzag Ge-Ge-chains. The hydrogen richest β-phase exhibits a three-fold superstructure perpendicular to the plane of the chains (*a* direction regarding to the Zintl phase). While the hydrogen poorest α-phase is similar to the parent Zintl phase, the intermediate α'-phase already shows the superstructure discussed above.

Upon hydrogenation tetrahedral Sr_4 voids are partially filled with hydrogen (deuterium) (α: 23%, α': 52%, β: 100% + additional H(D) position) . The *a* direction perpendicular to the chains is subsequently compressed (Zintl: 482 nm α: 476 nm α': 3x466 nm β: 3x397 nm). The *c* lattice parameter in direction of the chains only slightly changes from 417 nm to 406 nm. Since there is an overall increase of volume during the hydrogenation the second lattice parameter in the plane of the Ge-Ge-chains and perpendicular to the propagation of the chains needs to increase (Zintl: 1139 nm α: 1182 nm α': 1251 nm β: 1527 nm). The Figure shows *in situ* synchrotron diffraction data at 50 bar hydrogen pressure. The formation of three clearly distinct hydride phases can be seen as well as a decomposition step.

[1] H. Wu, W. Zhou, T. J. Udovic, J. J. Rush, T. Yildirim, *Phys. Rev. B: Condens. Matter Mater. Phys.* **2006**, *74*, 224101.
[2] M. Armbruster, M. Wörle, F. Krumeich, R. Nesper, *Z. Anorg. Allg. Chem.* **2009**, *635*, 1758-1766.
[3] H. Auer, P. Wenderoth, H. Kohlmann, *Z. Kristallogr. Supplement* **2015,** *S35*, 52.

Figure 1

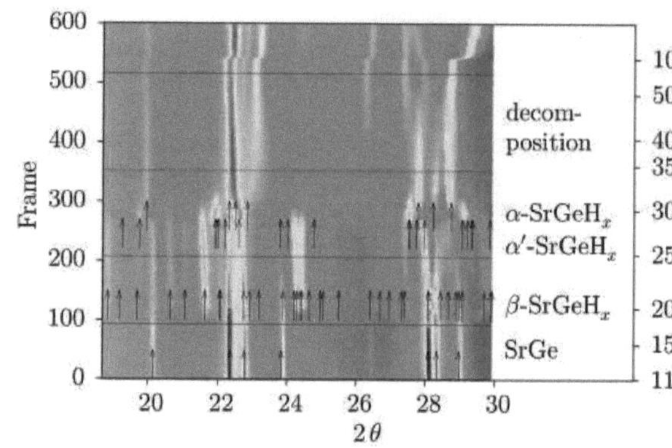

MS13-03
Simultaneous resonant x-ray diffraction measurement of polarization inversion and lattice strain in polycrystalline ferroelectrics

S. Gorfman[1], H. Simons[2], H. Choe[1], U. Pietsch[1], J. Jones[3]
[1]*University of Siegen, Siegen, Germany*
[2]*Technical University of Denmark, Department of Physics, Kopenhagen, Denmark*
[3]*North Carolina State University, Department of Materials Science and Engineering, Raleigh, United States*

Polycrystalline ferroelectrics offer a cost-effective and versatile route to such functionalities as large dielectric coefficients and strong electro-mechanical coupling. They are the backbone of many sensors, actuators, non-linear optics and power conversion devices. Understanding the microscopic mechanism of these functionalities is the key to the optimization and discovery of new materials with enhanced properties. However, existing techniques lack the capability to directly measure polarization-strain coupling across the relevant spatiotemporal regime.

This presentation shows the novel efficient technique [1]: it combines time-resolved and resonant synchrotron X-ray diffraction to directly and simultaneously measure the dynamic strain and ionic response of a polycrystalline ferroelectric to cyclic electric fields. The time-resolved X-ray diffraction implements a bespoke stroboscopic data acquisition system operating on the principle of a multi-channel analyser, and providing an ideal platform for the investigation of repetitive processes down to the nanosecond time scale [2]. The figure below shows the output of the stroboscopic data acquisition system, showing the time / field dependence of {111} powder diffraction profile under applied electric field. The structural inversion in individual crystallites of ferroelectric ceramics by the application of an electric field was observed through the Friedel pair contrast. This work also shows significant methodological innovation: because the hkl and -h-k-l powder rings exactly overlap with one another, the conditions at which the violation of Friedel's law can be observed using powder diffraction are very rare. These conditions can only be realized if: 1) a structure of individual powder grains can be actively inverted during the measurement and 2) the intensity of powder diffraction patterns is high enough to detect this small difference.

Our findings show that the investigation of the entire, structurally-derived functionality of ferroelectric powders and ceramics is now readily achievable. We demonstrate this approach using a high-energy (30 keV) X-ray beam on a material in which differentiating between multiple polarization reversal mechanisms remains a significant challenge: tetragonal $0.94{\cdot}BaTiO_3\text{-}0.06{\cdot}BiZn_{0.5}Ti_{0.5}O_3$

(BT-BZT) perovskite-based polycrystalline ferroelectrics. The approach offers significant opportunities for probing dynamics of intrinsic polarization using resonant X-ray scattering.

[1] S. Gorfman, H. Simons, T. Iamsasri, S. Prasertpalichat, D.P. Cann, H. Choe, U. Pietsch, Y. Watier, J.L. Jones. Scientific Reports, (2016) (under review)
[2] S. Gorfman, H. Choe, V. V Shvartsman, M. Ziolkowski, M. Vogt, J. Strempfer, T. Łukasiewicz, U. Pietsch, and J. Dec, Phys. Rev. Lett. 114, 097601 (2015)

Figure 1

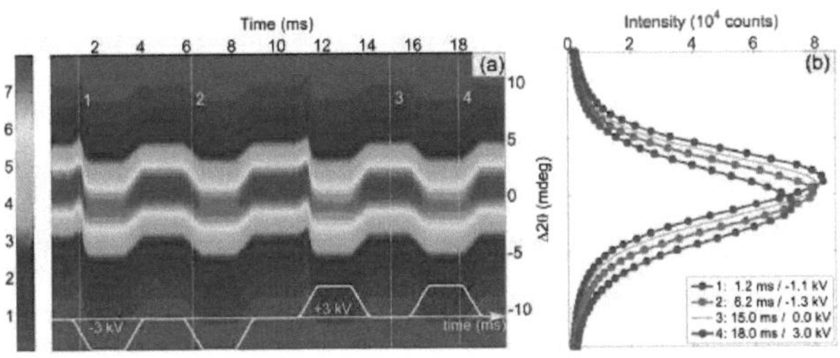

MS13-04
Time-resolved reciprocal space mapping probes the piezoelectricity in uniaxial $Sr_{0.5}Ba_{0.5}Nb_2O_3$ ferroelectric

H. Choe[1], U. Pietsch[1], J. Dec[2], U. Ruett[3], S. Gorfman[1]
[1]*University of Siegen, Department of Physics, Siegen, Germany*
[2]*University of Silesia, Institute of Materials Science, Katowice, Poland*
[3]*Deutsches Elektronen-Synchrotron DESY, Hamburg, Germany*

Ferroelectric crystals macroscopically deform under an applied electric field. It may be caused by two possible mechanisms. One is an intrinsic piezoelectricity which evolves from the atomic displacement and change of the bond lengths. The other one arises from the domain-wall motion or other mesoscopic phenomena. Many researches focused on pseudo-cubic ferroelectrics such as $PbZr_{1-x}Ti_xO_3$ (PZT), where domains share both ferroelectric and ferroelastic features. On the other hand, $Sr_xBa_{1-x}Nb_2O_6$ (SBN) crystals attract an interest as a model system of uniaxial ferroelectrics. Its symmetry (paraelectric and ferroelectric phase) allows for 180° ferroelectric domains only, while the formation of ferroelastic domains is forbidden. In this work, we studied the nature of relationship between ferroelectricity and piezoelectricity in SBN by using stroboscopic time-resolved X-ray diffraction [1]. This technique (described in [1,2]) enables to record a rocking curve of selected Bragg reflections as function of time and applied periodic electric field.

In previous measurements [1], we discovered the new hidden-in-the-bulk mechanism of piezoelectric activity, which suggests that it may arise from the correlation between average domain sizes and their lattice parameters. The observed field-induced rocking curves under sub-coercive fields showed the asymmetric behaviour. We suggested that the rocking curve may sum up for two components moving in the opposite directions. However, it was difficult to separate the rocking curve profiles into two components. For this reason, we tried to approach to the reciprocal space mapping. It can help us for separating the asymmetric Bragg peak and give more information.

In this work, we implemented stroboscopic technique to acquire time-resolved reciprocal space maps (RSM-STROBO) under an alternating electric field. The selected reciprocal space region was mapped by collecting the series of 2q scans at various values of w (rocking) angle. All RSMs were collected stroboscopically.

Figure 1 shows the observed RSMs on [-8 0 6]* point of the reciprocal space are asymmetric on opposite electric fields. We evaluated (1) strain-field hysteresis loops and (2) the Bragg peak separation in RSMs under sub-coercive fields. We will use these data to discuss the interaction between the field-induced 180° ferroelectric domains (whose intrinsic piezoelectric constants have opposite signs).

1. Time-resolved X-ray diffraction reveals the hidden mechanisms of high piezoelectric activity in uniaxial ferroelectrics, Gorfman S, Choe H, Shvartsman V.V., Ziolkowski M., Vogt M., Strempfer J., Lukasiewicz T., Pietsch U., Dec J., Phys. Rev. Lett., 114, 097601(2015)
2. Sub-microsecond X-ray crystallography: technique, challenges and applications for materials science, Gorfman S, Crystallography Reviews, 20(3), 210 (2014)

Figure 1 Top: Reciprocal space maps of -8 0 6 Bragg peak collected from $Sr_{0.5}Ba_{0.5}Nb_2O_6$ under applied sub-coercive field, showing dynamics of two peak components. Bottom: Dielectric P-E hysteresis loop.

Figure 1

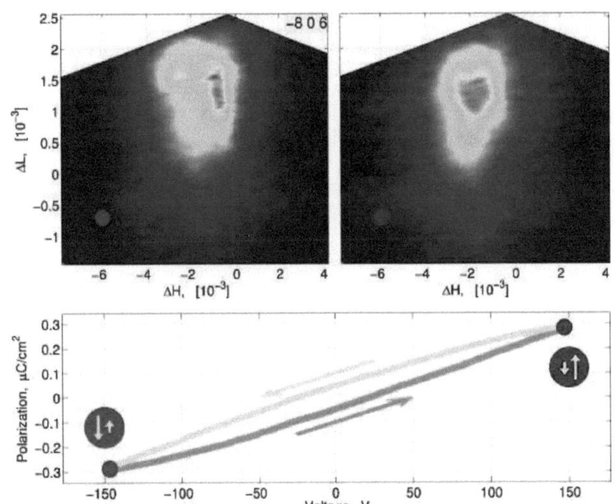

MS14 - Aperiodic and periodic complex materials

MS14-01
Structure and properties of refractory high-entropy alloys

W. Steurer[1], S. Maiti[1]
[1]*ETH Zürich, Zürich, Switzerland*

Refractory metal high-entropy alloys (HEAs) have potential applications as high-temperature materials. For that purpose they should show long-term stability at high temperatures without significant changes in their material properties. However, according to G = H-TS, entropy-stabilized phases will become unstable below a critical temperature leading to phase separation and the formation of characteristic microstructures. This may be welcome or not, in any case it will significantly change the properties of the HEA. Therefore, we studied the atomic ordering and mechanical properties of refractory HEAs (NbTaHfZr, NbTaHfZrTi, NbTaMoW, NbTaMoWV) as a function of annealing time and temperature. We applied temperatures up to eighty percent of the melting temperature and annealing times up to one week. The samples were characterized by X-ray and neutron diffraction, high-resolution electron microscopy and atom probe tomography. The resulting ordering phenomena were further modeled by molecular dynamics calculations. It was found that one class of samples (NbTaMoW) did not show any ordering phenomena as a function of time and temperature while the other (NbTaHfZr) exhibited local clustering, finally leading to phase separation. Consequently, due to its outstanding strength and significant ductility, which further increases by a factor of three for sample dimensions on the micro- and nano-scale [1], and its stability, NbTaMoW appears to be an excellent material for aerospace applications as well as for micro- and nano-electromechanical systems (MEMS and NEMS).

[1] Zou, Y., Maiti, S., Steurer, W., Spolenak, R., Acta Mater. **65** (2014) 85-97.

MS14-02
The oxidic dodecagonal quasicrystal and its approximant: X-ray analysis of the atomic structure.

S. Roy[1], M. Trautmann[2], R. Hammer[2], W. Adeagbo[2], S. Förster[2], F. Schumann[2], K. Mohseni[1], H. Meyerheim[1], W. Hergert[2], W. Widdra[2,1]
[1]*Max-Planck-Institut f. Mikrostrukturphysik, D-06120 Halle/S., Germany*
[2]*Institut f. Physik, Martin-Luther-Universität Halle-Wittenberg, D-06099 Halle, Halle/S., Germany*

Very recently a new type of quasicrystal has been discovered which is based on an oxide phase, closely related to Barium Titanate (BTO) [1]. The two-dimensional oxide quasicrystal (OQC) and its periodic approximant (AP) were grown in situ under ultra-high-vacuum conditions on Pt(111) and investigated by scanning tunneling microscopy (STM) which provided clear evidence for the presence of different tilings. Only one kind of protrusion could be imaged by STM, which are arranged in a 2D pattern identified as Stampfli-Gähler [2,3] tiling for the OQC and as Kepler tiling [4] for the AP, respectively. The tilings are composed of several elements, namely squares, triangles, and rhombs, which could be identified in the STM images and were fundamental to distinguish the two phases. Despite the precise analysis of the tilings, their metric and the approximate stoichiometry of the OQC and the AP, no detailed structural information is available on the atomic scale. For instance, a straightforward assignment of the protrusions to one atomic species is not possible. To this end we have carried out a surface x-ray diffraction (SXRD) study to elucidate the geometric structure of the AP which also allows the identification of the very closely related OQC structure.

Figure 1 shows the z-projected charge density contour plot ρ(x,y) calculated on the basis of 44 reflections of type (hk0). Four unit cells are displayed to indicate the tiling. Superimposed are the approximate positions of the atoms (labelled by 1-11) derived from the least squares fit (R=13%). We find that the protrusions observed in STM are related to titanium atoms (blue, 4, 5), which are surrounded by three oxygen atoms (red, 6-11) approximately forming a triangle. Furthermore, barium atoms (yellow, 1-3) are located at the origin of the unit cell with plane group p2 corresponding to the center of one type of rhomb. Further barium atoms are located in the vicinity of two out of the four edges of the squares (rectangles) leading to a stoichiometry which can be written as $Ba_4 Ti_4 O_{12}$ per unit cell, if no vacancies are present. This structure model also allows to develop a structure model for the very closely related OQC structure which is composed of identical tiling elements and which is derived from the AP by only minor changes of their arrangement. The experimental results are confirmed by ab-initio calculations.

Support by the German Science Foundation (DFG) through SFB 762 is gratefully acknowledged. HLM, RH, FS and KM thank the staff of the ESRF for hospitality and support during their visit in Grenoble.

References:

[1] S. Förster, K. Meinl, R. Hammer, and W. Widdra, Nature 502, 215 (2013)
[2] P. Stampfli, Helv. Phys. 59, 1260 (1986)
[3] F. Gähler in: Quasicrystalline Materials (World Scientific, 1988)
[4] J. Kepler: Harmonices Mundi (1619)

Figure 1: Calculated z-projected charge density contour plot ρ(x,y) showing four unit cells. Peaks in ρ(x,y) are identified with barium (green), titanium (blue) and oxygen atoms (red). The Kepler tiling is indicated by the solid black lines.

Figure 1

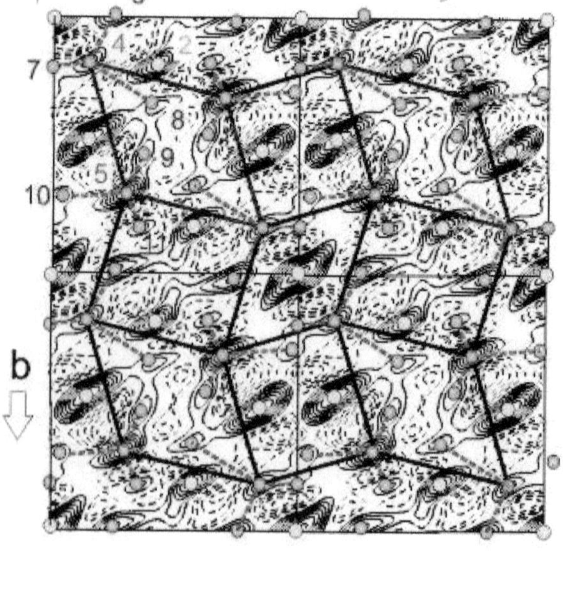

MS14-03

Simple Systems, Complex Structures

J. Dshemuchadse[1], M. Engel[1], P. F. Damasceno[2,3], C. L. Phillips[4], S. C. Glotzer[1,2,5]
[1]*University of Michigan, Department of Chemical Engineering, Ann Arbor, Michigan, United States*
[2]*University of Michigan, Applied Physics Program, Ann Arbor, Michigan, United States*
[3]*University of California, Department of Cellular and Molecular Pharmacology, San Francisco, California, United States*
[4]*Argonne National Laboratory, Argonne, Illinois, United States*
[5]*University of Michigan, Department of Materials Science and Engineering, Ann Arbor, Michigan, United States*

Complex structures can be found in all classes of materials. In macromolecular crystals they exist per definition, as the smallest building blocks are already large enough to form unit cells containing hundreds or thousands of atoms. Modulated structures are ubiquitous in inorganic systems such as oxides, and are described by either excessively large unit cells or in higher dimensional space. A more surprising class of giant unit cell structures is that of complex intermetallics, where as few as two components can form unit cells that contain hundreds, thousands, or even tens of thousands of atoms (*e.g.*, [1]). More recently, complex structures are also being discovered in soft-matter systems (*e.g.*, [2]), where previously only relatively simple structures had been observed.

We would like to understand why complex structures form and how they do it. Observing structure types in soft-matter systems that are well-known in intermetallics seems to defy our understanding of why they form - usually the main argument citing electronic factors. The process of how particles self-assemble into large unit cells whose dimensions far exceed their interaction ranges is still a mystery, let alone how they arrange in structures that can only be described as periodic in (3+*n*)-dimensional space.

In order to better understand the formation and stabilization of complex structures, we study different effective interaction potentials and analyse the structures that they self-assemble. We use the highly parallel molecular dynamics code HOOMD-blue [3] to simulate a wide variety of such systems, tuning different system parameters and observing the variation in the resulting structures. This information is then translated into phase diagrams that illustrate the stability regions of different crystal structure types, which are in turn evaluated with respect to specific factors that influence structure formation.

We have discovered a plethora of structures, ranging from the expected and well-known sphere packings and other simple structure types all the way to giant-unit cell structures, as well as aperiodic crystals [4]. We will present selected structure types - those that are ubiquitous in atomic systems, as well as previously unknown ones - and attempt to systematize the resultant phase diagrams, ultimately extracting the factors that govern the formation of these different structures and geometries.

[1] J. Dshemuchadse, D. Y. Jung, W. Steurer, Acta Crystallogr. B 67, 269-292 (2011).
[2] S. Lee, C. Leighton, F. S. Bates, Proc. Natl. Acad. Sci. U. S. A. 111, 17723-17731 (2014).
[3] J. A. Anderson, C. D. Lorenz, A. Travesset, J. Comp. Phys. 227, 5342-5359 (2008). J. A. Anderson, S. C. Glotzer, http://arXiv.org/abs/1308.5587 (2013). http://codeblue.umich.edu/hoomd-blue
[4] M. Engel, P. F. Damasceno, C. L. Phillips, S. C. Glotzer, Nature Mater. 14, 109-116 (2015).

MS14-04

FeOCl, magnetic order and crystal structure at low temperature

A. Schönleber[1], J. Zhang[1], S. van Smaalen[1], P. G. Reuvekamp[2], R. K. Kremer[2], A. Senyshyn[3]
[1]*University of Bayreuth, Laboratory of Crystallography, Bayreuth, Germany*
[2]*Max Planck Institute for Solid State Research, Stuttgart, Germany*
[3]*Forschungs-Neutronenquelle Heinz Maier-Leibnitz, FRM-II, Garching, Germany*

The metal(III) oxyhalide FeOCl is is built by slabs consisting of Fe_2O_2 bilayers enclosed by layers of Cl atoms, the interaction between those slabs is of van der Waals type. At room temperature the symmetry is orthorhombic, space group *Pmmn* [1,2]. A phase transition towards antiferromagnetic order was observed by means of Mössbauer spectroscopy [3] and other techniques [4-6] while the reported values for T_N vary in the range of 80 - 92 K. Two different models for the magnetic superstructure are proposed, an incommensurate cycloïdal one [4] and a commensurate spiral one [6]. Both models are based on orthorhombic symmetry.

To describe and better understand the magnetic and nuclear superstructures, we have performed low-temperature single crystal X-ray diffraction at beamline D3 (Hasylab/DESY, Hamburg, Germany) and low-temperature neutron powder diffraction at instrument SPODI (FRM2, Garching, Germany).

The magnetic phase transition occurs at T_N = 82.0(2) K and is of second order. It is accompanied by a symmetry reduction from orthorhombic at higher temperatures towards monoclinic at lower temperatures [5], indicating a strong magnetoelastic coupling. This monoclinic lattice distortion is also reported for the other *M*OCl compounds VOCl and CrOCl [7,8]. The magnetic and structural modulation wave vector varies in an intermediate temperature range as function of temperature, pointing to an incommensurate nature of the modulation. In our contribution we will propose models for the magnetic and nuclear structures applying magnetic superspacegroup symmetry [9,10].

[1] S. Goldsztaub (1934) *C. R. Hebd. Seances Acad. Sci.* **198** 667-669.
[2] M. D. Lind (1970) *Acta Crystallogr. B* **26** 1058-1062.
[3] R. W. Grant, (1971) *J. Appl. Phys.* **42** 1619-1620.
[4] A. Adam & G. Buisson (1975) *Phys. Stat. Sol. A* **30** 323-329.
[5] J. Zhang et al. (2012) *Phys. Rev. B* **86** 134428.
[6] S. R. Hwang et al. (2000) *Phys. Rev. B* **62** 14157-14163.
[7] A. Schönleber et al. (2009) *Phys. Rev. B* **80** 064426.
[8] J. Angelkort et al. (2009) *Phys. Rev. B* **80** 144416.
[9] V. Petříček, J. Fuksa & M. Dušek (2010) *Acta Crystallogr. A* **66** 649-655.
[10] J. M. Perez-Mato et al. (2012) *J. Phys.: Condens. Matter* **24** 163201.

MS15 - Materials: Properties and applications

MS15-01

Empirical electronic polarizabilities of ions in oxides and oxysalts for the determination of refractive indices

R. D. Shannon[1], R. X. Fischer[2]
[1]*University of Colorado, Geological Sciences, Colorado, United States*
[2]*Universität Bremen, Fachbereich Geowissenschaften, Bremen, Germany*

A set of electronic polarizabilities of ions is presented allowing the prediction of mean refractive indices (RI) with high accuracy showing deviations from observed values better than 2% for most oxides and oxysalts. The concept is based on the additivity rule for electronic polarizabilities of ions assuming that the total electronic polarizability of a compound can be calculated as a simple linear combination of individual ion electronic polarizabilities. Total polarizabilities of crystalline compounds are calculated from the mean RI's using the relationship $\alpha=(n^2-1)V_m/(4\pi+(4\pi/3-c)(n^2-1))$ from Eggleton (1991) with the total polarizability α, the refractive index *n*, the molar volume V_m and *c* as an electronic overlap factor empirically determined by Eggleton to be 2.26 for silicates.

We have calculated the polarizabilities for 80 cations in various coordinations representing 66 elements, for 4 anions (O, Cl, F, OH), and for NH_4^+ and 5 H_xO_y species by least squares procedures based on ~ 2500 data sets of RI's of ~1500 compounds. While the polarizabilities of the cations are represented by single parameters, the polarizabilities of the anions are calculated according to $\alpha_- = \alpha_-^o \cdot 10^{-N_o/X}$ with $X = V^e$, α_- = anion polarizability, α_-^o = free-ion polarizability, *V* = anion molar volume, and the exponent *e* empirically found to be 1.20 for a best fit.

In order to describe the relationship between cation polarizabilities and their coordination numbers (CN) we modified the light-scattering model by Jemmer et al. (1998) to give the expression $\alpha(CN) = (a_1+a_2CNe^{-a_3CN})^{-1}$, with a_i being least squares parameters (Shannon and Fischer (2006). Using $n = sqrt(4\pi\alpha/((2.26-4\pi/3)\alpha+V_m)+1)$ the mean refractive index can be calculated from the chemical composition and the polarizabilities of ions determined here. Thus, the RI of minerals even having a complex composition like elbaite ($Na_{0.74}Ca_{0.01}Al_{8.11}Li_{1.05}Fe_{0.01}Mn_{0.01}(BO_3)_3Si_{5.94}O_{18}(OH)_{3.63}O_{0.32}F_{0.05}$) can be estimated with high accuracy yielding n(calc) = 1.636 in excellent agreement with n(obs) = 1.635.

Systematic deviations of certain classes of compounds are discussed.

References

Eggleton, R.A. (1991) Can. Mineral. 29, 525-532.
Jemmer, P., Fowler, P.W., Wilson, M. and Madden, P.A. (1998).J.Phys.Chem. A102, 8377-8385.
Shannon, R.D. and Fischer, R.X. (2006). Phys. Rev. B73, 235111, 1-28.

MS15-02

Synthese und magnetische Eigen-schaften von 1D- und 2D-Thio- und Selenocyanat-Koordinationspolymeren

C. Näther[1], S. Wöhlert[1], J. Boeckmann[1], J. Werner[1], Z. Tomkowicz[2], M. Rams[2]
[1]*Universität Kiel, Institut für Anorganische Chemie, Kiel, Germany*
[2]*Universität Krakow, Krakau, Poland*

In den vergangenen Jahren wurden verstärkt Untersuchung-en zur Synthese und den magnetischen Eigenschaften von Einzelkettenmagneten (single chain magnets; SCMs) durchgeführt. Hierbei handelt es sich um Verbindungen, welche unterhalb der sog. Blocking-Temperatur eine langsame Relaxation der Magnetisierung zeigen und daher in der Lage sind ein magnetisches Moment dauerhaft zu speichern. Zur Darstellung derartiger Verbindungen müssen Kationen mit hoher magnetischer Anisotropie durch Liganden zu Ketten verbunden werden, bei denen ein großes Verhältnis zwischen den Intra- und Interketten-Wechselwirkungen auftritt. In diesem Zusammenhang haben wir eine Verbindung der Zusammensetzung $[Co(NCS)_2(Pyridin)_2]_n$ dargestellt. In dieser sind die Co(II)-Kationen oktaedrisch von zwei Pyridin-Liganden sowie von zwei S- und zwei N-koordinierenden Thiocyanat-Anionen umgeben und durch Paare der anionischen Liganden zu Ketten verbunden. Die Verbindung zeigt eine langsame Relaxation der Magnetisierung, welche auf Einzelkettenmagnetismus hindeutet.

Ausgehend von diesem Befund wurde der Einfluss einer chemischen und einer strukturellen Modifizierung auf die Parameter, welche die Leistungsfähigkeit derartiger Materialien beschreiben, systematisch untersucht. Dabei ergab sich, dass ungeachtet der Tatsache, dass die Co(II)-Verbindungen immer aus den gleichen Kettenstrukturen bestehen, sich alle hinsichtlich ihrer magnetischen Eigenschaften in zwei Gruppen einteilen lassen. In der Gruppe mit einem antiferromagnetischen Grundzustand sind die, in den Experimenten beobachteten Relaxationen auf die von Einzelketten zurückzuführen, während dies in den Verbindungen mit einem ferromagnetischen Grundzustand nicht der Fall ist. Dabei ergab sich kein einfacher Zusammenhang zwischen dem magnetischen Grundzustand sowie den Interketten-Abständen oder der Packung der Ketten im Kristall.

Weitere Untersuchungen belegen, dass der Austausch der Thio- gegen Selenocyanat-Anionen zu einer Erhöhung der Energiebarriere für die Spinrelaxation führt. Im Gegensatz zu den Co(II)-Thiocyanat-Koordinationspolymeren wird in den entsprechenden Verbindungen mit Fe(II)-Thiocyanat keine langsame Relaxation der Magnetisierung beobachtet. Diese kann jedoch induziert werden, wenn die Thio- gegen Selenocyanat-Anionen ausgetauscht werden. Ob dieses Verhalten auf eine Erhöhung der Intraketten-Wechselwirkungen oder einer Erhöhung der Anisotropie zurückgeführt werden kann, müssen weitere Untersuchungen zeigen.

MS15-03

Diffusion and Ion Transport in Oxide Glasses

H. Mrhrer[1]
[1]*Institut für Materiaklphysik, Physik, Münster, Germany*

Studies of ionic conduction and of radiotracer diffusion in oxide glasses from the Münster laboratory are summarized. In contrast to ionic conduction tracer diffusion is element-specific. The tracer experiments comprise measurements of Na and Rb diffusion in single and mixed alkali borate glasses and of Na and Ca diffusion in soda-lime silicate glasses as function of temperature and of composition. Diffusivities and ionic conductivities are Arrhenius activated and are compared via the Nernst-Einstein relation. Haven ratios for both silicate and borate glasses are deduced. Viscosity diffusion coefficients are deduced from viscosity data for a soda-lime silicate glass via the Stokes-Einstein relation. Viscosity diffusion is considerably slower than diffusion of the network modifiers Na and Ca indicating that the motion of the network formers and network modifiers are decoupled. The ionic conductivity in soda-lime glasses is dominated by Na ions. The mixed alkali effect of Na-Rb borate glasses was studied by conductivity measurements and Na and Rb tracer diffusion and permit a qualitative understanding of the mixed alkali effect.

The influence of hydrostatic pressure on diffusion and ionic conduction was studied for borate glasses. The activation volume of ionic conduction is relatively small on the Na rich side, it is larger on the Rb rich side. The activation volumes in the single alkali glasses correlate with the ionic volumes of Na or Rb ions. The activation volume increases to a shallow maximum in the mixed alkali range.The activation volumes of tracer diffusion are significantly larger than those of ionic conduction in both, single and mixed-alkali glasses. The tracer diffusivities have a stronger pressure dependence than the ionic conductivity. This is equivalent to a pressure-dependent Haven ratio. We attribute these observations to a varying degree of collectivity of the ionic jump process. Whereas ionic corresponds to a chain-like motion of several alkali ions, the transport of tracer atoms is almost random.

MS15-04
A structural perception about intrinsic point defects in kesterite type chalcogenides

S. Schorr[1,2], G. Gurieva[1], L. E. Valle Rios[2], K. Neldner[1]
[1]*Helmholtz-Zentrum Berlin für Materialien und Energie, Berlin, Germany*
[2]*Freie Universität Berlin, Institut für Geologische Wissenschaften, Berlin, Germany*

The understanding of the interplay between structural, chemical and electronic properties of kesterite type compound semiconductors, applied as absorber materials in photovoltaic devices, can give crucial informations aiding the continuous improvement of device efficiencies. Various reasons are discussed about efficiency losses in the bulk material of kesterite type based solar cells. These absorber layers, which are Cu2ZnSn(S, Se)4, exhibit in general a off-stoichiometric composition (Cu-poor, Zn-rich). The presence of secondary phases and deep defect levels caused by intrinsic point defects are efficiency limiting factors.

The present study reports a detailed crystallographic investigation about cationic point defects in off-stoichiometric Cu2ZnSnS4 and Cu2ZnSnSe4. Both compounds crystallize in the kesterite type structure (space group I-4). The basis of the systematic study are series of powder samples synthesized by solid state reaction. Their chemical composition has been determined by WDX spectroscopy (electron microprobe analysis). Due to the isoelectronic character of Cu+ and Zn2+ it is necessary to apply neutron powder diffraction to investigate the distribution of the three cations on the four cationic sites of the kesterite type structure. Rietveld refinements and the average neutron scattering length analysis method have been used to evaluate the neutron diffraction data.

The kesterite type structure can be described by a stacking sequence of cation layerd Cu/Sn - Cu/Zn - Sn/Cu - Cu/Zn - Cu/Sn perpendicular to the crystallographic z-direction. An off-stoichiometric composition, for instance Cu-poor, originates from the prospensity of the structure to stabilize copper vacancies, the charge balancees beeing commonly insured by appropriate substitutions on the cationic sites. If the oxidation states of the cations and anions are retained going from stoichiometric to an off-stoichiometric composition, a number of specific substitutions can be envisioned to account for the charge balance in the off-stoichiometric material. These substitutions lead to the formationof point defects in the crystal structure. The presented study shows, that the point defects, expected by this model, are found experimentally in the synthesized compounds. This demonstrates the ability of the kesterite type phase to tolerate large deviations from stoichiometric composition without collaps of the structure

MS16 - Spectroscopic methods in crystallography

MS16-01
Spectroscopic methods in Crystallography - A Supplementary benefit to solve crystallographic research questions - An overview

M. Fechtelkord[1]
[1]Ruhr-Universität Bochum, Institut für Geologie, Mineralogie und Geophysik, Bochum, Germany

A wide variety of analytical methods such as diffraction methods, electron microscopy, thermal analysis and spectroscopy is used in the different research fields of crystallography, each method contributing small pieces to the large puzzle of unsolved scientific questions. Often several methods fail in the investigation of structural aspects as in the case of amorphous materials (e.g., glasses and ceramics) or microcrystalline materials and for specific structural questions, e.g., the location and dynamics of hydrogen atoms in minerals. Spectroscopic methods can be of an ideal supplementary benefit here.

For example, in Nuclear Magnetic Resonance all nuclei, that possess a magnetic moment ($I > 0$), are able to provide detailed information about their local environments as local probe e.g., about bond angles, neighbouring atoms (1st and 2nd coordination sphere), the local symmetry, the coordination number, as well as being sensitive to dynamic processes. This is due to the fact that, beside the outside static magnetic field, there are small internal local fields that contain the appropriate structural information, which influence the effective magnetic field at the nucleus.

Similarly, Mössbauer spectroscopy has an extremely fine energy resolution and can detect even small deviations in the nuclear environment of the relevant atoms. Typically, there are three types of nuclear interactions that are detected, the isomeric shift, the quadrupole splitting and magnetic Zeeman interaction.

Infrared (IR) and Raman spectra result from dipolar transitions between vibrational energy states. Thus, vibrations of molecules, atomic groups and the crystal lattice can give structural information about the semi-range order of the atomic environment.

Finally, optical (UV-VIS) spectroscopy of transition metal ion bearing solids is a widespread experimental technique to investigate the coordination symmetry and crystal field strength of these transition metals.

In conclusion, spectroscopy is a powerful complementary method to X-ray structure analysis and to many other methods used in crystallography, probing short range structural effects rather than long range order.

MS16-02
Coherence spectroscopy as a tool for the characterization of local environments of rare-earth doped materials

N. Kunkel[1], P. Goldner[1]
[1]IRCP - Chimie ParisTech , Materials for Photonics and Opto-Electronics, Paris, France

Coherence spectroscopy techniques are usually applied in quantum information technologies and related fields. Here, materials doped with trivalent rare earth ions that show remarkable narrow homogeneous linewidth are of great interest for the application in, e.g., quantum storage and processing [1].

However, these techniques, such as photon echo measurements and spectral hole burning, could also complement the characterization of rare-earth doped materials in general. The homogeneous linewidths of the activator ions, which can be determined by photon echo experiments, can serve as a very sensitive probe for the local environment. Different pulse techniques and temperature and magnetic field dependence allow the observation of perturbations such as defects, fluctuating spins or low frequency vibrational modes that suggest the presence of local disorder.

In order to relate the coherence properties measured with high resolution laser spectroscopy with structural properties of the local environment we used Eu^{3+}-doped Y_2O_3 and the Eu^{3+} $^5D_0$ - $^5F_0$ transition as a model system [2]. The properties of different ceramic samples after different thermal treatments under different atmospheres were studied both by coherence spectroscopy and conventional methods such as, e.g., XRD, EPR, luminescence and thermoluminescence spectroscopy.

It was shown that the homogeneous linewidth as well as its temperature dependence is strongly related to the thermal treatment under different atmospheres as well as the purity of the starting materials. Thus, coherence spectroscopy can deliver information on the type and concentration of defects, such as colour centres, as well as impurities, such as d-metal or lanthanide ions and serve as a supplementary characterization tool.

References

[1] F. Bussières, C. Clausen, A. Tiranov, B. Korzh, V. B. Verma, S. W. Nam, F. Marsili, A. Ferrier, Ph. Goldner, H. Herrmann, C. Silberhorn, W. Sohler, M. Afzelius, N Gisin, Nature Photonics 2014, 8, 775.
[2] N. Kunkel, A. Ferrier, C. W. Thiel, M. O. Ramírez, L. E. Bausá, R. L. Cone, A. Ikesue, Ph. Goldner, Appl. Phys. Lett. Mat. 2015, 3, 096103 and unpublished results.

Figure legend: Temperature dependence of the homogeneous linewidth in different Y_2O_3:Eu^{3+} samples.

Figure 1

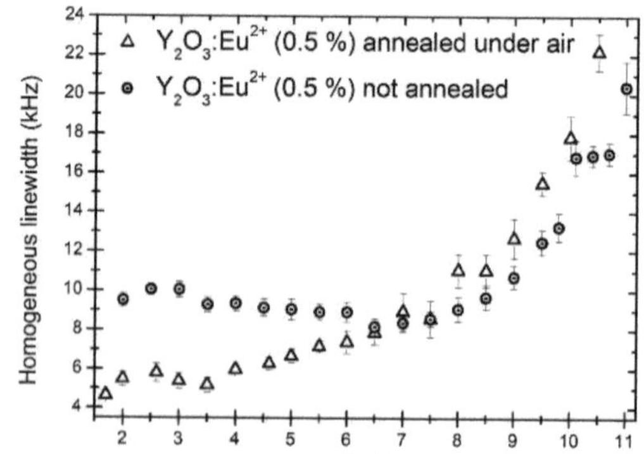

MS16-03

The incorporation of tungsten (VI) into hematite and goethite: a chemical and spectroscopic study combining natural and synthetic iron oxides developed from precursor ferrihydrite

S. Kreißl[1], R. Bolanz[2], J. Göttlicher[3], R. Steininger[3], M. Tarassov[4], G. Markl[1]
[1]*Universität Tübingen, Mineralogie & Geodynamik, Tübingen, Germany*
[2]*University of Jena, Institute of Geosciences, Jena, Germany*
[3]*Karlsruhe Institute of Technology, ANKA Synchrotron Radiation Facility, Eggenstein-Leopoldshafen, Germany*
[4]*Bulgarian Academy of Sciences, Institute of Mineralogy and Crystallography, Sofia, Bulgaria*

Hematite (hem; α-$Fe^{3+}{}_2O_3$) and goethite (goe; α-$Fe^{3+}OOH$) are common mineral phases in a wide variety of geo-environments e.g. primary hydrothermal vein formations or even supergene weathering products. Especially botryoidal hematite and goethite, crystallizing from the precursor phase ferrihydrite (fh; $Fe^{3+}{}_{10}O_{14}(OH)_2$), are known from hydrothermal systems. To date it is still unclear to what extent the decomposition of primary Fe-phases and the reprecipitation or even primary crystallization of Fe-oxi/-hydroxides link to either supergene or hypogene events. Tungsten (W) which frequently originate from late stage magmatic to early hydrothermal processes expects to be an indicator for a hypogene influence in hydrothermal systems. Since botryoidal hem and goe show a keen differentiated growth zonation with fluctuating W concentrations on macroscopic scale, they can record the influence of hypogene or supergene fluids on their hydrothermal systems.

If the nature and specification of W, to date only hypothetically present within the hem and goe crystal structures, has not approved, the time of its coprecipitation stays unclear and fluctuating concentrations cannot been linked to an either hypogene or supergene origin. The objective of this study was to verify the incorporation mechanism of W^{6+} and at which point of time the W species, adsorbed on the fh crystal surface, transfer into the hem and goe structures. In addition, the study shows the partitioning between W concentrations within the fluids and the resulting hem or goe.

Investigations on natural botryoidal hem and goe specimens with variable genetic connections to tungsten deposits are combined with synthetic W-spiked iron oxide samples. All samples formed from the same precursor phase ferrihydrite. Chemical data by electron microprobe (EMPA) and total reflection X-ray fluorescence (TXRF) is combined to spectroscopic data by X-ray absorption fine structure (XAFS) measurements, X-ray diffraction (XRD) and Raman spectroscopy (RS).

Natural samples, not associated to W ore deposits, reach higher W concentrations for hem (0.8 mol%) than for goe (0.12 mol%). Samples from a tungsten deposit and the synthetic samples, both grown in the presence of excess W, show the opposite trend: natural samples 5.5 mol% in goe 2.15 mol% in hem; synthetic samples 7 mol% in goe and 1.3 mol% in hem. Supporting the structural incorporation of W into the hem and goe lattices, XRD analyses indicate no presence of separate W-phases. With increasing W, refined lattice parameters indicate a rising structural disorder only in the synthetic goe' structures. However, Raman results show a rising structural disorder for synthetic hem and goe, indicating an increase of Fe vacancies and no evidence for a deprotonation of the goe structure. The natural and synthetic goe and hem show striking similarities at extended XAFS spectra, at which models suggesting W^{6+} resides in all samples on the Fe^{3+} position.

According to these spectroscopic and chemical approaches, hem and goe from synthetic, W adsorbed ferrihydrite incorporate W^{6+} by a direct isomorphic exchange against Fe^{3+}. Charge balance constitutes by the formation of two Fe^{3+}-vacancies in close proximity to the incorporated W^{6+} in addition to a protonation of the structures. Natural samples show the same exchange mechanism, which postulates an uninterrupted process of W adsorption to ferrihydrite and additionally a W signature preservation after its transformation at least for hematite.

MS16-04

Structural alteration across the morphotropic phase boundary in xBiMg$_{0.5}$Ti$_{0.5}$O$_3$-(1-x)PbTiO$_3$: combined pair-distribution-function and Raman-scattering analysis

K. Datta[1], R. Neder[2], J. Chen[3], B. Mihailova[1]
[1]*University of Hamburg, Department of Earth Sciences, Hamburg, Germany*
[2]*Friedrich-Alexander-Universitat Erlangen-Nurnberg, Department of Crystallography and Structure Physics, Erlangen, Germany*
[3]*University of Science and Technology Beijing, Department of Physical Chemistry, Beijing, China*

Morphotropic phase boundary (MPB) is an important theme in the research of ferroelectric materials since materials often exhibit enhanced physical properties at MPB. Although an established understanding of creating MPB behaviour relies on invoking instability of average crystal structure or a low-symmetry monoclinic phase in the system driven by composition [1], the actual mechanism is far more complex and heavily dependent on the local ordering of the cations as evidences found in recent diffuse scattering studies [2] on either lead-free or lead-based based systems. We have studied an important and novel ferroelectric system **xBiMg$_{0.5}$Ti$_{0.5}$O$_3$-(1-x)PbTiO$_3$** (BMT-PT), which was first reported in 2004 with MPB characteristics [3], through total scattering and Raman scattering method to understand its structural evolution at the atomic level as a function of composition and temperature in the vicinity of the MPB.

Total scattering technique which is essentially the analysis of the pair distribution function (PDF) of a system, provides crucial structural information at the microscopic level which are not easily available from conventional structural analysis, like Rietveld refinement. Neutron powder diffraction experiments were carried out on nine different compositions of xBMT-PT in the range $0.10 \leq x \leq 0.70$ and neutron PDFs were analysed through RMC simulations to extract the behaviour of individual cations. A categorical change in terms of favored local polarization directions of the cations as a function of composition was seen including an atomic ordering between Ti and Mg, and in particular, for the MPB compositions an enhancement in the randomness of the directions was detected which could be linked to the reported average structure as well as the properties of the system. Complementary Raman scattering data were collected at ambient and non-ambient temperature as a function of composition which further corroborates some of the results of the PDF analysis.

Reference: [1] R. Guo, L. E. Cross, S-E. Park et al Phys. Rev. Lett. 84, 5423-5426 (2000). [2] D. J. Goossens , ISRN Materials Science, 2013, 107178/1-17 (2013). [3] C. A. Randall et al, J. Appl. Phys. 95, 3633 (2004). [4] K. Datta, A. Richter, M. Goebbels, R. B. Neder and B. Mihailova Phys. Rev. B 90, 064112 (2014).

MS17 - Phase transitions and dynamic phenomena

MS17-01
The Power of Powder Diffraction in In Situ Studies of the Solid State

T. Runčevski[1]
[1]*University of California, Department of Chemistry, Berkeley, United States*

X-ray powder diffraction (XRPD) is as a powerful technique for structural analysis of crystalline compounds which do not grow as single crystals, but form a polycrystalline powder. If XRPD is applied under non-ambient conditions, or if data are collected at consecutive time intervals, it can be readily turned into a powerful method for in situ following, visualizing and ultimately explaining many chemical reactions and solid state processes. In the scope of the presented talk, the applicability of in situ XRPD in chemistry and physics is presented via several case studies of a) photo-induced [2+2] chemical reactions, b) thermo-induced phase transitions and c) gas adsorptions in porous materials. Finally, a perspective of this technique is given.

MS17-02
High-temperature PXRD studies of the phase transition kinetics of $CsGaSe_2$

D. Friedrich[1], M. Schlosser[1], A. Pfitzner[1]
[1]*Universität Regensburg, Regensburg, Germany*

Among the known group 13 chalcogenometallates with the composition MTQ_2 (M = alkali metal, T = triel, Q = chalcogen), $CsGaSe_2$ is the only example showing a reversible phase transition at elevated temperatures [1]. The most fascinating fact about this phase transition is a complete structural reconstruction from two-dimensional layers $^2_\infty[Ga_4Se_8^{4-}]$ in the low temperature phase $CsGaSe_2$-*mC*64 to one-dimensional chains $^1_\infty[GaSe_2^-]$ in the high-temperature phase $CsGaSe_2$-*mC*16 (Figure 1). Both polymorphs crystallize in the monoclinic space group *C*2/*c* (no. 15) with lattice parameters a = 11.046(1), b = 11.051(1), c = 16.827(1), β = 99.402(9), V = 2026.5(3), and Z = 16 for $CsGaSe_2$-*mC*64 ($KInS_2$ structure type) and a = 7.651(1), b = 12.555(2), c = 6.179(1), β = 113.53(2), V = 544.2(3), and Z = 4 for $CsGaSe_2$-*mC*16 ($KFeS_2$ structure type).

This reversible phase transition was studied *in situ* by high-temperature X-ray powder diffraction (Figure 2) using a STOE Stadi P diffractometer equipped with a high temperature capillary furnace (monochromatic Mo-$K_{\alpha 1}$ radiation). Diffraction experiments performed under isothermal conditions were used to study the phase transition kinetics, crystallization behavior and activation energies. The obtained data were analyzed by using the JMAK-theory of crystal growth [2] and different growth mechanisms found in literature [3]. A detailed analysis for both phase transitions revealed, that the initial stages of the crystallization are dominated by a preferred nucleation at phase boundaries, whereas the latter stages are dominated by a one-dimensional growth of $CsGaSe_2$-*mC*16 and a two-dimensional growth of $CsGaSe_2$-*mC*64, respectively.

Figure 1: Schematic representation of the reversible phase transition in $CsGaSe_2$ featuring the anionic substructures in both polymorphs and the crystal shape.

Figure 2: 3D plot of the progression of the phase transition from $CsGaSe_2$-*mC*64 (front) to $CsGaSe_2$-*mC*16 (back) on an arbitrary timescale.

[1] D. Friedrich, M. Schlosser, A. Pfitzner, *Z. Anorg. Allg. Chem.* **2014**, *640*, 2356.
[2] M. Avrami, *J. Chem. Phys.* **1939**, *7*, 1103; M. Avrami, *J. Chem. Phys.* **1940**, *8*, 212; M.

Avrami, *J. Chem. Phys.* **1941**, *9*, 177.
[3] Sharp, J. H.; Brindley, G. W.; Achar, N. N. *J. Am. Ceram. Soc.* **1966**, 49, 379-382.

Figure 1

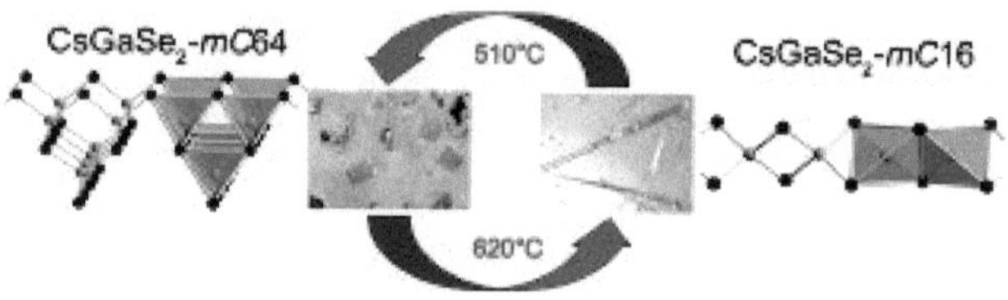

Figure 2

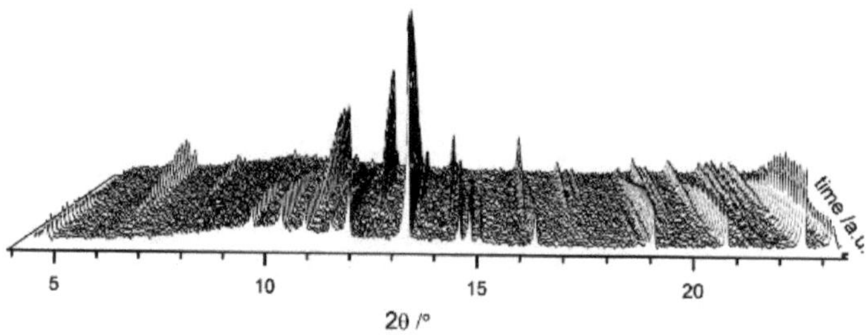

MS17-03
Effect of temperature and pressure on magnetocaloric compounds in the system $Mn_{5-x}Fe_xSi_3$

K. Friese[1], P. Hering[1], M. Maswada[2], Y. Chen[1], M. Hanfland[3], A. Grzechnik[4]
[1]*Forschungszentrum Jülich GmbH, Jülich Centre for Neutron Science-2, Jülich, Germany*
[2]*Al-Quds University, Jerusalem, Palestinian Territory*
[3]*European Synchrotron Radiation Facility, Grenoble, France*
[4]*Rheinisch-Westfälisch Technische Hochschule, Institut für Kristallographie, Aachen, Germany*

The magneto-caloric effect (MCE) describes the change of temperature of a magnetic material in an applied magnetic field. Magnetic refrigeration based on the MCE holds the potential to replace conventional vapor compression cooling [1]. Giant MCEs are observed when the magnetic phase transition involves large changes of the underlying structure.

Subject of our study is the system $Mn_{5-x}Fe_xSi_3$. The compounds undergo a variety of magnetic phase transitions at different temperatures depending on the iron content [2]. According to the literature [3], they crystallize in space group $P6_3/mcm$ with lattice parameter of $a \approx 6.8$Å and $c \approx 4.7$ Å, however, our structure refinement of the compound with x=4 using both neutron and x-ray single crystal data shows that the true symmetry is *P-6* [4].

Within the structure, M1 atoms (Mn or Fe) are interconnected to form distorted octahedra.These octahedra share common triangular faces and form chains of composition $\infty[\square(M1)_3]$ along the *c*-direction. The M2 site is surrounded by six silicon atoms in the form of distorted octahedra. Neighboring $[M2Si_6]$ octahedra share triangular faces and also form infinite chains of composition $\infty[(M2)Si_3]$ along the *c*-direction. Fe preferentially occupies the M2 site. For values of $x>2$ the M1 site shows mixed occupancy of Mn and Fe, while the M2 site is fully occupied by Fe. For the compound $x=4$ we observe an additional partial ordering on the M1 site which lowers the symmetry from $P6_3/mcm$ to $P\text{-}6$ [4].

We performed multiparametric studies in the system varying temperature, pressure and composition to elucidate the underlying mechanism of the MCE in multiple site driven magnetocaloric materials. Our studies show that the effect of hydrostatic pressure and "chemical pressure" (through substitution on Mn with Fe) seems to be equivalent at ambient temperature. However, if, in addition, temperature is simultaneously varied, the resulting behaviour is very complex and the analogy between hydrostatic and "chemical pressure" is no longer observed: Different phase transitions occur in the individual compounds as a function of temperature and pressure and the distortions of the unit cell induced by temperature or pressure are very different.

Surprisingly - and in contrast to what is observed in other systems - the effect of temperature on the structure of these magnetocaloric compounds seems to be far more drastic than the effect of pressure. As, to our knowledge, this is the first time that such a multiparametric study is carried out on a magnetocaloric material it remains to be seen whether this is a characteristic observation for caloric materials in general.

[1] Tegus et. al., Nature , 415, 150–152 (2002).
[2] Songlin, et al., J. Alloys Comp. 334, 249-252 (2002).
[3] Binczycka et. al., Phys. stat. sol. 19, K13–K17 (1973).
[4] Hering et. al., Chem. Mat. *27* , 7128-7136 (2015).

MS17-04

Pressure-induced phonon softening and phase transition of 5-aminotetrazole monohydrate

N. Schrodt[1], W. Morgenroth[1], L. Bayarjargal[1], B. Winkler[1]
[1]*Goethe University, Institute of Geoscience, Frankfurt am Main, Germany*

Chemical reactions at high pressures and temperatures are of interest because novel compounds with unusual structural features and interesting properties can be obtained [1]. In the framework of a DFG research project, we are particularly interested in understanding polymerisation reactions of small molecules with different N:C ratios. 5-aminotetrazole monohydrate (ATM, $CH_3N_5{\cdot}H_2O$) is an interesting candidate because of its high N:C ratio. A recent study of the anhydride found three different crystalline phases upon temperature increase at ambient pressure [2]. However, there was no phase transition of the anhydride reported upon pressure increase to around 12 GPa at ambient temperature [3]. The high pressure and temperature behaviour of the monohydrate and the influence of water in this system on possible phase transitions has not been investigated, yet. At ambient conditions ATM crystallises in the monoclinic space group $P2_1/c$ with 4 molecules per unit cell [4].

In this study, we investigated the high-pressure behaviour of ATM. We loaded Boehler-Almax [5] type diamond anvil cells with single crystals of ATM and performed micro-Raman spectroscopy up to 51 GPa and single-crystal X-ray diffraction up to 16 GPa at PETRA III, DESY. Theoretical Raman intensities were calculated with DFT-based model calculations using the program CASTEP [6].

Figure 1 shows the experimental Raman spectra of ATM. Between 9 and 11 GPa we observed discontinuities indicating a structural phase transition. There is also one phonon which broadened and softened upon pressure increase (dashed box in figure 1) before the phase transformation took place. DFT calculations confirmed this behaviour with the mode being predominantly an N-H stretching motion. The single crystal data (figure 2) confirmed the observed phase transition. The pattern at 15.9(3) GPa showed additional reflections which indicate a doubling of a cell parameter.

In conclusion, in contrast to the anhydrous form, we found a structural phase transition of ATM between 9 and 11 GPa. Further studies will now focus on inducing chemical reactions by heating the sample at high pressures.

Figure 1: Raman spectra of ATM from ambient pressure to 16 GPa showing discontinuities between 9 and 11 GPa and a softening of one mode (dashed box).

Figure 2: Diffraction patterns of ATM collected with an ω rotation of ±25°.

The authors gratefully acknowledge financial support from the DFG (project RA 2585/1-1) and the BMBF (project 05K13RF1).

[1] R. Bini, M. Ceppatelli, M. Citroni, and V. Schettino, Chem. Phys., 398, 262 (2012).
[2] S. Obata, S. Takeya, H. Fujihisa, K. Honda, and Y. Gotoh, J. Phys. Chem. B, 114, 12572 (2010).
[3] H. Fujihisa, K. Honda, S. Obata, H. Yamawaki, S. Takeya, Y. Gotoha, and T. Matsunaga, CrystEngComm, 13, 99 (2011).
[4] D. D. Bray, and J. G. White, Acta Cryst., B35, 3089 (1979).
[5] R. Boehler, Rev. Sci. Instrum., 77, 115103 (2006).
[6] S. Clark, M. Segall, C. Pickard, P. Hasnip, M. Probert, K. Refson, and M. Payne, Z. Kristallogr. 220, 567 (2005).

Figure 1

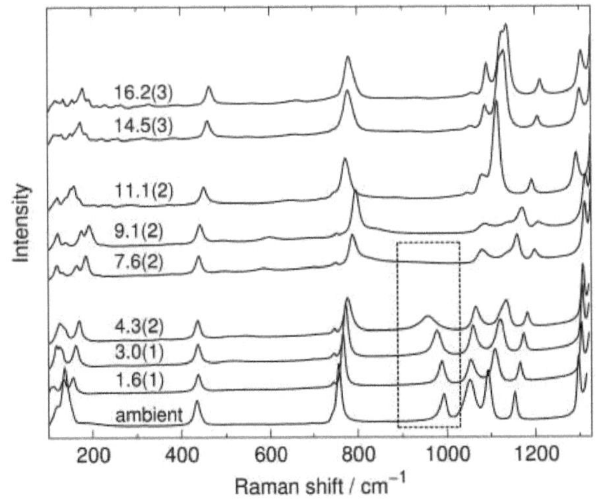

Figure 2

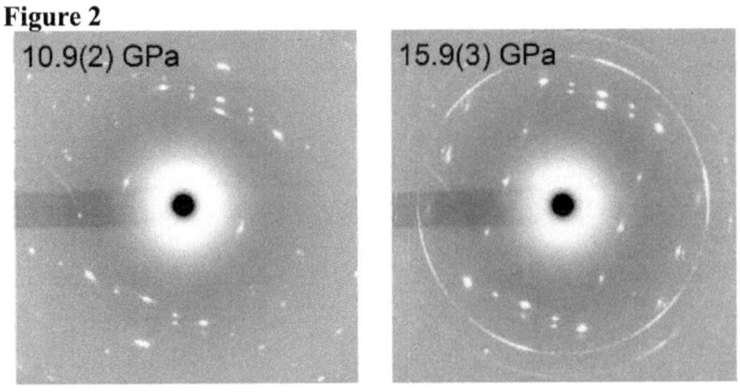

MS18 - Materials: Syntheses and structure II

MS18-01
Synthesis, growth and structural aspects of $RCa_4O(BO_3)_3$ (R=La, Sm, Gd, Y)

C. Reuther[1], R. Möckel[2], J. Schreuer[3], H. Schmidt[4], M. Hengst[1], J. Götze[1], G. Heide[1]
[1]*TU Bergakademie Freiberg, Mineralogie, Freiberg, Germany*
[2]*Helmholtz Institute Freiberg for Resource Technology, Germany, Germany*
[3]*Ruhr Universität Bochum, Institut für Geologie, Mineralogie und Geophysik, Bochum, Germany*
[4]*TU Bergakademie Freiberg, Institut für Anorganische Chemie, Freiberg, Germany*

In 1991 $SmCa_4O(BO_3)_3$ was discovered by Khamaganova et al. (1). Norrestam et al. (2) investigated structural details on small single crystals grown by a long-term sintering process. They found that $RCa_4O(BO_3)_3$ has the acentric space group *Cm*. In 1996 Aka et al. (3) have gown the first crystal of $GdCa_4O(BO_3)_3$ using the Czochralski method based on the congruent melting behaviour. Many growth experiments and optical studies followed because of the possible use for non-linear optical application. Only a few papers deal with structural aspects or the piezoelectric properties of the material. Jiang et al. (4), for example, have successfully tested $YCa_4O(BO_3)_3$ up to 1.000 °C as piezoelectric crystal. $RCa_4O(BO_3)_3$ is a possible high temperature piezoelectric material.

We synthesized and grew crystals of $RCa_4O(BO_3)_3$ (R=La, Sm, Gd, Y) using the Czochralski method. While the Gd-bearing oxoborate is relatively easy to grow, oxoborate crystals of La, Y and especially Sm tended to crack or show other macroscopic defects. The growth of crystals free of macrodefects is possible using a orientated seed. Pieces of the crystals were investigated by single crystal X-ray diffraction, thermal expansion and heat capacity measurements.

Single crystal measurements at room temperature provided lattice parameters of ca. a_0 = 8.0 Å, b_0 = 16.0 Å, c_0 = 3.6 Å, varying in dependence on the radius of the rare earth element. The structure consists of four different oxygen polyhedra: $Ca^{[6-8]}$, $Ca^{[6]}$, $R^{[6]}$ and $B^{[3]}$. Depending on the ionic radius of R, the $Ca^{[6]}$-position and $R^{[6]}$-position are mixed occupied in a different way. Small rare earth elements show a stronger disorder than larger ones. Single crystals, which were heated up to 1200 °C and then quenched, show a significant change in the occupation of the $R^{[6]}$- and $Ca^{[6]}$-positions.

The heat capacity as well as the thermal expansion reveals structural instabilities in form of a discontinuity at 700 °C - 1000 °C, also depending on the rare earth radius. The reason for the discontinuity is still unclear, but a possible explanation is the change from a statistic to a dynamic disorder of $R^{[6]}$- and $Ca^{[6]}$-positions.

(1) Khamaganova, T.; Trunov, V. & Dzhurinski, B. (1991): The crystal structure of calcium samarium oxide borates $Ca_8Sm_2O_2(BO_3)_6$, *Russian Journal of Inorganic Chemistry* 36, 484-485
(2) Norrestam, R.; Nygren, M. & O.-Bovin, J. (1992): Structural investigations of new calcium-rare earth (R) oxyborates with the composition $Ca_4RO(BO_3)_3$, *Chemistry of Materials* 4, 737-743
(3) Aka, G.; Kahn-Harari, A.; Vivien, D.; Benitez, J.-M.; Salin, F. & Godard, J. (1996): A new non-linear and neodymium laser self-frequency doubling crystal with congruent melting: $Ca_4GdO(BO_3)_3$ (GdCOB), *European Journal of Solid State and Inorganic Chemistry* 33, 727-736

(4) Jiang, X.; Kim, K.; Zhang, S.; Johnson, J. & Salazar, G. (2014): High-Temperature Piezoelectric Sensing, *Sensors* 14, 144-169

MS18-02
Structural investigation and large single crystal growth of (In,Na):MnWO4
U. Gattermann, S.-H. Park, C. Paulmann

U. Gattermann[1], S.- H. Park[1], C. Paulmann[2]
[1]*Ludwig-Maximilians-Universität München, Kristallographie, München, Germany*
[2]*Universität Hamburg, Geowissenschaften, Hamburg, Germany*

Manganese tungstate ($MnWO_4$, mineral name: huebnerite) is well known for its complex magnetic phase diagram [1] and particularly for its multiferroic phase AF2 between 7.6 K and 12.7 K. The magnetoelectric coupling in AF2 could be proved by a weak electric polarisation of 50-60 mC/cm² which is measurable only in a helical spin spiral order of Mn^{2+} in $MnWO_4$ [2, 3]. Therefore this magnetically induced electric polarisation has been widely accepted for the polar character of AF2. However, our recent investigation on a huebnerite sample clearly indicates that even its paramagnetic structure exhibits the polar space group P2 [4]. This new finding requests for reconsideration of the generally accepted centrosymmetric space group P2/c for the nuclear structure of $MnWO_4$. To understand the influence of spin-orbital interaction on magnetoelectric coupling in this one of the most interesting multiferroic compounds we have synthesized and characterized an In/Na-doped solid-solution system, $Mn_{1-2x}In_xNa_xWO_4$ ($0 < x < 0.25$) systematically.

All single crystal samples were grown from stoichiometric mixtures of MnO, WO_3, and In_2O_3 using high temperature solutions, such as Na_2WO_4 and $Na_2W_2WO_7$ in two steps:

(1) Synthesis of small single seed crystals via spontaneous nucleation;
(2) Czochralski and top seeded solution growth of several cm-sized large crystals using obtained seeds (**Fig. 1**).

X-Ray diffraction study confirmed the huebnerite topology of all crystals of $Mn_{1-2x}In_xNa_xWO_4$ even with a distinctive volume expansion with increasing substitution of Na^+ and In^{3+} for Mn^{2+}. More importantly, reconstruction of X-ray single crystal diffraction data confirmed the polar space group P2 not only for the synthetic pure $MnWO_4$ crystal, but also the whole $Mn_{1-2x}In_xNa_xWO_4$ crystals ($0 < x < 0.25$) (**Fig. 2**). We present at DGK 2016 structural distortion in $Mn_{1-2x}In_xNa_xWO_4$ and substitution effects on its magnetic properties. Subtle information about the introduction of Na^+ and In^{3+} into the $MnWO_4$ topology will be discussed based on data acquisition by means of electron microprobe and micro-Raman spectroscopy, as well.

[1] G. Lautenschläger , H. Weitzel, T. Vogt, R. Hock, A. Böhm, M. Bonnet, H. Fuess, Phys. Rev. B, 48(9) (1993) 6087-6098.
[2] A. H. Arkenbout, T. T. M. Palstra, T. Siegrist, T. Kimura, Phys. Rev. B, 74 (2006) 184431.
[3] O. Heyer, N. Hollmann, I. Klassen, S. Jodlauk, L. Bohaty, P. Becker, J. A. Mydosh, T. Lorenz, D. Khomskii, J. Phys.: Condens. Matter, 18 (2006) L471-L475.
[4] S.-H. Park, B. Mihailova, B. Pedersen, C. Paulman, D. Behal, U. Gattermann, R. Hochleitner, J. Magn. Magn. Mater., 394 (2015) 160-172.

Fig. 1 Large single crystals of synthetic $MnWO_4$ (left) and $Mn_{0.94}In_{0.03}Na_{0.03}WO_4$ (right)

Fig. 2 Reciprocal h0l plane for synthetic $MnWO_4$ (left) and $Mn_{0.88}In_{0.06}Na_{0.06}WO_4$ (right) from X-ray single crystal diffraction using a rotating Cu tube without monochromator. Weak h0l-reflections forbidden for P2/c (l = odd) are marked with arrows.

Figure 1

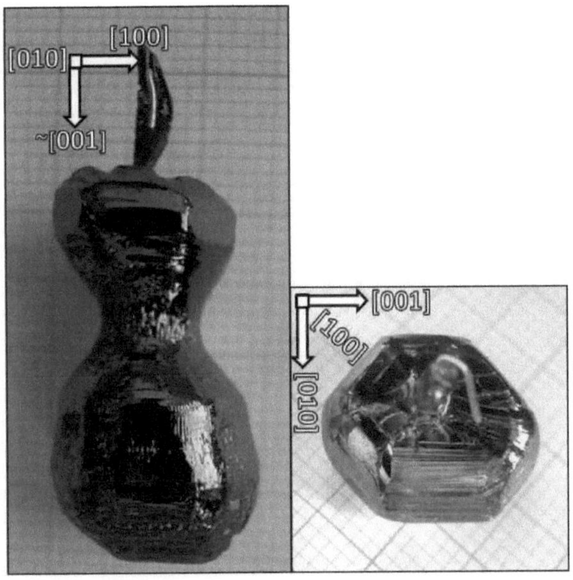

Figure 2

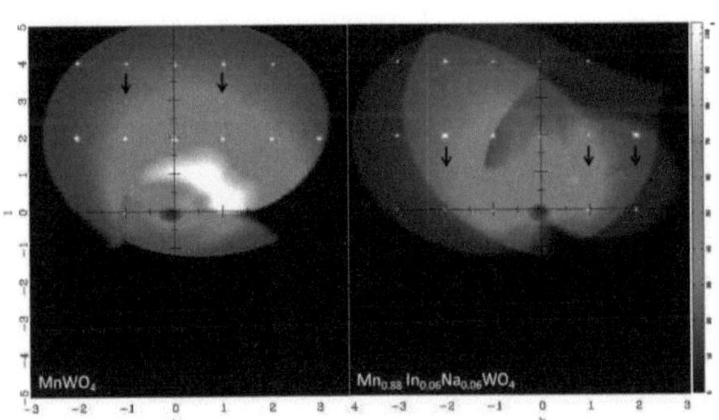

MS18-03
Synthesis of Perovskite-type Oxides and Oxynitrides Located at the Stability Border by Pechini Method as Materials for Energy Conversion

M. Widenmeyer[1], W. Xie[1], A. Weidenkaff[1]
[1]*University of Stuttgart, Institute for Materials Science, Stuttgart, Germany*

The conversion of “waste” heat into electricity by using a thermoelectric (TE) converter is one promising way to reduce CO_2 emission and the dependence on fossil fuels. Another alternative is the photoelectrocatalytic splitting of H_2O into H_2 and O_2 by a semiconducting photocatalyst(PC).[1,2]

Perovskite-type oxides ABX_3 are suitable TE materials at higher $T > 1000$ K, since the perovskite structure tolerates a variety of cationic substitution allowing for significant modification of the band structure and hence the TE performance. Especially the cubic $EuTiO_{3\pm x}$ offers promising TE properties due to the localized $4f$ states near the VB edge. The electrical resistivity can be controlled by the oxygen partial pressure resulting in different amounts of Ti^{3+}. Further cationic substitution can result in a lowered thermal conductivity due to enhanced phonon scattering.[3] The required substitutions were often difficult to achieve in single phase by classical solid state reactions (SSR). In contrast soft chemistry methods (SCM) such as the Pechini method easily allows for even extended substitution and the smaller initial particle size provided a nearly doubled sintering density up to 90 %.

Since perovskites cannot be only modified on the cationic site, but also on the anionic part. The introduction of N into ABX_3 forming oxynitrides $AB(O,N)_3$ by ammonolysis is known to reduce the optical band gap (oBG) by an expansion of the VB, which is crucial for an application as PC. Further narrowing can be achieved by expansion of the CB by partial B cation substitution with B' bearing a higher electronegativity value than B. Charge compensation of the additional negative charge (N^{3-} vs. O^{2-}) can be achieved by partial A-site substitution to avoid an enhanced formation of O vacancies.[2,4] The control of the thereby introduced distortion of the octahedral network is crucial since this can result in an opening of the oBG. Enhanced catalytic activity is often observed at the borders of the respective stability field. A single phase synthesis of such materials is highly interesting. SCM approaches are typically superior compared to SSR for such tasks by avoiding thermodynamic reaction control. Double anionic substitution leading to oxyfluoronitrides $AB(O,N,F)_3$ showing typically a reduced degree of distortion of the octahedral network is an alternative.

Figure 1. *A*-, *B*-, *X*-site substitutions of $SrTiO_3$ and their effect on the crystal structure.

[1] W. Xie, A. Weidenkaff, X. Tang, Q. Zhang, J. Poon, T. Tritt, Nanomater. 2012, 4, 379
[2] S. Yoon, A. Maegli, L. Karvonen, A. Shkabko, S. Populoh, K. Galazka, L. Sagarna, M. Aguirre, P. Jakes, R. Eichel, S. Ebbinghaus, S. Pokrant, A. Weidenkaff, Z. Anorg. Allg. Chem. 2014, 640, 797
[3] L. Sagarna, K. Rushchanskii, A. Maegli, S. Yoon, S. Populoh, A. Shkabko, S. Pokrant, M. Lezaic, R. Waser, A. Weidenkaff, J. Appl. Phys. 2013, 114, 033701
[4] M. Widenmeyer, C. Peng, A. Baki, W. Xie, R. Niewa, A. Weidenkaff, Solid State Sci. 2015, in press

Figure 1

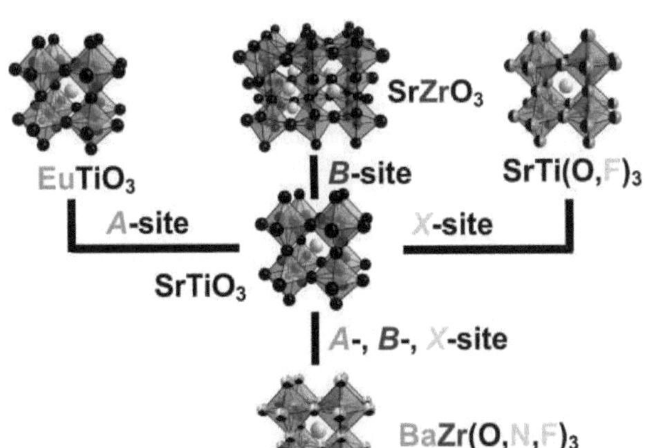

MS18-04

Thermal dependence of Cu/Zn ordering in CZTSe kesterites by anomalous diffraction

D. Többens[1], G. Gurieva[1], S. Schorr[1]
[1]*Helmholtz-Zentrum Berlin für Materialien und Energie , Abteilung Struktur und Dynamik von Energiematerialien, Berlin, Germany*

We demonstrate by means of direct determination of the site occupancies from anomalous X-ray powder diffraction data at the Cu- and Zn absorption edges the ordering of Cu^{+} and Zn^{2+} in B-type $Cu_2ZnSnSe_4$ (CZTSe) kesterite upon annealing at temperatures below 180°C.

$Cu_2ZnSn(S,Se)_4$ (CZTSSe) semiconductor material is a promising alternative for absorber layers in thin film solar cells. One structural property of particular interest is the distribution of the cations Cu^{1+} and Zn^{2+} in the crystal structure. In a recent paper [1] ordering of Cu and Zn in the CZTS structure was reported at temperatures below 533 K; similar results were found for CZTSe below 473 K [2]. The implications of such a low critical temperature would be significant. CZTSSe is normally grown at 720-830 K, far above the critical temperature. The exact extent of ordering by the end of the synthesis will depend only on the part of the cooling that occurs below the critical temperature, in a temperature region in which cooling history often is neither controlled nor reported. Thus the concentration of Cu_{Zn} and Zn_{Cu} antisite defects in CZTSSe and from these the electrical properties of CZTS could be modified by relatively low temperature processes. Indeed, low temperature postdeposition annealing has been shown to increase photovoltaic power conversion efficiency significantly [3]. However, the low temperature ordering effects were only deducted indirectly, from changes in the Raman spectrum [1], from anomalous lattice parameter expansion [4], and from kinetics simulations [2].

Cu^{1+} and Zn^{2+} cannot be distinguished by conventional X-ray diffraction, since they have essentially the same scattering factor for X-rays (same number of electrons). Anomalous diffraction has been used to overcome this, but currently published papers either restrict themselves to qualitative changes [5,6] or used single crystal data [7] with limited explanatory power for realistic samples. We used anomalous X-ray diffraction on the Cu- and Zn-K absorption edges to determine the distribution of Cu and Zn over the crystallographic sites in a B-type CZTSe kesterite ($Cu_{1.949}Zn_{1.059}Sn_{0.983}Se_4$) powder. Rietveld refinement allowed the quantitative determination of both Cu- and Zn-occupancy for all relevant sites. From this, the temperature dependency of a structure-based, quantitative order parameter could be determined. The critical temperature of the phase transition was confirmed at 460±10 K. The ordering mechanism is in agreement with a transition from disordered to ordered kesterite.

[1] J. J. S. Scragg *et al.*, Appl. Phys. Let. **2014**, 104, 041911
[2] G. Rey *et al.*, Appl. Phys. Let. **2014**, 105, 112106
[3] M. Neuschitzer *et al.*, Chem. Mater. **2015**, 27, 5279–5287
[4] S. Schorr, G. Gonzalez-Aviles, Physica Status Solidi A **2009**, 206(5), 1054
[5] H. Nozakia *et al.,* J. Alloys Comp. **2012**, 524, 22
[6] T. Washio *et al.*, J. Appl. Phys. **2011**, 110, 074511
[7] A. O. Lafond *et al.*, Acta Crystallographica B **2014**, 70, 390

MS19 - Crystal physics

MS19-01
Mechanical properties of natural radiation damaged minerals and temperature-induced structural reorganization

T. Beirau[1,2], W. D. Nix[3], R. C. Ewing[1], G. A. Schneider[4], L. A. Groat[5], U. Bismayer[2]
[1]*Stanford University, Department of Geological Sciences, Stanford, United States*
[2]*University of Hamburg, Department of Earth Sciences, Hamburg, Germany*
[3]*Stanford University, Department of Materials Science and Engineering, Stanford, United States*
[4]*Hamburg University of Technology, Institute of Advanced Ceramics, Hamburg, Germany*
[5]*University of British Columbia, Department of Earth, Ocean and Atmospheric Sciences, Vancouver, Canada*

We present new insights into the relation between thermally induced structural reorganization and the macroscopic mechanical properties of radiation-damaged titanite (Beirau et al. 2016). Low-temperature annealing affects only slightly the sample stiffness and leads to a softening resulting from the defect annihilation in crystalline regions. In the high-temperature annealing regime, amorphous domains recrystallize and this leads to further recovery of defects, reduction of interfaces, grain growth, and, in general, an increase in the long-range order. The thermally induced recrystallization is accompanied by massive dehydration leading to considerable stiffening and hardening. This interpretation of the recrystallization process in titanite based on the correlation of new results from nanoindentation and Raman-spectroscopic measurements complementing previous investigations using thermogravimetric and gas analyses by Hawthorne et al. (1991) and infrared spectroscopy by Zhang et al. (2001). Additionally, we present the results of recent alpha-radiation dose dependent nanoindentation measurements of zircon. With increasing radiation dose, the defect concentration and the amount of amorphous fraction increase and the density decreases which has a complex influence on the materials elastic modulus and hardness.

Beirau, T., Nix, W.D., Ewing, R.C., Schneider, G.A., Groat, L.A. and Bismayer, U. (2016) *Mechanical properties of natural radiation damaged titanite and temperature induced structural reorganization: A Nanoindentation and Raman spectroscopic study*. American Mineralogist, in press.

Hawthorne, F.C., Groat, L.A., Raudsepp, M., Ball, N.A., Kimata, M., Spike, F., Gaba, R., Halden, N.M., Lumpkin, G.R., Ewing, R.C., and others. (1991) *Alpha-decay damage in titanite*. American Mineralogist, 76, 370-396.

Zhang, M., Groat, L.A., Salje, E.K.H., and Beran, A. (2001) *Hydrous species in crystalline and metamict titanites*. American Mineralogist, 86, 904-909.

MS19-02
Microstructure characterization of HTVPE GaN layers grown on sapphire substrates

M. Barchuk[1], C. Röder[2], G. Lukin[3], F. Zimmermann[4], J. Kortus[2], O. Pätzold[3], J. Heitmann[4], D. Rafaja[1]
[1]*TU Bergakademie Freiberg, Institute of Materials Science, Freiberg, Germany*
[2]*TU Bergakademie Freiberg, Institute of Theoretical Physics, Freiberg, Germany*
[3]*TU Bergakademie Freiberg, Institute of Nonferrous Metallurgy and Purest Materials, Freiberg, Germany*
[4]*TU Bergakademie Freiberg, Institute of Applied Physics, Freiberg, Germany*

Gallium nitride (GaN) is nowadays one of the most common wide-bandgap semiconductors, which is used for optoelectronic applications such as light emitting diodes, blue lasers etc. The growth of GaN on the foreign substrates (Al_2O_3, SiC) is economically profitable. However, a large lattice mismatch between the substrates and the GaN coatings is compensated by structural defects mainly misfit dislocations, which propagate towards the GaN surface as threading dislocations and decrease the efficiency of the GaN-based devices [1]. Therefore, new technologies are developed permanently in order to design microstructure features facilitating the annihilation of the unwanted defects.

In this work, we investigated thin (up to 5 mm) c-plane GaN samples grown in a vertical HTVPE reactor on MOVPE GaN templates or directly on sapphire substrates [2]. We have produced 3 series of samples by varying a single selected growth parameter in each batch. The parameters that were varied in individual sample series were the V/III ratio and the growth rate. The V/III ratio was controlled by the ammonia flow (for HTVPE samples on MOVPE GaN or sapphire), the growth rate by the enhanced diffusion in the gas phase due to reduced reactor pressure. The aim of this study was to determine the most appropriate growth conditions that enable to reduce the density of threading dislocations and to control the emergence of residual stresses and incorporation of impurities coming from the reactor environment. The samples were characterized by means of high-resolution X-ray diffraction (HRXRD), transmission electron microscopy (TEM), micro-Raman spectroscopy and photoluminescence (PL) techniques. HRXRD in conjunction with the Monte Carlo simulation [3] that includes the description of the dislocation bunching allowed us to determine the density of edge and screw threading dislocations. It was found that the density of threading dislocations (TDs) strongly depends on the growth rate, on the presence of the planar defects, on the stoichiometry of the material and on the layer thickness. The TEM observations revealed a high degree of dislocation bunching and a possible termination of TDs on the basal stacking faults [4]. The presence of bunched dislocations was correlated with residual stresses, which were quantified from the shift of the E_2(high) Raman mode [4]. A low ammonia flow caused a deficiency of nitrogen atoms during the layers growth, which subsequently resulted in formation of partial stacking faults, where the N positions are occupied by O or C unintentional impurities. Complementary, photoluminescence measurements revealed the highest PL intensity for the samples with the lowest ammonia flow, which confirms the substitution of lacking N atoms by O or C atoms.

References

- [1] S. Nakamura, G. Fasol, The Blue Laser Diode, *Springer*, New York, 1997.
- [2] G. Lukin, C. Röder, M. Barchuk, G. Schreiber, O. Pätzold, J. Kortus, D. Rafaja, M. Stelter, *Phys. Stat. Solidi C*, **11**, (2014), 491-494.
- [3] M. Barchuk, V. Holý, B. Miljevic, B. Krause, T. Baumbach, J. Hertkorn, F. Scholz, *J. Appl. Phys.*, **108**, (2010), 43521.
- [4] M. Barchuk, C. Röder, Y. Shashev, G. Lukin, M. Motylenko, J. Kortus, O. Pätzold, D. Rafaja, *J. Cryst. Growth*, **386**, (2014), 1-8.

MS19-03

Anisotropy of Oxygen Vacancy Migration in $SrTiO_3$

M. Zschornak[1], J. Hanzig[1], E. Mehner[1], F. Hanzig[1], W. Münchgesang[1], T. Leisegang[1,2], H. Stöcker[1], D. C. Meyer[1]
[1]*TU Bergakademie Freiberg, Institut für Experimentelle Physik, Freiberg, Germany*
[2]*Samara State Aerospace University, Samara, Germany*

High ionic mobility in solids is of fundamental interest not only for chemical sensors and solid electrolytes but also for various microstructural changes and chemical reactions. Oxygen migration in perovskites is well known to occur via vacancies along the TiO_6 octahedron edges. Ionic conduction depends further on the orientation of the crystal in the electric field.

Electroformation of pure $SrTiO_3$ crystals has recently been correlated to the redistribution of intrinsic oxygen vacancies caused by the applied electric field. The here presented temperature-dependent electroformation studies give evidence of such a vacancy migration process, following an Arrhenius behavior with activation energies in the order of 0.7 eV. The experimental barriers are close to energy barriers from first-principles calculations with density functional theory.

Anisotropy of vacancy migration has been studied for representative crystallographic directions. Mobility of oxygen vacancies has been found enhanced by up to half an order of magnitude in <001> directions compared to <110> and <111> directions. A sophisticated migration model based on atomistic migration paths and their multiplicities will be presented, which accounts for these experimental variations in mobility. The significant changes of ionic mobility due to the crystal's orientation have to be taken into account for optimized performance of related devices.

MS19-04

Elastic and inelastic properties of a novel single-crystal Co-base superalloy

C. Betzing[1], K. Demtröder[1], J. Schreuer[1], C. Zenk[2]
[1]*Ruhr-University Bochum, Institute for Geology, Mineralogy and Geophysics, Bochum, Germany*
[2]*Friedrich-Alexander University, Department Materials Science and Engineering, Erlangen, Germany*

Question

Currently, the most used high-temperature materials in aircraft engines and power generation systems are Ni-base superalloys. They are primarily used for critical units, like turbine blades and selected parts of an engine [1]. Co-based superalloys in general are known to feature better corrosion, oxidation and wear resistance than Ni-based superalloys and exhibit a higher melting point [2]. The main reason for the continuous high-temperature strength of the new Co-based alloys is the existence of a stable microstructure consisting of $L1_2$-ordered Co_3(Al, W) coherent precipitates (γ'-phase) embedded in the fcc-type matrix (γ-phase) [3]. Though, a drawback of the current Co-Al-W-based superalloys is their high density compared to the Ni-based superalloys which limits the applicability in critical situations [4]. Thus, the development of strength and ductility in Co-Al-W-based alloys at room temperature and high temperatures in combination with hot corrosion and oxidation resistance is now a primary issue [5].

Methods

The focus of our work is on the elastic and inelastic properties of a novel single-crystal Co-based superalloy. To this end its elastic stiffnesses, coefficient of thermal expansion and specific heat capacity were determined between room temperature and about 1500 K employing resonant ultrasound spectroscopy (RUS), dilatometry (DIL) and differential scanning calorimetry (DSC), respectively. Additionally, the effective lattice misfit and the microstructure were studied by a combination of high resolution X-ray diffraction, electron probe micro analysis and scanning electron microscopy.

Results & Conclusions

The results of our RUS measurement at ambient conditions (c_{11} = 258 GPa, c_{12} = 158 GPa and c_{44} = 141 GPa) are in reasonable agreement with literature data [6]. A small but significant change in the temperature dependence of the elastic stiffnesses at about 770 K is probably connected with the beginning dissolution of the γ'-phase. A corresponding anomaly is also visible in the evolution of the thermal expansion coefficient. Most remarkably is a hitherto unknown fully reversible discontinuity in thermal expansion at approximately 1470 K, which resembles the characteristic behavior of thermal expansion often observed in the vicinity of second order phase transitions.

[1] R. C. Reed (2006): The Superalloys - Fundamentals and Applications, Cambridge University Press.
[2] S. K. Makineni, B. Nithin, K. Chattopadyay (2015): Synthesis of a new tungsten free γ-γ' cobalt-based superalloy by tuning alloying additions, in: Scripta Materialia 85, 85-94.
[3] J. Sato, T. Omori, K. Oikawa, I. Ohnuma, R. Kainuma, K. Ishida (2006): Cobalt-Base High-Temperature Alloys, in: Science, Vol. 312, 90-91.
[4] S. K. Makineni, B. Nithin, K. Chattopadyay (2015): A new tungsten-free γ-γ' Co-Al-Mo-Nb-based superalloy, in: Scripta Materialia 98, 36-39.
[5] G. Feng, H. Li, S.S. Li and J.B. Sha (2012): Effect of Mo additions on microstructure and tensile behavior of a Co-Al-W-Ta-B alloy at room temperature, in: Scripta Materiala, Vol. 67, 499-502.
[6] H. Inui, T. Oohashi, N. L. Okamoto, K. Kishida, K. Tanaka (2011): Mechanical Properties of the Ternary $L1_2$ Compound Co_3(Al,W) in Single and Polycrystalline Forms, in: Advanced Materials Research, Vol. 278, 1-6.

MS20 - Biostructures V: Structure based drug design

MS20-01
Specific Antidotes For Dabigatran: Structure-Guided Affinity Optimisation and Functional Characterisation

H. Nar[1]
[1]*Boehringer Ingelheim GmbH & Co KG, LIOS, Biberach, Germany*

Dabigatran etexilate is a direct thrombin inhibitor and widely used as an anticoagulant for the prevention of stroke in patients with atrial fibrillation (1 & 2). Despite clear advantages in efficacy and safety above traditional treatment options, anticoagulation therapy with dabigatran is associated with a risk of bleeding.

Here, we present data on the identification, humanization and in vitro pharmacology of an antibody fragment for dabigatran, idarucizumab. The X-ray crystal structure of dabigatran in complex with the antidote reveals many structural similarities of dabigatran recognition compared to thrombin. By a tighter network of interactions, the antidote achieves an affinity for dabigatran that is ~350 times stronger than its affinity for thrombin. Despite the structural similarities in the mode of dabigatran binding, the antidote does not bind known thrombin substrates and has no activity in coagulation tests or platelet aggregation. In addition we demonstrate that the antidote rapidly reversed the anticoagulant activity of dabigatran in vivo in a rat model of anticoagulation. This is the first specific antidote for a next-generation anticoagulant which may become a valuable tool in patients that require emergency procedures (3).

Further, we present a distinct, novel antibody fragment against dabigatran that was identified in a search for backup clinical candidates to idarucizumab. By structure-guided protein design we were able to improve the affinity and residence time of the original FAB significantly. The optimized mutant exhibits prolonged and more effective neutralization of dabigatran's anticoagulative effects in vivo compared to the parental antibody fragment (4).

References:

(1) van Ryn, Joanne; Goss, Ashley; Hauel, Norbert; Wienen, Wolfgang; Priepke, Henning; Nar, Herbert; Clemens, Andreas, Frontiers in Cardiovascular and Smooth Muscle Pharmacology (2013), 4, 12-19
(2) Hauel, Norbert; Clemens, Andreas; Nar, Herbert; Priepke, Henning; van Ryn, Joanne; Wienen, Wolfgang, The Discovery of Dabigatran Etexilate, in Analogue-based Drug Discovery III, First Edition. Edited by Janos Fischer, C. Robin Ganellin, and David P. Rotella. (2013) Wiley-VCH Verlag GmbH & Co. KgaA, p. 243-267
(3) Schiele, Felix; van Ryn, Joanne; Canada, Keith; Newsome, Corey; Sepulveda, Eliud; Park, John; Nar, Herbert; Litzenburger, Tobias, Blood (2013), 121(18), 3554-3562
(4) Schiele, Felix; van Ryn, Joanne; Litzenburger, Tobias; Ritter, Michael; Seeliger, Daniel; Nar, Herbert, mAbs (2015) 7:5, 871

Figure 1

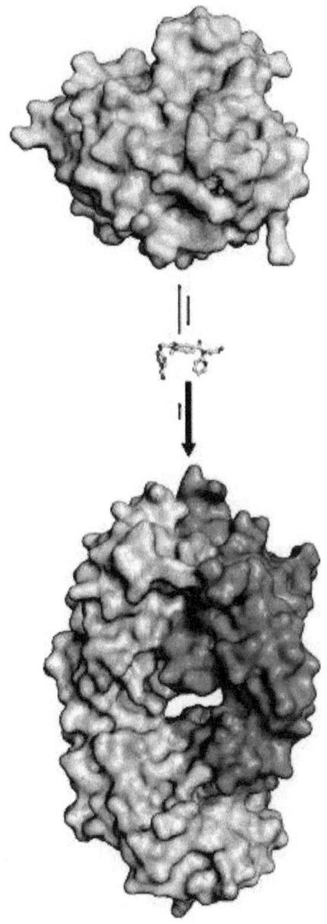

MS20-02
Griselimycins, Novel Anti Tuberculosis Agents: Structural Insights into the Mode of Action and Resistance

P. Lukat[1], A. Kling[2], N. Zaburannyi[2], J. Herrmann[2], S. Wenzel[2], A. Bauer[3], E. Fontaine[4], S. Lagrange[4], M. Brönstrup[5], L. Fraisse[4], D. Heinz[5], W. Blankenfeldt[1], R. Müller[2]
[1]*Helmholtz-Zentrum f. Infektionsforschung, SFPR, Braunschweig, Germany*
[2]*Helmholtz-Institut f. Pharmazeutische Forschung Saarland, MINS, Saarbrücken, Germany*
[3]*Sanofi-Aventis R&D, LGCR/Chemistry, Frankfurt a. M., Germany*
[4]*Sanofi-Aventis R&D, Infectious Disease Therapeutic Strategic Unit, Toulouse, France*
[5]*Helmholtz-Zentrum f. Infektionsforschung, Braunschweig, Germany*

Introduction
Tuberculosis remains a major global health problem that caused an estimated 1.5 million deaths in 2013, whereby a growing percentage is multidrug resistant tuberculosis. Novel anti TB agents with distinct mechanism of action are urgently needed. Griselimycins (GMs) from *Streptomyces caelicus* were found to exhibit bactericidal effects almost exclusively on mycobacteria and were thus developed as promising anti-TB agents showing exceptional potential in *in vivo* mice models.

Objectives
Aiming to identify the mode of action of these novel antibacterial lead substances, we analyzed the biosynthetic gene cluster of GMs and found that the GM producing strain *S. caelicus* harbors an additional variant of the *dnaN* gene (encoding the sliding clamp of DNA polymerase) in the cluster. Upon heterologous overexpression, this variant confers GM resistance to GM sensitive *Streptomyces*. While the common DnaN of the producer strain seems to be inhibited by GMs, this seems not to be the case for the additional variant, which thus represents a mechanism of GM self-resistance and hence lead to the identification of bacterial DnaN as a novel target for a natural product.

Material & Methods
Mycobacterial DnaN and both sliding clamp variants of the GM producer strain were heterologously expressed, purified and crystallized in the presence of GMs. The crystal structures were solved by molecular replacement. In addition, binding of GMs to the proteins was characterized by SPR.

Results
The crystal structures clearly show the binding of GMs to the bacterial sliding clamps and give insight into their mode of action. In the case of the *S. caelicus* proteins, the crystal structures also point to a potential mechanism of GM self-resistance. SPR data show that GMs bind to mycobacterial and *S. caelicus* DnaN with high affinity, to *E. coli* DnaN and *S. caelicus* GriR with low affinity. No binding could be detected for the human sliding clamp.

Conclusion
GMs are promising leads for therapy of drug resistant TB. GMs bind to the bacterial sliding clamp, thereby inhibiting the recruitment of enzymes involved in DNA replication & repair. GMs do not interfere with the eukaryotic sliding clamp. Self-resistance of the producer strain is mediated by the resistance protein GriR, while resistance development mycobacteria occurs at low frequency and is accompanied by a high fitness loss.

MS20-03
Structural basis of proline-proline peptide bond specificity of the secreted protease PPEP-1 implicated in motility of *Clostridium difficile*.
M. Schacherl[1], C. Pichlo[1], I. Neundorf[1], U. Baumann[1]
[1]*University of Cologne, Biochemistry, Köln, Germany*

Introdution
Clostridium difficile is a pathogenic bacterium causing gastrointestinal diseases from mild diarrhea to severe colitis and toxic megacolon. As other pathogenic bacteria, *C. difficile* secretes proteins involved in adhesion, colonization and dissemination. A subset of these constitutes extracellular proteases. The recently identified PPEP-1 is a novel metalloprotease showing a unique specificity for Pro-Pro peptide bonds [1,2]. It is able to cleave fibronectin, fibrinogen, Hsp90b and IgA2. The endogenous substrates of PPEP-1 are two peptidoglycan-attached surface proteins implicated in adhesion of *C. difficile* to surface proteins of human cells. Thus, PPEP-1 is believed to be involved in the regulation or the adhesion-motility balance of *C. difficile*.

Objectives
The aim of this study was to determine the structural basis of the strikingly high specificity of PPEP-1 for Pro-Pro bonds and to determine important residues for proteolytic activity.

Methods
We have crystallized the wild-type form and several mutant forms of PPEP-1 alone and in complex with short peptides resembling the natural substrate. We solved the PPEP-1 crystal structures by Zn-SAD and molecular replacement. We also determined kinetic parameters of wild-type protein vs. mutant protein with a fluorogenic model substrate.

Results
Here we present crystal structures of PPEP-1 from *C. difficile* in its apo- and a peptide-bound form at atomic resolution [3]. The structure analysis revealed a fold similar to the metalloprotease domain of *B. anthracis* lethal factor. Crystal structures in the open and closed conformation of the large S-loop shed light on the mode of binding of the substrate. The peptide-bound structure reveals important residues for substrate recognition and the strict specificity of PPEP-1 for Pro-Pro peptide bonds. A unique aromatic-aliphatic side-chain stack at the upper rim of the active-site cleft, together with the diverting loop and the mobile S-loop seem to act together to make binding of the double-kinked Pro-Pro-peptides possible (Fig. 1). In kinetic analysis the wild-type and mutant forms of PPEP-1 differ drastically in the *kcat* value.

Conclusions
The unique shape of PPEP-1 active-site cleft enables PPEP-1 to bind proline-rich proteins and cleave Pro-Pro bonds efficiently.

References

[1] V. Cafardi, M. Biagini, M. Martinelli, R. Leuzzi, J. T. Rubino, F. Cantini, N. Norais, M. Scarselli, D. Serruto, M. Unnikrishnan, Identification of a novel zinc metalloprotease through a global analysis of *Clostridium difficile* extracellular proteins., PLoS One 8 (2013) e81306.
[2] P. J. Hensbergen, O. I. Klychnikov, D. Bakker, V. J. van Winden, N. Ras, A. C. Kemp, R. A. Cordfunke, I. Dragan, A. M. Deelder, E. J. Kuijper, J. Corver, J. W. Drijfhout, H. C. van Leeuwen, A novel secreted metalloprotease (CD2830) from *Clostridium difficile* cleaves specific proline sequences in LPXTG cell surface proteins., Mol Cell Proteomics 13 (2014) 1231-1244.
[3] M. Schacherl, C. Pichlo, I. Neundorf, U. Baumann, Structural Basis of Proline-Proline Peptide Bond Specificity of the Metalloprotease Zmp1 Implicated in Motility of *Clostridium difficile.*, Structure 23 (2015) 1632-1642.

Figure legend

Figure 1: PPEP-1 is a proline-proline specific metalloprotease that displays an aliphatic-aromatic network of amino acid side chains and the unique diverting loop to shape its active site for substrate binding.

Figure 1

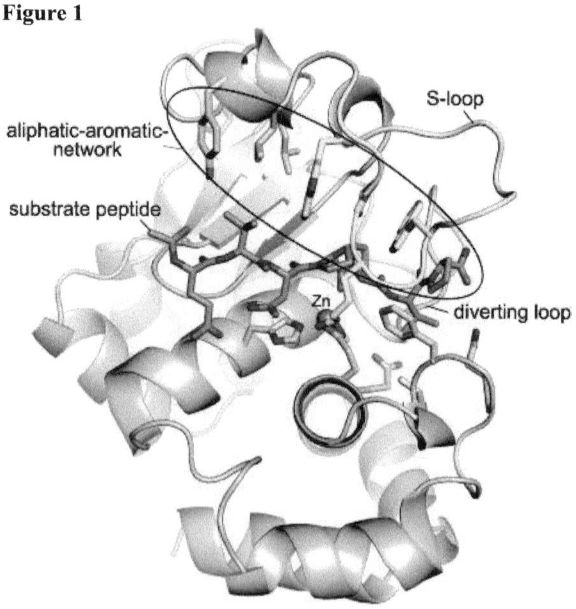

PPEP-1-substrate peptide complex

MS20-04

MCPIP3 (ZC3H12c) regulate the innate immune response by acting as a Ribonuclease

A. Garg[1], U. Heinemann[1]
[1] *Max-Delbrück-Centrum für Molekulare Medizin (MDC), Molecular Crystallography, Berlin, Germany*

Inflammation is a protective response by the body to ensure removal of detrimental stimuli, as well as a healing process for repairing the damaged tissue (1). Different innate pattern recognition receptors like Toll-like receptors (TLR) and RIG-1 like receptors sense the infection, and then the inflammatory response is orchestrated by pro-inflammatory cytokines such as tumor necrosis factor (TNF), interleukin (IL)-1 and IL-6 which ultimately result in elimination of pathogens. Cytokine levels are tightly controlled at transcriptional and post-transcriptional level by various proteins, because a cytokine storm could lead to immunodeficiency and autoimmune disorders (2).

The MCPIP (macrophage chemotactic protein-induced proteins) family (also known as ZC3H12 family) of proteins are ribonucleases containing a CCCH-type zinc finger. These proteins play an important role in innate immune response by regulating cytokine levels through targeting their mRNAs (3). MCPIP1 (Regnase1) has recently been shown to recognize specific stem-loop structures in the 3' untranslated region (UTR) of translationally active mRNAs for IL6 and other immunological factors and lead to their degradation by utilizing the helicase activity of UPF1 (4).

MCPIP3 (ZC3H12C) is another member of the family which has also been reported to be involved in cytokine regulation by mRNA degradation. Moreover, it is highly expressed in precursor B-lymphocytes and is also involved in lymphocyte differentiation, as MCPIP3 knock-out mice show severe defects in lymphocytes.

A crystal structure for the RNase domain of ZC3H12C has been determined at a resolution of 1.9 Å which shows a Rossmann-fold like structure with an Mg^{2+} ion bound in the catalytic site. Initial RNase assays show that the protein variants used in structure analysis are catalytically active. Further biochemical analyses of these proteins are required.

Overall, we are trying to understand the structure-function relation of MCPIP3 which will enable us to understand how it recognizes its target mRNA and how it coordinates with other family members in order to regulate the lymphocyte maturation and inflammatory responses.

References;

1) Medzhitov, R. Origin and physiological roles of inflammation. Nature, 454, 428-435 (2008).
2) Takeuchi, O., Akira, S. Pattern recognition receptors and inflammation. Cell, 140, 805-820 (2010).
3) Xu, J., Peng, W., Sun, Y., Wang, X., Xu, Y., Li, X., Gao, G., Rao, Z. Structural study of MCPIP1 N-terminal conserved domain reveals a PIN-like RNase. Nucleic Acids Research, 40, 6957-6965 (2012).
4) Mino, T., Murakawa, Y., Fukao, A., Vandenbon, A., Wessels, H.H., Ori, D., Uwata, T., Tartey, S., Akira, S., Suzuki, Y., Vinuesa, C.G., Ohler, U., Standley, D.M., Landthaler, M., Fujiwara, T., Takeuchi, O. Regnase-1 and roquin regulate a common element in inflammatory mRNAs by spatiotemporally distinct mechanism. Cell, 141, 1-16 (2015).

Poster

Biostructures

P001

BioMAX - The First Macromolecular Crystallography Beamline of MAX IV

U. Mueller[1], T. Ursby[1], R. Appio[1], J. Nan[1], J. Unge[1], M. Eguiraun[1], C. Ward[1], D. Logan[1], A. Nardella[1], F. Blomsten[1], M. Thunissen[1]
[1]*Lund University, MAX IV Laboratory, Lund, Sweden*

BioMAX is the first macromolecular crystallography beamline under development at the 3 GeV ring of MAX IV in Lund (Sweden). It is designed to become a world-class resource accommodating the majority of XRD experiments in structural biology that are being performed by the international user community. The experimental end-station will make use of advanced instrumentation for goniometry, sample delivery and diffraction detection. Nevertheless BioMAX is designed as well to be a stable and reliable research instrument that is easy to use, even by unexperienced users. The beam focus will be 20 x 5 μm^2 with a photon flux of 2 x 10^{13}ph/s. Smaller beam-crossections at the sample position are possible by aperture overfilling. A parallel beam option to resolve crystals with hugh unit cells will be made available too. The main commissioning and testing with beam will take place in 2016. We expect that the beamline can open to users early 2017 but it is of uttermost importance that the beamline is fully tested before being handed over to users. Some capabilities will be added later. BioMAX does not aim at the smallest beam or crystal sizes since it is foreseen that it will be complemented with the MicroMAX beamline, aiming at a beam size of 1 μm. The presentation will summarize the current state of the project and its future potential for the user community.

P002

MeshAndCollect: A New Synchrotron Serial Crystallography Method

U. Zander[1], G. Bourenkov[2], A. Popov[1], D. De Sanctis[1], O. Svensson[1], A. McCarthy[3], E. Round[4], V. Gordeliy[4], C. Mueller-Dieckmann [1], G. Leonard[1]
[1]*ESRF, Grenoble, France*
[2]*EMBL, Hamburg, Germany*
[3]*EMBL, Grenoble, France*
[4]*IBS, Grenoble, France*

The recent development in macromolecular crystallography on X-ray Free electron Lasers (XFELs) gave a new rise to serial crystallography methods on synchrotron sources. By making use of high-flux microfocus beamlines and fast readout photon counting detectors, the possibility of collecting high-resolution data from many micrometer-sized crystals, either by injecting them into the beam or by mounting them on a sample holder has already been demonstrated. However, processing data that has been collected in such a way style is challenging: the hit rate can be relatively low only very few diffraction images per crystal are collected which makes the determination of structure factors difficult. With MeshAndCollect we have developed a synchrotron serial crystallography method that allows protein crystallographers to collect data in a different way [1]: microcrystals are mounted on a standard sample holder and rastered in a low-dose X-ray mesh scan. The collected images are inspected automatically to identify protein diffraction patterns and well diffracting positions on the target are ranked. Subsequently, Data wedges over a small rotation range (e.g. 10°) are collected at each determined position. After the initial data processing, the best datasets for merging can be identified by hierarchical cluster analysis. So far, this method has been successfully used on different systems with crystal sizes down to 5um. MeshAndCollect is available to users on all MX beamlines of the ESRF.

[1] MeshAndCollect: an automated multi-crystal data-collection workflow for synchrotron macromolecular crystallography beamlines. U. Zander, G. Bourenkov, A.N. Popov, D. de Sanctis, O. Svensson, A.A. McCarthy, E. Round, V. Gordeliy, .C Mueller-Dieckmann and G. A. Leonard

Acta Cryst. (2015). D71, 2328-2343

P003

Pushing the limits of experimental phasing in the laboratory

A. Förster[1]
[1]*DECTRIS, Baden, Switzerland*

Success in experimental phasing, in particular when relying on the weak signal from light atoms like sulfur, phosphorous, potassium and calcium, critically depends on the accuracy of the measured anomalous differences. Concerted upgrades to radiation sources, beamline optics and detector instrumentation as well as software improvements have enabled the collection of more accurate data at synchrotron facilities, which drives the current resurgence of the native single anomalous dispersion (SAD) method [1]. Hybrid photon counting (HPC) detectors like the PILATUS3 and the recently introduced EIGER collect noise-free data at frame rates in the hundreds of hertz, allowing the user to collect better data in a shorter time.

This technology is also available for laboratory sources. One of the most exciting new data collection strategies for experimental phasing is the low-intensity, high-multiplicity method developed by Wang and colleagues [2]. By its very nature, this method is perfectly suited for laboratory instruments, where beam stability is high, radiation intensity is low and four-axis goniometers allow the collection of datasets with high multiplicity. Even if the measured anomalous differences are not sufficiently strong to allow for the phasing of data, they will facilitate or improve a molecular replacement solution and add information (e.g. on the identity of metal ions) to the final model.

A brief introduction of EIGER and PILATUS3 detectors will highlight key aspects of the underlying technology and compare their differences. The focus of the presentation will then be twofold. First, basic strategies to maximize the accuracy of the recorded diffraction data will be explored. In favorable cases and with carefully collected data, protein structures can be solved by experimental phasing with very little, but highly accurate, anomalous data. Second, advanced protocols for native SAD experiments will be outlined. These concern the collection of highly redundant low-dose data and the merging of data from multiple crystals [3]. These novel approaches to experimental phasing take crystallography beyond the previous state of the art set by fine phi slicing on PILATUS [4]. There is now no reason not to measure anomalous differences with every dataset and obtain better structural models as a result.

1. J. P. Rose, B.-C. Wang, M. S. Weiss, *IUCrJ.* 14, (2015), 431-40.
2. T. Weinert, V. Olieric, S. Waltersperger, E. Panepucci, L. Chen, et al., *Nature Methods*, 12, (2015), 131-133.
3. Q. Liu, Q. Liu, W. A. Hendrickson, *Acta Cryst.* D69, (2013), 1314-1332.
4. M. Mueller, M. Wang, C. Schulze-Briese. *Acta Cryst.* D68, (2012), 42-56.

P004

Automated software solutions for fragment-screening at the HZB MX beamlines

K. Sparta[1], J. Schiebel[2], F. Huschmann[1,2], M. Ühlein[1], A. Heine[2], G. Klebe[2], M. S. Weiss[1], U. Mueller[1]
[1]*Helmholtz-Zentrum Berlin, Macromolecular Crystallography, Berlin, Germany*
[2]*University of Marburg, Department of Pharmaceutical Chemistry, Marburg, Germany*

Thanks to continuous improvements of macromolecular crystallography (MX) beamlines at synchrotron sources worldwide, complete single-crystal X-ray diffraction data sets can now be routinely collected within a few minutes, using exposure times below one second per image. Two of the three beamlines managed by the HZB MX group are equipped with Pilatus detectors and sample changers: our beamline BL14.2 has recently been upgraded for fragment-screening experiments [1].

We have developed the expert software XDSAPP [2] to support users in the analysis of their diffraction data during measurements at the beamline with minimal effort and time. It mainly uses the diffraction data processing program XDS [3], along with additional software like POINTLESS from the CCP4 suite [4], XDSSTAT [5], SFCHECK [6] and PHENIX.XTRIAGE [7] for automatic decision making.

Furthermore, an independent refinement pipeline for automated fragment search based on PHENIX and COOT [8] has been developed for use after processing with XDSAPP at the HZB MX beamlines. Using a given PDB model, a refinement strategy dependent on data quality is applied to each data set. At the end of the pipeline, all unmodelled blobs in the 2Fo-Fc maps of all data sets are compared in order to select the blobs used to check the potential hits.

XDSAPP is available free of charge for academic users from www.helmholtz-berlin.de/bessy-mx. The refinement and fragment search pipeline is accessible to users working with our fragment-screening facilities [9].

References

[1] Mueller, U. *et al.* (2015). *Eur. Phys. J. Plus* **130**, 141-150.
[2] Krug, M. *et al.* (2012). *J. Appl. Cryst.* **45**, 568-572.
[3] Kabsch, W. (2012). *Acta Cryst.* D66, 125-132.
[4] Collaborative Computational Project (1994). *Acta Cryst.* **D50**, 760-763.
[5] Diederichs, K. (2006). *Acta Cryst.* **D62**, 96-101.
[6] Vaguine, A. *et al.* (1999). *Acta Cryst.* **D55**, 191-205.
[7] Adams, P.D. *et al.* (2010). *Acta Cryst.* **D66**, 213-221.
[8] Emsley, P. *et al.* (2010). *Acta Cryst.* **D66**, 486-501.
[9] Huschmann, F. *et al.* (2016). In preparation.

P005

Macromolecular data collection and processing solutions for home laboratory from Rigaku Oxford Diffraction.

T. Skarzynski[1], M. Meyer[2], P. Stec[2]
[1]*Rigaku Oxford Diffraction, Sevenoaks, Great Britain*
[2]*Rigaku Polska, Wroclaw, Poland*

While the majority of macromolecular X-ray data are currently collected using highly-efficient beam lines on an ever-increasing number of synchrotrons, there is still a need for low-maintenance, reliable systems for in-house experiments. In addition to crystal screening and optimization of x-ray experiments before a synchrotron trip, the home system allows collection of data as soon as the crystals are produced to get the initial solution of novel structures and is invaluable in the quick turnover often required in ligand-binding studies.

We will describe how the combination of the Rigaku Oxford Diffraction X-ray systems with the new version of fully automated CrysAlis^{Pro} data collection and processing software, optimized for macromolecular crystallography, results in ideal home lab solutions, complementing synchrotron data collection.

New unique features of CrysAlis^{Pro} software and examples of high quality results will be presented.

Figure 1. Rigaku Oxford Diffraction microfocus sealed-tube system with HPAD detector.

Figure 2. Multiple lattice de-convolution using CrysAlis^{Pro}

Figure 1

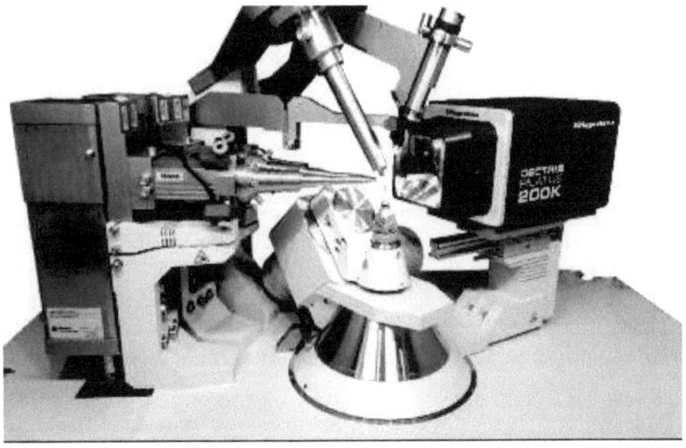

Figure 2

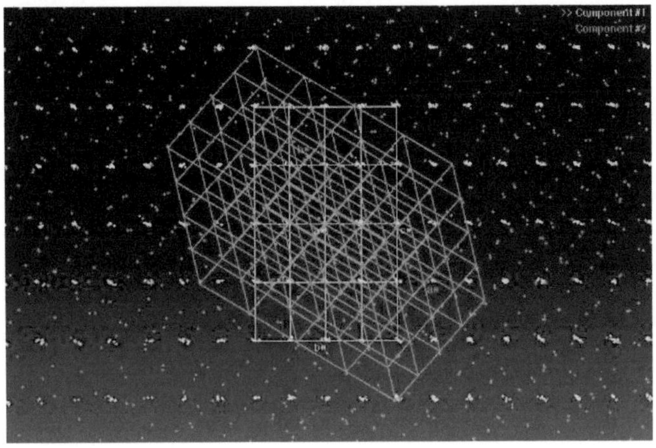

P006
HUMIDITY INDUCED PHASE TRANSITIONS OF HEW LYSOZYME INVESTIGATED BY MICROCRYSTALLINE POWDER DIFFRACTION ON A LABORATORY XRD SYSTEM

D. Beckers[1], T. Degen[1], G. Nénert[1], F. Karavassili[2], A. Valmas[2], S. Logotheti[2], S. Saslis[2], I. Margiolaki[2], S. Trampari[3]
[1]*PANalytical B.V., Almelo, Netherlands*
[2]*University of Patras, Patras, Greece*
[3]*Kapodistrian University of Athens, Athens, Greece*

Introduction
Proteins often crystallize in microcrystalline precipitates. The protein molecules are then surrounded by solvent and their packing arrangement is retained by limited intermolecular contacts. A change in the crystal environment first affects the bulk solvent that fills the intermolecular space, with resulting changes in the crystal structure. In literature it is reported that protein crystals in controlled humidity environments show a large change in unit-cell parameters when the humidity is decreased [1-2]. When a protein crystal is carefully dehydrated, it is in a metastable state in which the crystal initially still retains the original packing structure [2]. Further dehydration may cause the collapse of the crystal lattice; however in some crystals the dehydration induces a molecular arrangement change resulting in a new crystal structure [3].

Objectives
We studied dehydration and hydration processes of microcrystalline tetragonal and monoclinic HEWL (**H**en **E**gg **W**hite **L**ysozym) in the relative humidity range 55% - 95% by in situ measurements on a laboratory X-ray powder diffractometer. Phase transitions and cell parameter variations are compared for HEWL prepared with different crystallization conditions.

Materials & Methods
Several batches of microcrystalline HEWL samples were investigated on a laboratory X-ray powder diffractometer equipped with a temperature controlled humidity chamber. The samples were crystallized with precipitant solutions partly with and partly without PEG4000. After crystallization the samples were prepared by centrifugation and pipetting the concentrated precipitates onto the sample holders of the humidity chamber. *In situ* measurements were performed at room temperature under variable relative humidity conditions. The observed gradual structural changes as well as phase transitions upon dehydration and hydration were analyzed in the relative humidity range 55% - 95% for different stabilization times after humidity change.

Results
Dehydration and hydration processes of tetragonal HEWL are reversible in humidity cycles in the range of 95% rH to ~75% rH. Without stabilizing PEG the lower limit for dehydration of HEWL is between 71% rH and 80% rH. With PEG the HEWL samples remain crystalline below 75% rH, but show phase transitions and larger variations of the cell parameters. Below 75% rH a new tetragonal polymorph was discovered.

The dehydration and hydration behavior of monoclinic HEWL was in contrast more sensitive to the variation of stabilization times in the humidity cycle.

Conclusions
The flexibility and sensitivity of modern multi-purpose X-ray laboratory systems allows the crystallographic study of polycrystalline proteins by X-ray powder diffraction (XRPD). Polymorph identification and the analysis of structural changes of protein precipitates upon dehydration and hydration using a humidity chamber show its potential for troubleshooting and process analysis in the production of protein based pharmaceuticals.

[1] Kiefersauer, R., et al. (2000). J. Appl.Cryst. 33, 1223-1230.
[2] Dobrianov, I., et al.(1999).J. Cryst. Growth, 196, 511-523.
[3] K. Harata, T. Akiba, Acta Cryst.(2007). D63, 1016-1021

P007
Tailor-made beams for macromolecular crystals

A. Polyakova[1], G. Bourenkov[1], I. Karpics[1], S. Fiedler[1], F. Liu[1], T. Schneider[1]
[1]*European Molecular Biology Laboratory, Hamburg outstation, Hamburg, Germany*

Introduction
Recently, a transfocator system (Vaughan et al, 2011) encompassing a series of Compound Refractive Lenses (CRLs) has been installed at the EMBL beamline P14 at PETRA III. Acting as an additional focusing device, the CRLs produce a beam with different focal properties across a wide energy range. The transfocator allows the amount of CRLs in the beam path to be easily changed, adjusting the focal distance and hence the beam size to match the dimensions of protein crystals of a varying sizes.

Objectives
The introduction of CRLs will further increase the range of available beam properties at P14. The main objective of this study is to establish the conditions required to adjust the beam size to crystals in the size range of 10-500 μm and making the CRLs available to P14 users through the beamline control software MxCuBE (Gabadinho et al, 2010), enabling rapid and easy adjustment of the beam size.

Materials & Methods
Following characterisation of the beam size and flux as obtained by employing different combinations of CRLs, the quality of the diffraction data from several macromolecular crystal systems is evaluated in order to establish optimum conditions. Diffraction data is collected from standard test systems (lysozyme, insulin, trypsin), as well as more challenging crystals from collaborators from the Karolinska Institute in Sweden.

Results
Initial experiments have shown that the CRLs can be successfully used to adjust the size of the beam at P14 in the 20 to 150 micron range for X-ray energies between 6.1-12.7 keV. Diffraction data acquired at various energies have also shown that high quality native and anomalous diffraction data can be collected with CRLs in the beam path.

Conclusions
CRLs have been integrated and demonstrated to work well at the crystallography beamline P14. Initial beam characterisation and diffraction data collection on several crystal systems have shown that the beam size can be adjusted over a wide energy range while allowing the collection of high quality diffraction data. These results show promise of the CRLs soon to be made available to P14 users, enabling them to optimise data quality for given crystals by matching the beam properties to the crystal.

The project is performed in a collaboration with the groups of Gunter Schneider and Robert Schnell (Karolinska Institute, Stockholm) and is funded by the RÅC programme.

References
Gabadinho, J., et al. (2010). MxCuBE: a synchrotron beamline

control environment customized for macromolecular crystallography experiments. *Journal of Synchrotron Radiation*, *17*(5), 700-707.

Vaughan, G. B. M., et al. (2011). X-ray transfocators: focusing devices based on compound refractive lenses. *Journal of Synchrotron Radiation*, *18*(Pt 2), 125-133.

P008

Low multiplicity Cadmium SAD phasing at 12 keV

S. Panneerselvam[1], A. Burkhardt[1], A. Meents[1]
[1]*DESY, Hamburg, Germany*

SAD phasing becoming a fast and reliable method for macromolecular crystal structure determination due to the recent developments in light sources and beam line instrumentation to collect a high quality data (1). Still most of the SAD phasing is routinely carried out by using X-ray beam of longer wavelength which in turn leads to the radiation damage due to absorption, beam instability and requires additional instrumentation to avoid air absorption related issues.

We have studied the capability of Cadmium SAD phasing of macromolecular crystal structures at the typical wavelength of 1Å, where the most of the synchrotron beamlines were optimized for beam stablity and data collection. Cadmium ions are shown to improve the growth and diffraction quality of macromolecular crystals (2).

Hen egg-white lysozyme crystals were obtained in the presence of 25 mM Cadmium sulfate and a high resolution diffraction data set was collected at a wavelength of 1.033Å at the beam line P11, PETRA III, DESY, Germany. The complete diffraction dataset comprised of 9000 frames (360°) was collected by using a fine oscillation increment of 0.04°, exposure time of 60 ms and recorded with a Pilauts 6MF detector.

The dataset was integrated by XDS(3) and SAD phasing analysis was carried out with SHELX programs (4). The SAD phasing was successful with automatically tracing of 126 residues from the complete 129 residues. Inorder to check the minimum multiplicity needed for SAD phaising, the complete data set was split in to smaller wedges and analyzed for successful SAD phasing. The minimum data necessary for a successful SAD phasing from this data set was comprised of 800 frames(32°) with a relatively low anomalous completeness (44.8%) and low anomalous multiplicity (1.1). We have further analyzed the dependence between the anomalous data multiplicity and the critical resolution limit for the correct substructure determination.

References:

1. Rose JP, Wang BC, Weiss MS. Native SAD is maturing. **IUCrJ. 2015** May 14;2(Pt4):431-40.
2. Trakhanov S, Kreimer DI, Parkin S, Ames GF, Rupp B. Cadmium-induced crystallization of proteins: II. Crystallization of the Salmonella typhimurium histidine-binding protein in complex with L-histidine, L-arginine, or L-lysine. **Protein Sci. 1998** Mar;7(3):600-4.
3. Kabsch W. XDS. **Acta Crystallogr D Biol Crystallogr. 2010** Feb;66(Pt 2):125-32.
4. Sheldrick GM. Experimental phasing with SHELXC/D/E: combining chain tracing with density modification. **Acta Crystallogr D Biol Crystallogr. 2010** Apr;66(Pt 4):479-85.

P009

New beam lines for structural biology at the ESRF

D. von Stetten[1]
[1]*ESRF, Structural Biology, Grenoble, France*

Six beam lines for macromolecular crystallography are currently operational at the ESRF, of which three have recently been opened to users on the canted ID30 straight section. Three of the ESRF Structural Biology beam lines are operated at a fixed energy near the selenium edge (12.8 keV), while the other three are tunable over a large energy rang (6-20 keV). The latest hard- and software developments on these beam lines will be presented. The fixed-energy MASSIF-1 (Massively Automated Sample Selection Integrated Facility) beam line is equipped with a newly developed sample-changing robot (RoboDiff), which also functions as goniometer, and focuses on fully automatic high-throughput sample evaluation and data collection based on user-defined criteria (*e.g.*, resolution, presence of ligands or heavy atoms). The fixed-energy MASSIF-3 beam line offers a very intense ($1.5x10^{13}$ ph/s) and small (15 μm diameter) beam, and is thus very well suited for small crystals. It is now permanently equipped with an online microspectrophotometer which allows to easily record UV/vis absorption and fluorescence spectra. On the tunable beam line ID30B a newly developed robotic sample changer (FlexHCD) is being commissioned, which can handle SPINE standard and UniPuck samples in a high-capacity dewar as well as crystallisation plates. In addition, offline high-pressure cooling setups allow to cryo-cool crystals under up to 2000 bar of noble gases or 40 bar of oxygen, which allows avoiding the use of cryoprotecting agents and/or exploring gas and water channels in the protein. Advanced automated workflows (*e.g.*, X-ray centring to find small crystals, helical data collection for needle-shaped crystals, and multi-crystal strategies for serial crystallography) help the users to optimally exploit their samples to obtain the best possible diffraction data sets. In combination with automatic processing of all collected data and the ISPyB database for keeping track of samples and results, collecting diffraction data of a large number of samples on the ESRF structural biology beam lines is easy, fast, and efficient.

P010

Status of the Crystallography Endstation at Beamline P11 at PETRA III

A. Burkhardt[1], S. Panneerselvam[1], O. Lorbeer[1], B. Reime[1], T. Pakendorf[1], N. Stübe[1], P. Fischer[1], J. Meyer[1], P. Rödig[1], D. Göries[1], M. Warmer[1], A. Meents[1]
[1]*Deutsches Elektronen-Synchrotron DESY, Photon Science, Hamburg, Germany*

The "Bio-Imaging and Diffraction Beamline" P11 at PETRA III provides two state-of-the-art endstations for structural investigations of biological samples at different length scales: an X-ray microscope utilizing tender X-rays between 2.4 and 10 keV [1, 2, 3] and a crystallography experiment operated between 5.5 and 30 keV [4].

Basis of beamline design was to make full use of the excellent source properties of PETRA III and to deliver most of the photons from the source into a very small focal spot at the sample position.

The P11 optics concept involves an LN_2 cooled double crystal monochromator, followed by a first Kirkpatrick-Baez (KB) system which is composed of two horizontal and one vertical deflecting mirror. All KB mirrors can be dynamically bent and, in the case of crystallography experiments, are used to generate a secondary source at 65.50 m with a size of 25×210 μm^2 (FWHM, v × h). A second refocusing KB system which is installed in the P11 experimental hutch is used for refocusing the X-ray beam with full flux from the source (2×10^{13} ph/s at 12 keV). With this, beam sizes between 4 × 9 µm² (2nd KB system fully bent) and 300 × 300 µm² (2nd KB system flat, FWHM, v × h) can be generated at the sample position at 73.30 m. Smaller beam sizes down to 1 × 1 µm² with more than 2 × 10^{11} ph/s in the focus can be obtained by slitting down the secondary source at the cost of flux.

The flexible X-ray optics allow to tailor the beam properties to the needs of the experiment: A large parallel beam is available for investigations of large unit cell systems, such as large molecular complexes [5] and virus crystals. A highly intense microbeam allows for structure determinations from microcrystals including serial crystallography experiments using liquid delivery systems [6] or silicon microchips [7, 8].

The P11 crystallography endstation (see Fig. 1) is equipped with a high precision single axis goniostat with a combined sphere of confusion of less than 100 nm. Crystals can be mounted in less than 10 s using an automatic sample changer. The robot is combined with a large storage Dewar which provides space for 368 samples. In addition, a Pilatus 6M-F detector allows for fast data collection with frame rates of up to 25 Hz. This makes P11 ideally suited for high-throughput crystallography and fast crystal screening.

Figure 1: Overview of the P11 crystallography endstation.

[1] A. Meents, B. Reime, N. Stübe et al., Proceedings of SPIE 8851, 88510K (2013).
[2] I. Vartiainen, M. Warmer, D. Göries et al., J. Synchrotron Radiat. 21, 790-794 (2014).
[3] K. Stachnik, I. Mohacsi, I. Vartiainen et al., Appl. Phys. Lett. 107, 011105 (2015).
[4] A. Burkhardt, T. Pakendorf, B. Reime et al., Eur. Phys. J. Plus, accepted.
[5] K. Schulte, N. Pawlowski, K. Fälber et al., submitted.
[6] F. Stellato, D. Oberthür, M. Liang et al., IUCrJ 1, 204-212 (2014).
[7] P. Rödig, I. Vartiainen, R. Duman et al., Scientific Reports 5, 10451 (2015).
[8] P. Rödig, R. Duman, J. Sanchez-Weatherby et al., submitted.

Figure 1

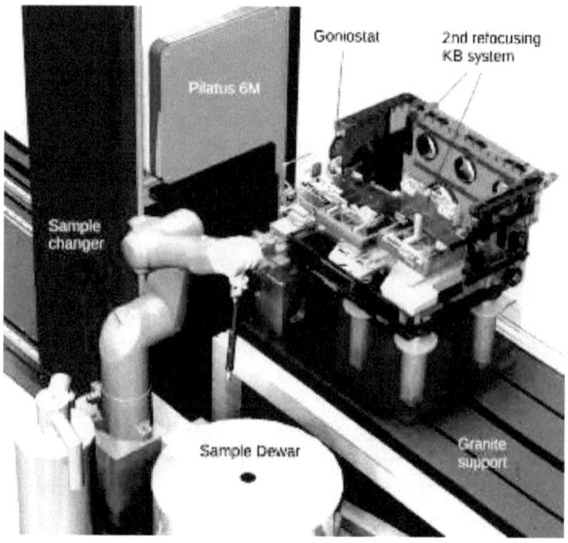

P011

BioDiff - a neutron diffractometer optimized for crystals with large unit cell dimensions

T. E. Schrader[1], A. Ostermann[2], M. Monkenbusch[1,3], B. Laatsch[4], P. Jüttner[2], W. Petry[2], D. Richter[1,3]
[1]Forschungszentrum Jülich GmbH, Jülich Centre for Neutron Science (JCNS), Garching, Germany
[2]Technische Universität München, Heinz Maier-Leibnitz Zentrum (MLZ), Garching, Germany
[3]Forschungszentrum Jülich GmbH, Institute for Complex Systems ICS, Jülich, Germany
[4]Forschungszentrum Jülich GmbH, Engineering and Technology (ZEA-1), Jülich, Germany

The newly built neutron single crystal diffractometer BIODIFF at the research reactor Heinz Maier-Leibnitz (FRM II) is especially designed to collect data from crystals with large unit cells. The main field of application is the structural analysis of proteins, especially the determination of hydrogen atom positions. BIODIFF is a joint project of the Jülich Centre for Neutron Science (JCNS) and the FRM II. Typical scientific questions addressed are the determination of protonation states of amino acid side chains and the characterization of the hydrogen bonding network between the protein active centre and an inhibitor or substrate. BIODIFF (see picture below) is designed as a monochromatic instrument with a narrow wavelength spread of less than 3 %. By using a highly orientated pyrolytic graphite monochromator (PG002) the diffractometer is able to change its central wavelength in the range between 2.4 Å and 5.6 Å. Contaminations of higher order wavelengths are removed by a neutron velocity selector. To cover a large solid angle the main detector of BIODIFF consists of a neutron imaging plate in a cylindrical geometry with online read-out capability. With a radius of 200 mm and a height of 450 mm it covers a solid angle of approximately 2π with a spatial resolution of up to 125 µm. A fast Li/ZnS scintillator CCD camera is available for additional detection abilities. An optical CCD-camera pointing at the sample position is used to quickly align the sample with respect to the neutron beam. The main advantage of BIODIFF is the possibility to adapt the wavelength to the size of the unit cell of the sample crystal while operating with a cleanmonochromatic beam that keeps the background level low. BIODFF is equipped with a standard Oxford Cryosystem "Cryostream 700+" which allows measurements in the temperature range from 90 K up to 500 K. A standard closed cycle cryostat is also available for measurements down to 4 K in vacuum. This can be used for the determination of magnetic unit cells of inorganic crystals in zero magnetic field. Recent applications of BIODIFF also include clay powder diffraction and diffraction from lipid membranes in a controlled humidity environment.

Figure 1

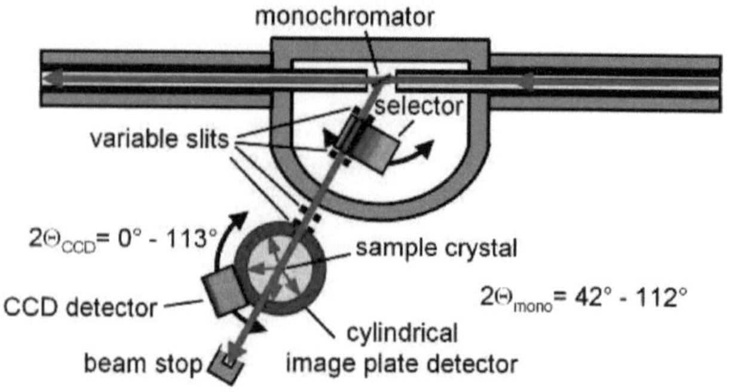

Figure 2

P012

UsnRNP assembly machinery of *Schizosaccharomyces pombe*

J. Veepaschit[1], C. Grimm[1], U. Fischer[1]
[1]*University of Wuerzburg, Biochemistry, Wuerzburg, Germany*

Eukaryotic protein coding genes are discontinuous in nature. The coding sequences (exons) of an eukaryotic gene are interrupted by non-coding sequences (introns). Removal of introns from a primary mRNA transcript and subsequent ligation of exons to generate a mature mRNA is carried out by the spliceosome. The spliceosome consists of several protein-RNA complexes called **u**ridine-rich **s**mall **n**uclear **r**ibo**n**ucleo**p**roteins (UsnRNPs). During the biogenesis of UsnRNPs, assembly of the heptameric Sm core (common to all UsnRNPs; composed of 7 Sm proteins and an UsnRNA) is mediated by protein arginine methyltransferase 5 (PRMT5) complex and survival of motor neuron (SMN) complex. The PRMT5 complex symmetrically dimethylates Sm proteins which are then recognized by the SMN component of the SMN complex. The SMN complex acts as a catalyst to generate the mature Sm core. The human SMN (hSMN) complex consists of 9 components namely SMN, Gemins 2-8 and Unrip. Atomic structures of the C-terminal oligomerisation domain (YG-box), middle tudor domain and an N-terminal helix (in complex with Gemin2) of the SMN protein have been solved independently. In addition, structure of Gemin7:Gemin6 heterodimer has also been solved. The atomic structures of Gemin8 and the whole SMN complex, however, remain elusive. For our study we chose the SMN complex of *Schizosaccharomyces pombe* (ySMN complex) which is rather minimalistic, consisting of only ySMN, yGemin2, yGemin8, yGemin7 and yGemin6. The ySMN and yGemin2 have been established to be true orthologs of hSMN and hGemin2. So far there is no biochemical data for the newly discovered yGemin8, yGemin7 and yGemin6. We set out to establish an interaction map for the ySMN complex by domain mapping and interaction studies. Protein complexes were co-expressed in *E. coli* and purified by Ni-NTA and size exclusion chromatography. yGemin7:yGemin6 exists as a dimer in solution. ySMN binds to yGemin2 and oligomerises via its YG-box. yGemin8 serves as the link between ySMN:yGemin2 and yGemin7:yGemin6. The YG-box is necessary for the soluble expression of yGemin8 and is therefore likely the binding site for yGemin8. Two predicted N-terminal helices of yGemin8 are necessary for its binding to ySMN:yGemin2 complex and are likely the binding site for YG-box. yGemin7:yGemin6 heterodimeric complex yielded crystals. *In situ* limited proteolysis experiments during the crystallization trials of ySMN:yGemin2 have produced several hits. A crystal hit has also been obtained for the yGemin8ΔN-terminus:yGemin7:yGemin6 heterotrimeric complex. In order to remove flexible parts that might hinder crystallization, protein complexes were subjected to limited proteolysis, purified via gelfiltration and crystallization trials were setup. In conclusion, our results indicate that the ySMN complex has an interaction map similar to the hSMN complex. The ySMN complex proteins could be expressed in *E. coli* and purified with high purity and yield unlike their human counterparts. The initial hits obtained for yGemin7:yGemin6, ySMN:yGemin2 and yGemin8ΔN-terminus:yGemin7:yGemin6 suggest that the crystallization propensity of the complexes can be improved by further construct optimization. Due to its smaller size, ySMN complex is a suitable system for structural studies.

Figure 1

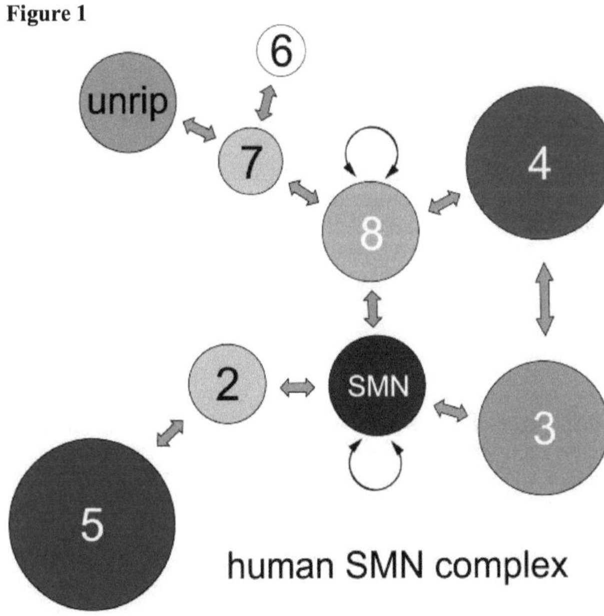

Figure 2

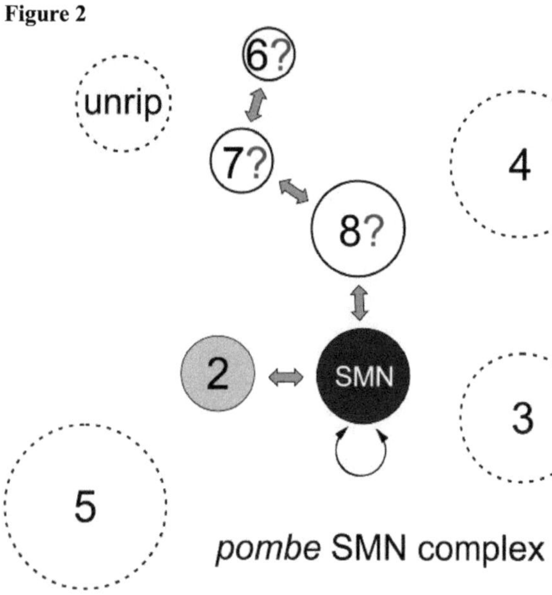

P013

Synthesis, crystal structure and physical properties of Bi3TeBO9

M. Daub[1], M. Krummer[1], H. Hillebrecht[1]
[1]*Albert-Ludwigs-Universität Freiburg, Institut für Anorganische und Analytische Chemie, Freiburg, Germany*

Borates are a well-known class of compounds for interesting optical properties.[1] The preferred coordination polyhedra are trigonal-planar BO_3 units and BO_4 tetrahedra, which both favor non-centrosymmetric structures. Furthermore, bismutates(III) frequently show low-symmetric surroundings of Bi^{3+} because of the ns^2 electron pair. Therefore, Bi(III)-containing borates are promising materials for new non-centrosymmetric structures. In continuation of our work on borosulfates [2] and boroselenates [3] we now report on the first borate tellurates of bismuth.

Pale-yellow single crystals of Bi_3TeBO_9 were obtained by heating stoichiometric amounts of B_2O_3, H_6TeO_6, and Bi_2O_3 in a Pt crucible up to 780 °C followed by slow cooling to RT. The crystal structure of Bi_3TeBO_9 ($P6_3$, $Z = 2$, $a = 8.7451(16)$Å, $c = 5.8917(11)$Å), 575 refl, 44 param. $R_1 = 0.033$, $wR_2 = 0.089$) represents a new structure type. It is characterized by isolated trigonal-planar BO_3 units and slightly distorted TeO_6 octahedra. The Bi^{3+} cations are localized between these polyhedra. The resulting coordination of Bi^{3+} is an octahedron with three shorter and three longer distances (2.25/2.50Å). Alternatively, the crystal structure can also be described as a 3D-net of corner-sharing BiO_6 octahedra with Te in octahedral and B in trigonal-planar voids. Overall, the oxygen atoms form a haexagonal closest packing with Bi and Te in octahedral voids. According to the topopogy of *hcp* voids which are linked by a common face must be occupied if the share of occupied voids exceeds one half. This is avoided in Bi_3TeBO_9 because B is in trigonal-planar voids.

Fig.1: Crystal structure of Bi_3TeBO_9 with BO_3 and TeO_6-polyhedra (left) and BiO_6 octahedra (right).

The vibrational spectra can be described as a superposition of the modes of isolated trigonal-planar BO3 units, isolated TeO6 octahedra and additional Bi-O modes. The isotopic splitting 10B/11B of the antisymmetric valence mode (IR: 1282/1248cm-1) is well resolved. The optical spectra confirm the color of the crystals as they show a slight absorption edge around 400 nm. The signals of the MAS-NMR spectra of 11B and 125Te are in line with the expectations for trigonal-planar and octahedral surroundings, respectively.

[1] P. Becker, Adv. Mater. 1998 10, 979-997.
[2] M. Daub, K. Kazmierczak, H. Höppe, H. Hillebrecht Chem. Eur. J., 2013, 19, 16954-16962.
[3] M. Daub, H. Hillebrecht, Chem. Eur. J., 2015, 21, 298-304.

Figure 1

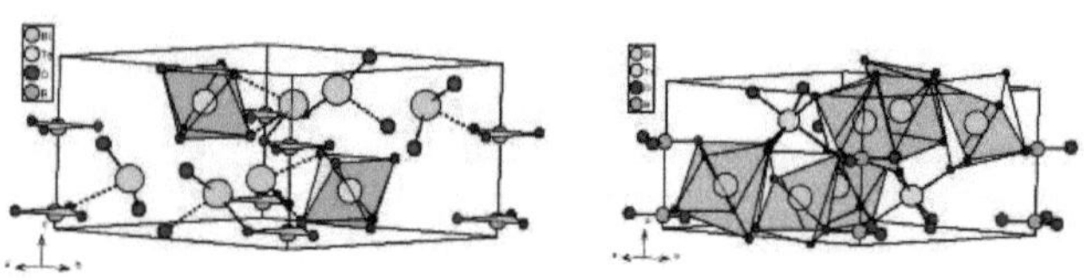

P014

Structure of the Leucine-rich Repeat of the *Salmonella*-recognizing E3 ubiquitin Ligase LRSAM1

C. Behrens[1], A. Scrima[1]
[1]*Helmholtz Centre for Infection Research, Braunschweig, Germany*

Introduction

In autophagy, damaged organelles and aggregated proteins are targeted to the lysosome via the enclosure in a *de novo* formed autophagosome. Autophagy is also known to function in innate immunity to restrict bacterial growth and survival in the host cell cytoplasm. To this end, free or vacuole-contained bacteria are marked with ubiquitin. Polyubiquitin coats around bacteria are detected by autophagy adaptors that consecutively recruit the autophagic machinery, resulting in lysosomal degradation of bacterial pathogens. The Leucine-rich repeat (LRR) containing RING E3 Ligase LRSAM1 was shown to function as E3 Ligase in anti-*Salmonella* autophagy (Huett et al 2012). LRSAM1 is binding to the pathogen through interaction of its N-terminal LRR with an unknown molecular pattern on the pathogen. This activates the C-terminal RING domain which subsequently generates a polyubiquitin coat around the bacterium. The autophagy adaptor NDP52 recognizes ubiquitin around the bacterium, but can also bind to LRSAM1 directly. This mechanism leads to bacterial capture in an autophagosome.

Objectives

Our current knowledge of LRSAM1 leaves several questions to be answered: What is the molecular pattern on the bacterium that is recognized by LRSAM1 LRR? How does binding of the LRR to this pattern change the activity of the catalytic RING domain? To gain insight into LRSAM1 function, structure and its activation, we set out for recombinant purification and crystallization of LRSAM1 and/or its domains.

Materials & Methods

LRSAM1 was recombinantly expressed and purified. A fragment of LRSAM1 was successfully crystallized by vapor diffusion method. Crystals were soaked with an Yb derivative and the anomalous signal was employed for SAD phasing.

Results

The structure of LRSAM1 LRR was solved to 1.8 Å resolution. The asymmetric unit contains two molecules related by NCS, which overlap with an rms of 0.098 Å. Seven Leucine rich repeats are enwrapped by helical elements at the N- and C-terminus. The entire LRR spans about 50 Å in width and 25 Å in height. In the dimer, the two inner surfaces of the LRRs are packed against each other in an almost perpendicular fashion. Next, we tested LRSAM1 function in ubiquitination assays. LRSAM1 shows little intrinsic activity when supplied with ubiquitin, E1/E2 enzymes and ATP. We also investigated whether LRSAM1 can be activated by pathogen associated recognition patterns (PAMP) and identified one, which lead to robust activation of the enzyme resulting in autoubiquitination. We confirmed this result by mass spectrometry analysis. Tryptic peptides of LRSAM1 were found to be ubiquitinated on several lysines covering all domains of LRSAM1 except the LRR and RING domain.

Conclusion
We were able to solve the structure of the LRR of the *Salmonella*-recognizing E3 ubiquitin ligase LRSAM1 at 1.8 Å resolution. The LRR includes seven repeats and is enwrapped by capping helices. In in vitro functional assays, we observed activation and specific autoubiquitination of LRSAM1 upon the addition of a PAMP. Further studies will be necessary to give detailed insight into LRSAM1 interaction and activation.

References
Huett, A. et al., 2012. The LRR and RING Domain Protein LRSAM1 Is an E3 Ligase Crucial for Ubiquitin-Dependent Autophagy of Intracellular Salmonella Typhimurium. *Cell Host & Microbe*, 12(6), pp.778-790.

P015
Structural and functional characterization of Invasin-like proteins
P. Sadana[1], P. Dersch[1], A. Scrima[1]
[1]*Helmholtz centre for Infection Research, Structural Biology of Autophagy, Braunschweig, Germany*

INTRODUCTION
Yersinia spp. and other enteric bacteria armour themselves with variety of adhesins that help them to adhere and invade the host tissues, escape and modulate the host immune system. Invasin-like proteins represent a subgroup of the adhesin family that specifically interacts with receptors on the host cell plasma membrane. Invasin-like proteins share a common architecture: they consist of i) an N-terminal beta-barrel domain, ii) a central region of varying length consisting of highly homologous repeats of Ig-like domains and iii) the C-terminal part, which varies significantly among Invasin-like proteins. The C-terminal domain provides specificity to Invasins for interaction with host-derived factors.

OBJECTIVE
The aim of this work is to get insight into the structure and function of the C-terminal domain of two Invasins, Invasin D (InvD) and Invasin E (InvE). The C-terminal part is significantly different, not only to other members of the Invasin-like proteins, but also between InvD and InvE. These studies will improve our understanding of the receptor-specificity of Invasins and their role in the internalization and infection process.

MATERIALS & METHODS

To achieve this, both the variants were cloned and expressed in *E. coli* and were purified by affinity chromatography and gel filtration. Both the proteins were successfully crystallized. InvD crystals diffracted to 2.6Å resolution and the structure was solved by combining molecular replacement and single wavelength anomalous dispersion (MR-SAD) while InvE crystals diffracted to 1.8Å resolution and the structure was solved by SAD phasing.

RESULTS
The N-terminal domains of crystallized InvD and InvE reveal similarities to the Immunoglobulin Superfamily (IgSF) fold. In contrast, the C terminal domain of InvD adopts a novel fold and the InvE C-terminal domain adopts a C-type lectin like fold.

CONCLUSION
We have solved the structure of two novel adhesins from *Yersinia pseudotuberculosis.*

The C-terminal domain of these two adhesins share weak structural similarity with proteins of different functions suggesting a role in targeting InvD/InvE to a specific, yet unidentified receptor on the host cell plasma membrane.

P016
Crystallization of *Streptomyces mobaraensis* transglutaminase substrates
A. Scrima[1]
[1]*Helmholtz Centre for Infection Research, Braunschweig, Germany*

Stefan Schmelz[1], David Fiebig[2], Vanessa Siegmund[3], Harald Kolmar[3], Hans-Lothar Fuchsbauer[2] and Andrea Scrima[1]

[1]Arbeitsgruppe Strukturbiologie der Autophagie, Abteilung Struktur und Funktion der Proteine, Helmholtz-Zentrum für Infektionsforschung, Inhoffenstr. 7, 38124 Braunschweig (Germany)

[2]Fachbereich Chemie- und Biotechnologie, Hochschule Darmstadt, Schnittspahnstraße 12, 64287 Darmstadt (Germany)

[3]Clemens-Schöpf-Institut für Organische Chemie und Biochemie, Technische Universität Darmstadt, Alarich-Weiss-Straße 4, 64287 Darmstadt (Germany)

Streptomyces mobaraensis DSM 40847 produces transglutaminase (MTG, P81453), which cross-links proteins irreversibly through N-γ-glutamyl-lysine isopeptide bonds. Although transglutaminases (TGases) are ubiquitously distributed among animals, plants and lower organisms, the MTG is unique. Structure analysis does not reveal any relationship to other TGases, not even to the intracellular TGase from *Bacillus subtilis*. One substrate of MTG is the *Streptomyces* subtilisin inhibitor (SSTI). Three so far unknown additional MTG substrates from *S. mobaraensis* culture supernatant have been identified: the dispase-autolysis-inducing protein (DAIP), the *Streptomyces* papain inhibitor (SPI), and an unpublished β-lactamase. Although these MTG substrates do not share any structural homology, all show antimicrobial activity. Structural analysis of MTG substrates is essential to get insights into their individual function as antimicrobial/protecting agent as well as their specific interactions as MTG substrates. We thus set out to determine the structures of the individual MTG substrates and characterize the structural properties mediating efficient modification by MTG. Alongside the structural characterization, we wanted to shed light onto each molecular mechanism to achieve antimicrobial activity. We were able to solve the structures of all four substrates by X-ray crystallography: SSTI, DAIP, SPI and *β*-lactamase (Figure 1). One interesting MTG substrate, the 7-bladed β-propeller DAIP, is known to induce autolysis of neutral bacterial metalloproteases such as bacillolysin and thermolysin, thereby reducing pathogenicity of invading bacteria. The mechanism however is still unclear. The crystal structure of DAIP was solved to 1.7 Å resolution using Yb-SAD phasing, with two Yb-ions, which are partially occupying potential Ca^{2+}-binding sites. The structure provides first insights into the orientation of the potential MTG binding sites. Interestingly, most glutamines are located on the same side, which correlates with biochemical and mutational studies. Although preferred MTG binding sites have been located and one site was used for efficient conjugation to a reporter cargo (recently published in Angewandte Chemie [1]), it is not yet clear how DAIP acts on dispase. Two possible mechanisms are likely: the β-propeller acts as structural scaffold to recruit the dispase and induce autolysis, or DAIP chelates Ca^{2+} ions to prime auto-catalytic degradation. We favor the latter mechanism and have first results to show that DAIP is likely to scavenge crucial Ca^{2+} ions.

[1] Siegmund V, Schmelz S, Dickgiesser S, Beck J, Ebenig A, Fittler H, Frauendorf H, Piater B, Betz UA, Avrutina O, Scrima A, Fuchsbauer HL, Kolmar H (2015) Angew Chem. doi: 10.1002/anie.201504851

Figure 1

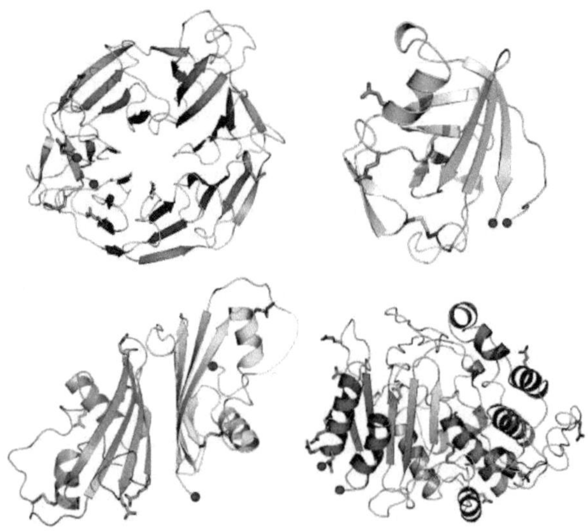

Materials: Syntheses and structure

P017

Synthesis and Crystal Structure of $In_{4.64}Ho_{3.36}S_{12}$

I. Hartenbach[1], T. Lebherz[1], T. Schleid[1]
[1]*University of Stuttgart, Institute for Inorganic Chemistry, Stuttgart, Germany*

An attempt to synthesize indium(I) derivatives of holmium(III) sulfide thiosilicate(IV) with the formula $InHo_3S_3[SiS_4]$ by fusing elemental indium, holmium, and sulfur together with indium trichloride and silicon disulfide in evacuated silica ampoules at 600 °C for 10 days did not result in the desired compound, but among other by-products, red, needle-shaped single crystals of $In_{4.64}Ho_{3.36}S_{12}$ emerged from this reaction mixture. The title compound crystallizes monoclinically in the space group $P2_1/m$ (a = 1094.70(4), b = 387.26(1), c = 2117.81(7) pm, β = 96.421(2)°; CSD-430425) with two formula units per unit cell, and in principle isotypically to the indium terbium sulfide $In_5Tb_3S_{12}$ [1]. The In^{3+} cations are surrounded octahedrally (In1 – In4) and tetrahedrally (In5), respectively, by eleven of the twelve crystallographically independent sulfide anions. The polyhedra around the In^{3+} cations build up a three-dimensional network by sharing common edges and vertices. The *non*-indium connected $(S12)^{2-}$ anions are surrounded exclusively by five Ho^{3+} cations each in the shape of square pyramids, which form chains along [010] via mutual edges. These chains reside within the voids of the aforementioned network (Figure 1). The three Ho^{3+} cations exhibit an either monocapped (Ho4, CN = 7) or bicapped (Ho2 and Ho3, CN = 8) trigonal-prismatic environment of sulfide anions. During the crystal structure refinement unusual small atomic displacement factors for the position of the $(In1)^{3+}$ cations occurred, which led to the conclusion of a mixed occupation with Ho^{3+} cations. The respective refinement of this position with In^{3+} against Ho^{3+} resulted in improved residual and atomic displacement values as well as in an (In1) : (Ho1) ratio of 64 : 36. Since the isotypic terbium derivative [1] also displays one of the indium positions with very small atomic displacement parameters, it is assumed that the possibility of a mixed occupation of this position was overlooked and the real formula is better described as $In_{5-x}Dy_{3+x}S_{12}$.

[1] D. Carré, *Bull. Soc. Chim. Fr.* **1977**, *33*, 1163-1166.

Figure 1. View at the crystal structure of $In_{4.64}Ho_{3.36}S_{12}$ along [001].

Figure 1

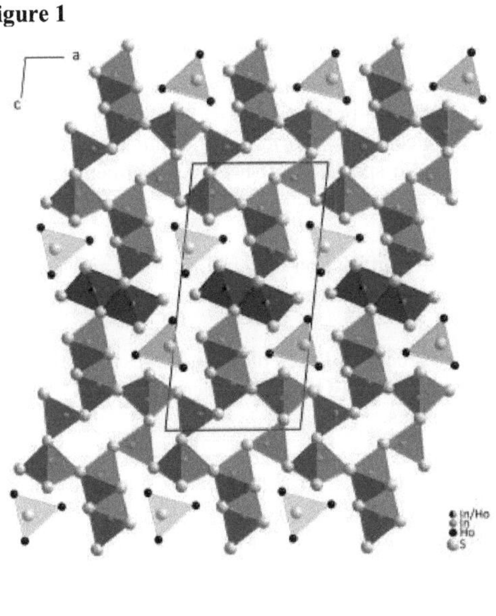

P018

2D coordination polymer employing deprotonated 3,3'-bipyridine-2,2' dicarboxylate as a ligand

A. Kochel[1]
[1]*University of Wrocław, Faculty of Chemistry, Wrocław, Poland*

The synthesis, structure, and properties of $[Mn(3,3'\text{-bpdc})]_n$ **1** will be presented. The ligand is coordinated bidentately through the N and O atoms and monodentately through its carboxylic O atoms to create a 2D coordination polymer. In the $[Mn(3,3'\text{-bpdc})]_n$, the polymeric layers are parallel to the [110] direction. Within these layers, O-H···O and weak C-H···O/N hydrogen bonds as well as weak C-H···π contacts can be found. Additionally, TGA studies of decomposition compound under the nitrogen atmosphere as well as studies of magnetic properties and topological analysis of $[Mn(3,3'\text{-bpdc})]_n$ were performed. Magnetic properties of the manganese(II) complex show weak antiferromagnetic interactions.

Figure 1.

Coordination sphere of the Mn^{2+} ion. Hydrogen atoms are omitted for clarity. Thermal ellipsoids are plotted at the 50% probability level.

Project was supported by Wroclaw Centre of Biotechnology, program The Leading National Research Centre (KNOW) for years 2014-2018.

Figure 1

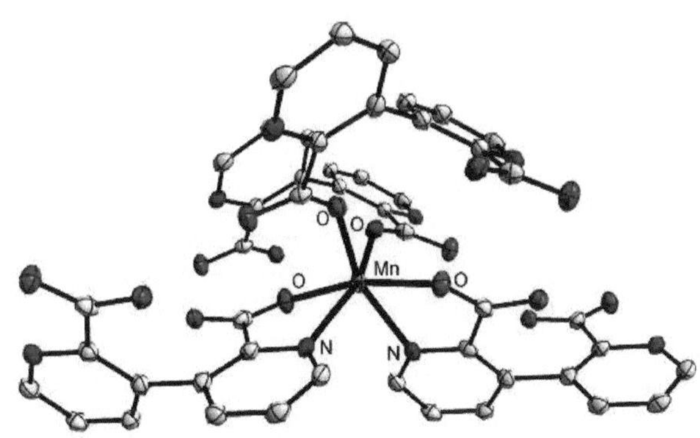

P019

The order-disorder transition in Cu_2ZnSnS_4

A. Ritscher[1,2], M. Hoelzel[3], M. Lerch[1]
[1]*Technische Universität Berlin, Institut für Chemie, Berlin, Germany*
[2]*Helmholtz Zentrum Berlin, Abteilung Kristallographie, Berlin, Germany*
[3]*Technische Universität München, Heinz Maier-Leibnitz Zentrum (MLZ), Garching, Germany*

Cu_2ZnSnS_4 is a potential compound for future thin film photovoltaic applications. It crystallizes in the kesterite type, space group .[1] In the fully ordered kesterite structure one Cu occupies the 2*a* (0, 0, 0) position, the remaining Cu is located at position 2*c* (0, ½, ¼), Zn at 2*d* (0, ½, ¾). In addition, kesterite-type phases exhibiting a statistical distribution of Cu/Zn on the two latter sites are also known.[2] Previous thin film near-order resonant Raman studies on CZTS proposed a critical temperature for a transition from ordered to disordered kesterite of 533 ± 10 K.[3]

Recently we developed a new route leading to single phase (CZTS) powders.[4] In order to follow the transformation from ordered to disordered kesterite, Cu_2ZnSnS_4 samples were synthesized and afterwards annealed at and quenched from different temperatures in the range of 473 - 623 K. As Cu and Zn are not distinguishable with standard X-ray techniques neutron diffraction measurements (SPODI, FRM II) were carried out (Fig 1).

The results of the first Rietveld refinements predict that Cu fully occupies the 2a and Sn the 2b position within the whole temperature range. When looking at Zn and the remaining Cu on sites 2d and 2c, a clear change from ordered to disordered kesterite is visible. An order parameter η as function of temperature was defined (fully ordered: $\eta = 1$, statistical distribution: $\eta = 0$) which was fitted to the following equation, commonly used in phase transition theories:

$\eta \propto ((T_c-T)/T_c)^{\beta}$ (eq. 1).

The refined value of the critical exponent β is 0.57 ± 0.06 which is in agreement with a Landau-type behavior ($\beta = 1/2$). The critical temperature T_c was determined as 552 ± 1 K (Fig.2). With these results a deeper understanding of the correlation between cation distribution and electrical properties seems to be possible.

Figure 1 Selected neutron diffraction pattern (513 K) with the results of the Rietveld refinements

Figure 2 Temperature variation of the order parameter η and fit to eq. 1

Financial support from the MatSEC graduate school of the Helmholtz Zentrum Berlin (HZB) is gratefully acknowledged.

[1] S.R. Hall, J.T. Szymanski, J.M. Stewart, Can. Mineral., 16 (1978) 131

[2] S. Schorr, H.-J. Hoebler, M. Tovar, Eur. J. Mineral., 19 (2007) 65

[3] J.J.S. Scragg et al., Appl Phys. Lett. 104, (2014) 041911

[4] A.Ritscher, J. Just, O. Dolotko, S. Schorr, M. Lerch, *submitted to* J. Alloys Comp.

Figure 1

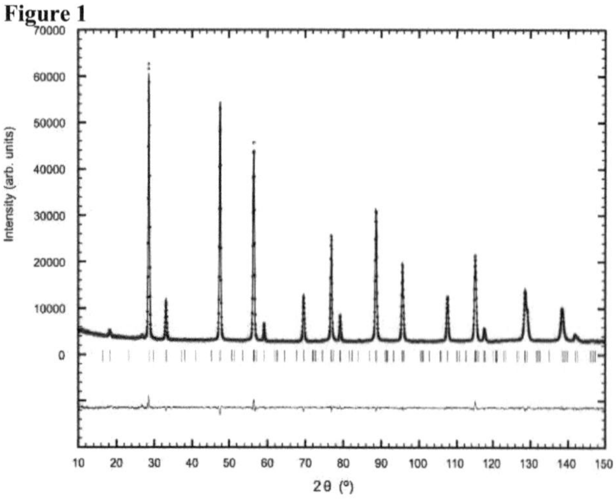

Figure 2

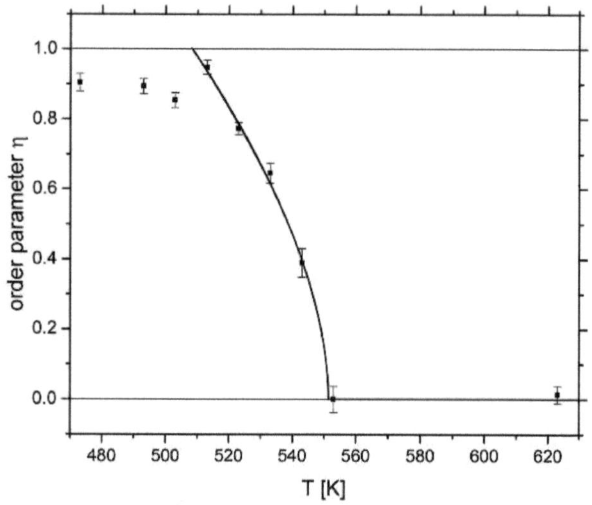

P020

Decreasing the Optical Band Gap in Photocatalysts: From $Ba_3Ta_5O_{14}N$ to $LaBa_2Ta_5O_{13}N_2$

B. Anke[1], M. Rohloff[1,2], A. Fischer[2], M. Lerch[1]
[1] *Technische Universität Berlin, Institut für Anorganische und Analytische Chemie, Berlin, Germany*
[2] *Albert-Ludwigs-Universität Freiburg, Institut für Anorganische und Analytische Chemie, Freiburg, Germany*

Barium tantalum compounds are well-known photocatalysts for water splitting.[1] Interestingly, also mixed-valent $Ba_3Ta^{IV}Ta^{V}_4O_{15}$ exhibits significant photocatalytic activity.[2] An approach to optimize the activity is band-gap narrowing by exchange of oxide by nitride ions and simultaneous oxidation of the tantalum(IV) ions leading to $Ba_3Ta^{V}_5O_{14}N$ [3]. A further increase of the nitrogen content in in this type of materials may be realized by co-substitution: Ba^{2+}/O^{2-} --> La^{3+}/N^{3-}.

We successfully synthesized light yellow $LaBa_2Ta_5O_{13}N_2$ as phase-pure material crystallizing isostructurally to $Ba_3Ta_5O_{15}$ and to a recently reported $Ba_3Ta_5O_{14}N$ oxide nitride (see Fig. 1). Nitrogen and oxygen contents were determined by the hot gas extraction method (exp.: 14.1 wt% O, 1.8 wt% N; theo: 13.4 wt% O, 1.8 wt% N).

By incorporating nitride, the band gap decreases from 2.80 eV for $Ba_3Ta_5O_{14}N$ to 2.77 eV for $LaBa_2Ta_5O_{13}N_2$, giving rise to an absorption band well in the visible-light region (see Fig. 2).

The photoelectrocatalytic properties for water oxidation are investigated using electrodes fabricated by an electrophoretic deposition method and a specific post-necking treatment [4]. First results show significant photocurrents for $LaBa_2Ta_5O_{13}N_2$.

Figure 1: X-ray powder diffraction pattern of $LaBa_2Ta_5O_{13}N_2$ with the results of the Rietveld refinement.

Figure 2: UV/Vis spectra (Tauc plot, indirect transition) of $LaBa_2Ta_5O_{13}N_2$ and $Ba_3Ta_5O_{14}N$ as derived from diffuse reflectance measurements.

This work is supported by the German Science Foundation (DFG) in the frame of the priority program 1613.

References

[1] R. Marschall, J. Soldat, M. Wark, *Photochem. Photobiol. Sci.* **2013**, *12*, 671-677
[2] C. Feger, R. Ziebarth, *Chem. Mater.* **1995**, *7*, 373-378.
[3] B. Anke, T. Bredow, J. Soldat, M. Wark, M. Lerch, *J. Solid State Chem.* **2015**, 233, 282-288.
[4] R. Abe, M. Higashi, K. Domen, *J. Am. Chem. Soc.* **2010**, *132*, 11828-11829.

Figure 1

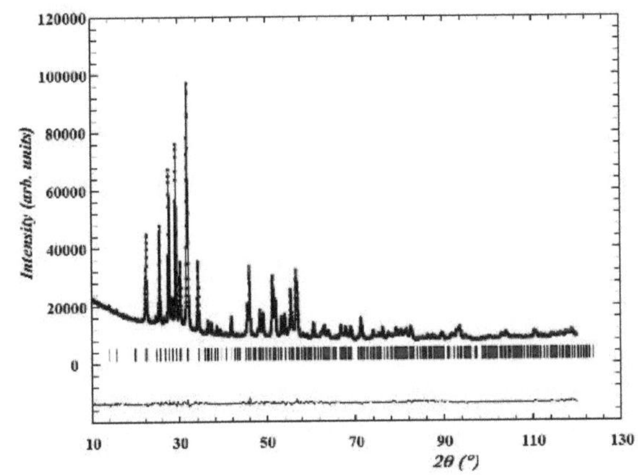

Figure 2

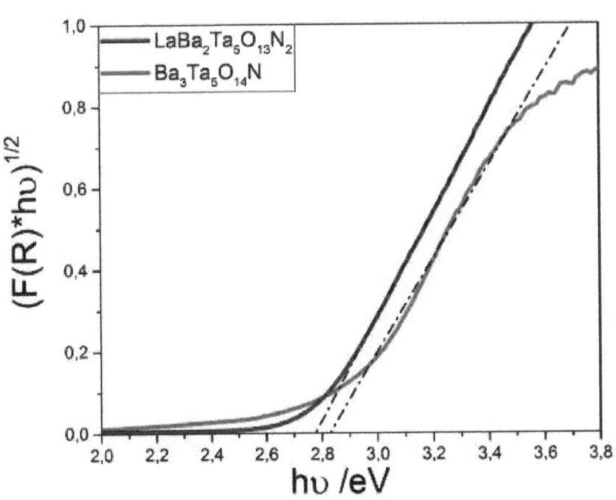

P021

Mixed Sr and Ba stannides/plumbides $A(Sn_{1-x}Pb_x)_3$

M. Langenmaier[1], M. Jehle[1], C. Röhr[1]
[1]*Universität Freiburg, Institut für Anorganische und Analytische Chemie, Freiburg, Germany*

Similar to the variation of the alkaline-earth elements (*A*) in tristannides [1] and triplumbides [2], a continuous substitution of tin by lead allows to explore geometric and electronic (Sn/Pb distribution) criteria for the stability of different stacking variants in the tritetrelides $A^{II}M^{IV}_3$. A series of ternary Sr and Ba mixed tri-stannides/plumbides $A(Sn_{1-x}Pb_x)_3$ (*A* = Sr, Ba) was synthesized from stoichiometric mixtures of the elements. Their structures were determined by means of single crystal X-ray data. All structures exhibit close packed ordered AM_3 layers containing *M* kagomé nets. Depending on the stacking sequence, the resulting *M* polyanion resembles the oxygen substructure of the hexagonal [face sharing octahedra, *h* stacking, Ni_3Sn-type, border compound $BaSn_3$, fig. 1 (a)] or the cubic [corner sharing octahedra, *c* stacking, Cu_3Au-type, border compound $SrPb_3$, fig. 1 (f)] perovskite. In $BaSn_3$ [3] (Ni_3Sn-type) up to 28 % of Sn can be substituted against Pb (*hP*8, $P6_3/mmc$, $x = 0.28(4)$: $a = 726.12(6)$, $c = 556.51(6)$ pm, $R1 = 0.0264$). A further increased lead content of 47 to 66 % causes the formation of the $BaSn_{2.57}Bi_{0.43}$-type with a $(hhhc)_2$ stacking
[*hP*32, $P6_3/mmc$, $x = 0.47(3)$: $a = 726.80(3)$, $c = 2235.78(14)$ pm, $R1 = 0.0437$, (b)]. The stability range of the $BaPb_3$-type sequence $(hhc)_3$ [(c)] starts at a lead proportion of 78 % (*hR*36, *R*-3*m*, $a = 728.77(3)$, $c = 2540.59(15)$ pm, $R1 = 0.0660$) and reaches up to the pure plumbide $BaPb_3$ [2,4]. A second new polymorph of $BaPb_3$ forms the Mg_3In-type with a further increased amount of cubic sequences [$(hhcc)_3$; *hR*48, $a = 728.7(2)$, $c = 3420.3(10)$ pm, $R1 = 0.0669$, (d)] and is thus isotypic with the border phase $SrSn_3$ [5] of the respective strontium series. In the Sr series, a Pb content of 32 % causes a small existence region of the $PuAl_3$-type structure [*hP*24, $P6_3/mmc$, $a = 696.97(6)$, $c = 1675.5(2)$ pm, $R1 = 0.1182$, fig. 1 (e)] with a $(hcc)_2$ stacking. The series is terminated by the pure *c* stacking of $SrPb_3$, the stability range of which starts at 75 % Pb (*cP*4, *Pm*-3*m*; $a = 495.46(9)$ pm, $R1 = 0.0498$).

As easily noted from this series and figure 2, the stacking sequence depends on the ratio of the atomic radii of the contributing atoms(*M*)/ions(*A*). Covalent as well as metallic aspects of the chemical bonding in the tri-tetrelides are discussed on the basis of the results of FP-LAPW calculations (density of states, band structure, v.e. densities).

[1] M. Wendorff, C. Röhr, Z. Anorg. Allg. Chem., **637**, 1013-1023 (2011).
[2] M. Wendorff, C. Röhr, Z. Naturforsch., **63b**, 1383-1394 (2008).
[3] Th. F. Fässler, Ch. Kronseder, Angew. Chem., **109**, 2800-2803 (1997).
[4] D. E. Sands, D. H. Wood, W. J. Ramsey, Acta Crystallogr., **17**, 986-989 (1964).
[5] Th. F. Fässler, S. Hoffmann, Z. Anorg. Allg. Chem., **626**, 106-112 (2002).

Fig 1: Crystal structures and lead contents *x* of ternary tritetrelides $A(Sn_{1-x}Pb_x)_3$.
Fig. 2: Amount of hexagonal stacking related to the radius ratio r_A/r_M.

Figure 1

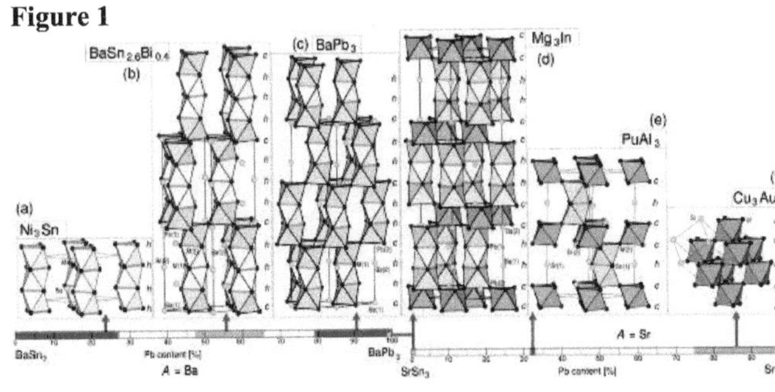

Figure 2

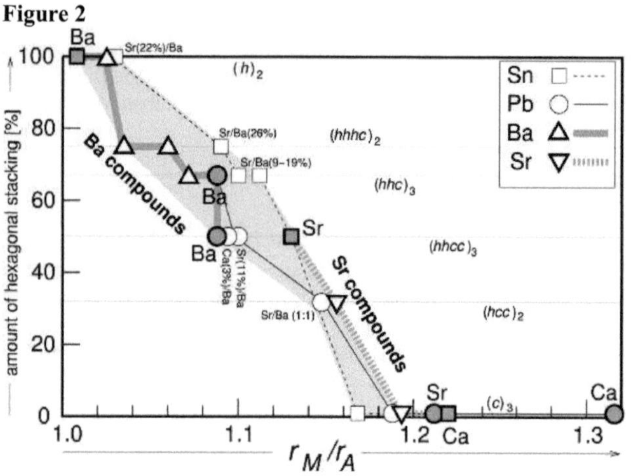

P022

$Cs_4Pr_2As_4S_{11}$: A New Quaternary Thioarsenate(III) According to $Cs_4Pr_2[AsS_3]_2[As_2S_5]$

F. Ledderboge[1], T. Schleid[1]
[1]*University of Stuttgart, Institute for Inorganic Chemistry, Stuttgart, Germany*

Introduction

In attempts to synthesize lanthanoid(III) thioarsenates(III) with the simple composition *Ln*[AsS_3], which are hitherto still unknown, yellow needles of $Cs_4Pr_2As_4S_{11}$ (≡ $Cs_4Pr_2[AsS_3]_2[As_2S_5]$) as first member of a new class of cesium-containing lanthanoid(III) thioarsenates(III) could be observed. Compared to the derivatives known from the literature, such as $Pr_4S_3[AsS_3]_2$ [1], $Pr_3S_2Cl_2[AsS_3]$ [2], $La_3S_2Cl_2[AsS_3]$ [3], $Cs_2CeCl_2[AsS_3]$ [4] and $Ce_5S_4Cl[AsS_3]_2$ [5], the new compound is the first one, which includes ψ^1-tetrahedral $[AsS_3]^{3-}$ monomers and $[As_2S_5]^{4-}$ dimers.

Materials and Methods

$Cs_4Pr_2[AsS_3]_2[As_2S_5]$ was prepared by using elemental praseodymium (Pr) and sulfur (S), arsenic(V) sulfide (As_2S_5) and cesium chloride (CsCl) as fluxing agent and potential reactant in sealed silica vessels at 750 °C for three days.

Results

The title compound crystallizes monoclinically in the *non-centrosymmetric* space group $P2_1$ with a = 1202.91(8) pm, b = 683.16(4) pm, c = 1526.81(7) pm and β = 94.202(4)° for Z = 2. It contains two crystallographically different Pr^{3+} cations, both showing an eightfold coordination of S^{2-} anions in the shape of distorted square antiprisms $[PrS_8]^{13-}$, which are connected by alternating edges and square faces to undulated chains running along [010] (Figure 1). Both polyhedra are built up by the sulfur atoms of three monomeric ψ^1-tetrahedral $[AsS_3]^{3-}$ anions and one $[As_2S_5]^{4-}$ dimer as vertex-connected fusion product of two of them (Figure 2), which are linking these chains to layers within (001). The space between the layers is filled with four crystallographically unique charge-balancing Cs^+ cations in seven- to ninefold sulfur coordination.

Conclusion

The systems of lanthanoid(III) thioarsenates(III) with fluxing alkali-metal halides tend to create more complex structures than without by fitting both components (A^+ and X^-) partially or completely into the resulting structures.

References

[1] D.-H. Kang, Th. Schleid, *Z. Anorg. Allg. Chem.* **2009**, *635*, 2170-2176.

[2] D.-H. Kang, F. Ledderboge, H. Kleinke, Th. Schleid, *Z. Anorg. Allg. Chem.* **2015**, *641*, 322-326.
[3] D.-H. Kang, Th. Schleid, *Z. Kristallogr.* **2009**, *Suppl. 28*, 46-46.
[4] D.-H. Kang, Th. Schleid, *Z. Anorg. Allg. Chem.* **2008**, *634*, 2050-2050.
[5] F. Ledderboge, Th. Schleid, *Z. Kristallogr.* **2015**, *Suppl. 35*, 87-87.

Figure 1. View at the chains of fused $[PrS_8]^{13-}$ polyhedra in the crystal structure of $Cs_4Pr_2As_4S_{11}$ along [010].

Figure 2. The different thioarsenate(III) anions $[(As1)S_3]^{3-}$ (*top, left*), $[(As2)S_3]^{3-}$ (*top, right*) and $[(As3)(As4)S_5]^{4-}$ (*bottom*) in the crystal structure of $Cs_4Pr_2As_4S_{11}$ with their cationic Cs^+ and Pr^{3+} surrounding.

Figure 1

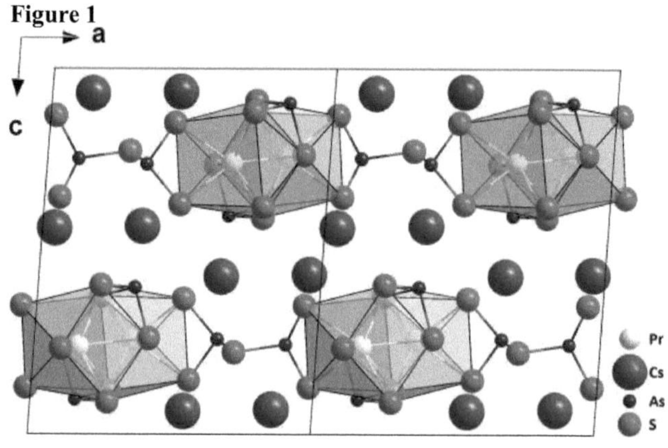

Figure 2

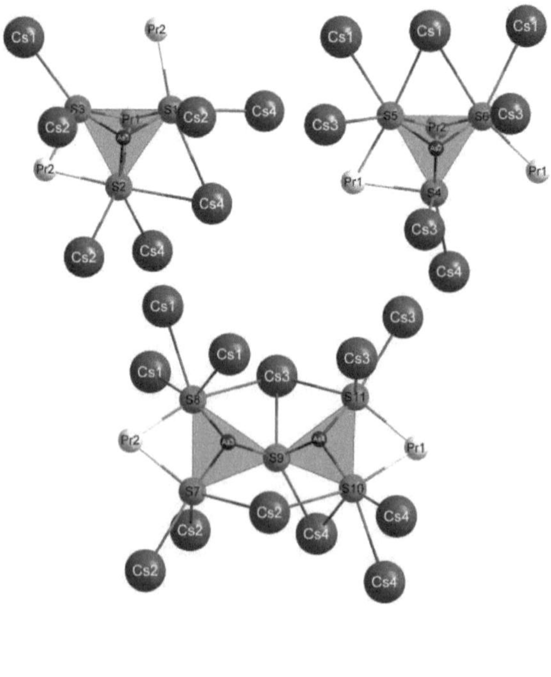

P023

Expanding the SrCu*RE*S_3 Series with the Rare-Earth Metals Scandium and Yttrium

M. Eberle[1], T. Schleid[1]
[1]*University of Stuttgart, Institute for Inorganic Chemistry, Stuttgart, Germany*

Quaternary strontium- and copper-bearing sulfides with trivalent lanthanoid cations that yield in the formula SrCu*Ln*S_3 have been synthesized with several lanthanoid(III) cations already. Here, we report on the expansion of this SrCu*RE*S_3 series with scandium and yttrium, which belong to the rare-earth metal family as well. Not surprisingly, the scandium compound $SrCuScS_3$ is crystallizing isotypically with $SrCuLuS_3$ [1], as both trivalent metal cations have the smallest radius among the rare-earth metals [2]. However, $SrCuYS_3$ is *not* isotypic with for example $SrCuErS_3$ [1] any more, despite the fact of the very similar ionic radii of Y^{3+} and Er^{3+} [2]. The needle-shaped compounds $SrCuScS_3$ and $SrCuYS_3$ were synthesized using Sr, Cu, Sc or Y and S in molar ratios of 1 : 1 : 1 : 3 along with CsBr as fluxing agent. The reactions took place in evacuated glassy silica ampoules for eight days at 850 °C. The results were a dark yellow sample of $SrCuScS_3$ and an orange one for $SrCuYS_3$ with single crystals big enough for single-crystal X-ray measurements. $SrCuScS_3$ is crystallizing in the orthorhombic space group *Cmcm* with a = 383.16(3), b = 1285.04(9) and c = 971.53(7) pm, whereas $SrCuYS_3$ adopts the orthorhombic space group *Pnma* with a = 1018.45(7), b = 393.78(3) and c = 1294.26(9) pm (further informations on quoting CSD-430506 for $SrCuScS_3$ and CSD-430508 for $SrCuYS_3$ at the FIZ Karlsruhe: crysdata@fiz-karlsruhe.de). In both crystal structures, the rare-earth metal trications are surrounded by six S^{2-} anions in the shape of octahedra, which are fused together via common edges and vertices resulting in 2D-{[*RE*S3]$^{3-}$} layers. Within these layers, notches occur, where the Cu^+ cations fit in and then are tetrahedrally surrounded by S^{2-} anions. The only difference between the two structures are the coordination spheres of the Sr^{2+} cations (CN = 8) that reside in between the anionic 2D-{[Cu*RE*S_3]$^{2-}$} layers (Figure 1). While trigonal prisms are visible in both structures (d(Sr-S) = 295 - 309 pm for $SrCuScS_3$, 298 - 309 pm for $SrCuYS_3$), the two capping S^{2-} anions in the crystal structure of $SrCuScS_3$ have the same distance to the prism-centering Sr^{2+} cation (d(Sr-S) = 332 pm). In contrast, there is just one close capping S^{2-} anion in the crystal structure of $SrCuYS_3$ (d(Sr-S) = 326 pm), while the second one is already 383 pm apart from Sr^{2+} (Figure 2).

[1] N. V. Sikerina, O. V. Andreev, I. P. Leven, *Russ. J. Inorg. Chem.* **2008**, *53*, 455-459.
[2] R. D. Shannon, *Acta Crystallogr.* **1976**, *A 32*, 751-767.

Figure 1. Sequence of the 2D-{[Cu*RE*S_3]$^{2-}$} layers with Sr^{2+} cations in between occurring in the crystal structures of both $SrCuScS_3$ and $SrCuYS_3$.

Figure 2. Interconnection of the [SrS_6]$^{10-}$ prisms; for $SrCuScS_3$ the dashed and straight bonds have the same distances from S^{2-} to Sr^{2+}, for $SrCuYS_3$ they are considerably different.

Figure 1

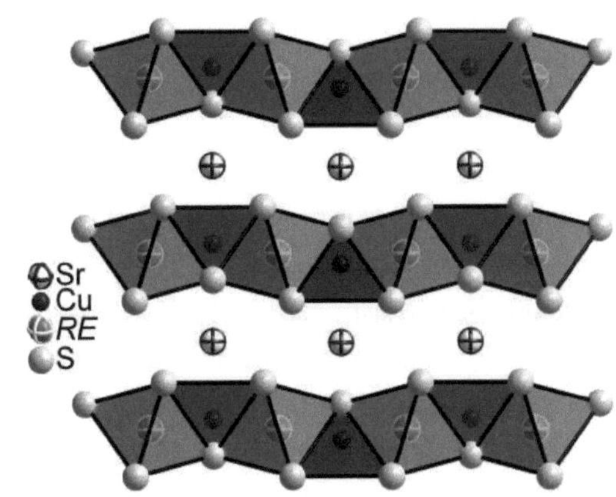

Figure 2

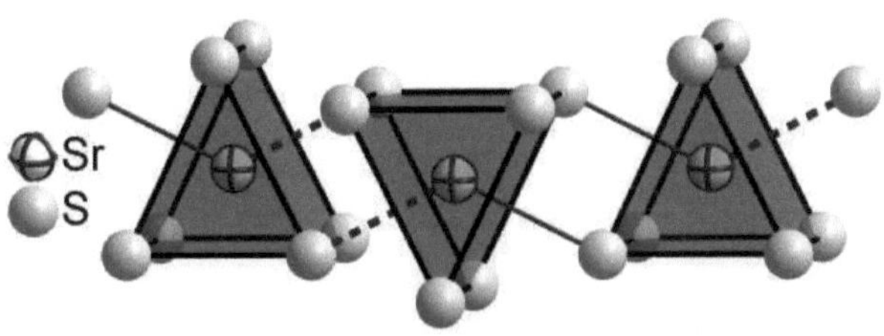

P024

Structural characterisation of $Cu_2ZnSn(S_{1-x}Se_x)_4$ by anomalous X-ray and neutron diffraction.

G. Gurieva[1], D. M. Többens[1], S. Zander[1], S. Schorr[1,2]
[1]*Helmholtz-Zentrum Berlin, Berlin, Germany*
[2]*Freie Universität Berlin, Berlin, Germany*

In spite of previously reported stannite type structure for CZTSe, CZTS and CZTSe crystallize both in the kesterite type structure (space group)[1]. A differentiation between the isoelectronic cations Cu^+ and Zn^{2+} and consequently kesterite and stannite type structures is not possible using X-ray diffraction due to their similar scattering factors. But neutrons diffraction can solve this problem; the coherent scattering lengths are sufficiently different for these cations[2]. It was shown by this method that both Cu_2ZnSnS_4 and $Cu_2ZnSnSe_4$ occur in the kesterite structure.[1] Another method which can solve the problem is anomalous X-ray diffraction on the Cu-K(8979 eV) and Zn-K(9659 eV) edges. The usage of multiple wavelengths above, below and between the absorption edges of Cu and Zn ensures significant over determination, so that the Cu-, Zn-, and vacancy concentrations can be refined reliably for the independent crystallographic sites.

A detailed structural analysis of stoichiometric $Cu_2ZnSn(S_{1-x}Se_x)_4$, grown by solid state reaction, was performed by both neutron diffraction and anomalous X-ray diffraction at the fine resolution neutron powder diffractometer E9 at BER II (λ=1.7986Å, RT) and the Diffraction station at KMC-2 beamline at BESSY II respectively. Rietveld refinement of both sets of diffraction data using the FullProf suite software [3] lead to accurate values of lattice constants, and site occupancy factors. The latter have given insights into the cation distribution within the crystal structure of $Cu_2ZnSn(S_{1-x}Se_x)_4$ solid solutions with different x values. The correlated information about changes in lattice parameters and cation site occupancies, details on the existing intrinsic point defects and their amounts obtained by both methods will be discussed.

This research was supported by projects IRSES PVICOKEST 269167 and KESTCELLS 316488, FP7-PEOPLE-2012 ITN, Multi-ITN.

[1] S.Schorr, Solar Energy Materials and Solar Cells, 95 (2011)1482.
[2] V.F. Sears, Neutron News 3(3) (1992) 26.
[3] Juan Rodriguez-Carvajal and Thierry Roisnel, www.ill.eu/sites/fullprof

P025

Eu_2O_2I: A New Mixed-Valent Europium Oxide Halide

D. Rudolph[1], T. Schleid[1]
[1]*University of Stuttgart, Institute for Inorganic Chemistry, Stuttgart, Germany*

After Eu_2O_2Br [1] and Sm_2O_2I [2] we report with Eu_2O_2I a further mixed-valent lanthanoid oxide halide. Like the two known compounds before, Eu_2O_2I was obtained by some serendipity while trying to synthesize europium(II) oxide hydride iodides, such as $Eu_5O_2H_2I_4$, resembling the analogous barium compound $Ba_5O_2H_2I_4$ [3]. Red brown, needle-shaped single crystals of the title compound emerged from the reaction of europium dihydride, europium diiodide, europium sesquioxide and europium metal in a molar ratio of 3 : 6 : 2 : 2 at 900 °C for two days in sealed niobium ampoules.

The crystal structure of Eu_2O_2I (CSD-429195) was determined by single-crystal X-ray diffraction and is isotypic to the two above-mentioned compounds Eu_2O_2Br and Sm_2O_2I. It crystallizes in the monoclinic space group *C*2/*m* with a = 1256.81(9) pm, b = 409.24(3) pm, c = 975.96(7) pm and β = 117.638(3)° for Z = 4.

In the crystal structure of $(Eu^{3+})(Eu^{2+})(O^{2-})_2(I^-)$, there are two crystallographically and functionally different europium cations. They can be assigned comfortably to sites containing only Eu^{2+} and Eu^{3+}, respectively, according to their distances to the surrounding oxide anions. $(Eu1)^{3+}$ is coordinated by two iodide and five oxide anions with Eu1-O distances ranging from 226 to 236 pm, resulting in an average distance of 230 pm. This value corresponds very well to the four Eu^{3+}-O^{2-} distances in PbFCl-type EuOI [4] with 228 pm, leading to the conclusion that Eu1 represents just trivalent europium. In contrast, $(Eu2)^{2+}$ is surrounded by five iodide and three oxide anions. The three slightly different Eu2-O distances are with 241 and 242 pm similar to the average europium(II)-oxygen bond lengths in oxide iodides with exclusively divalent europium, such as Eu_2OI_2 [5] (239 pm) and Eu_4OI_6 [6] (240 pm).

Central building blocks of the crystal structure are *cis*-edge connected $[(O1)(Eu1)_3(Eu2)]^{9+}$ tetrahedra as well as $[(O2)(Eu1)_2(Eu2)_2]^{8+}$ tetrahedra sharing common vertices and edges. Both structural features result in infinite strands running along the *b* axis (Figure 1), which are interconnected to each other via common edges and vertices to form layers parallel to the *ab* plane. These cationic $\{[O_2Eu_2]^+\}$ layers are separated and charge-balanced by the iodide anions (Figure 2).

Figure 1. Interconnection of the oxygen-centered europium tetrahedra in the crystal structure of Eu_2O_2I.

Figure 2. View at four unit cells of Eu_2O_2I along [001].

References

[1] S. Hammerich, I. Pantenburg, G. Meyer, *Z. Anorg. Allg. Chem.* **632** (2006) 1487-1490.
[2] M. Ryazanov, C. Hoch, Hj. Mattausch, A. Simon, *Z. Anorg. Allg. Chem.* **632** (2006) 2385-2588.
[3] O. Reckeweg, F. J. DiSalvo, *Z. Naturforsch.* **66 b** (2011) 21-26.
[4] H. Bärnighausen, *J. Prakt. Chem.* **14** (1961) 313-322.
[5] S. Hammerich, G. Meyer, *Z. Anorg. Allg. Chem.* **632** (2006) 1244-1246.
[6] L. Wuping, R. Dronskowski, *Acta Crystallogr.* **C 60** (2004) i23-i24.

Figure 1

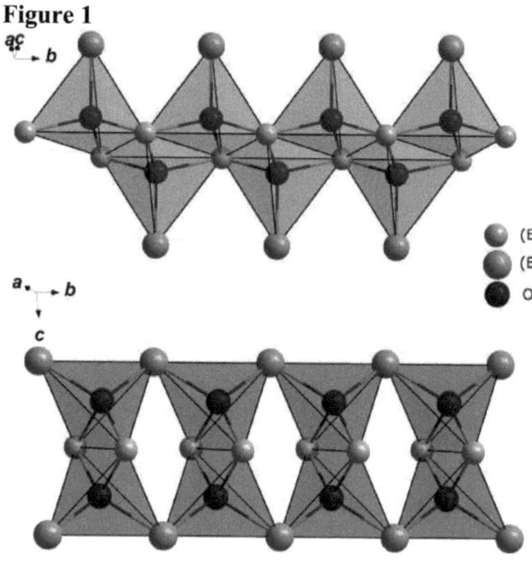

Figure 2

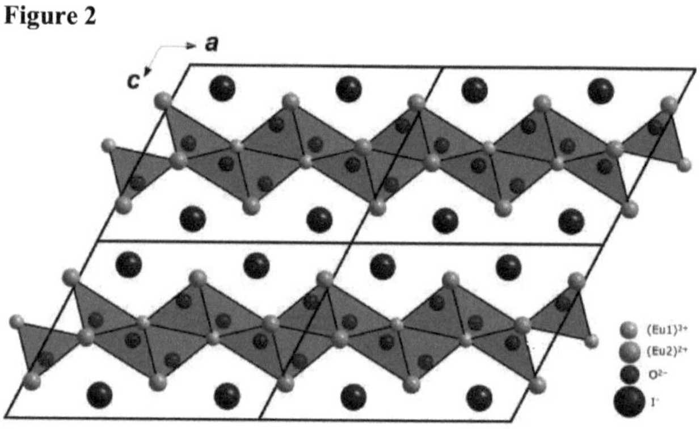

P026

The crystal structure of $MAPbI_3$: a complementary neutron and synchrotron x-ray diffraction study

A. Franz[1], D. Többens[1], S. Schorr[1]
[1]*Helmholtz-Zentrum Berlin für Materialien und Energie, Abteilung Struktur und Dynamik von Energiematerialien, Berlin, Germany*

Perovskites with ABX_3 - structure show a huge possibility on element substitutions on A-,B- and C-site which leads to a broad variety of physical properties. In recent years the interest become focused on hybrid perovskites as a future photovoltaic material. Our field of interest lays in lead methylammonium triodide in which A is the organic unit $[CH_3NH_3]^+$, B=Pb^{2+} and X=I_3^-.

The different perovskite crystal structures can be classified by their octahedral coordinated B-cation. The aristotype-structure (P m-3m) is symmetry lowered due to tilting, distortion of $[BX_6]$ - octahedra or displacement of B-cation from center of octahedron. These different types of displacements are shown in the 'Bärnighausen-Stammbaum' [1] a tree diagram which was extended in 2002 by Bock & Müller [2] who presented a tree diagram where all group-subgroup relations of octahedral tilted space groups are taken into account, based on the work of Bärnighausen and 'Glazer's tilt systems' [3].

First structure analysis and Rietveld refinements confirmed the $MAPbI_3$ - perovskite belongs to the tetragonal space group I 4/mcm. Further investigations showed that $MAPbI_3$ shows a highly disordered structure with large atomic displacement factors of the anions. The analyses furthermore revealed a distortion and statistical distributed tilting of the octahedra. This in turn takes influence on the organic unit and therefore influences photovoltaic properties.

The polycrystalline perovskite samples were synthesized from a equimolar mixture of methylammonium iodide and PbI_2, described by Im et al. [4].

For structural investigations Synchrotron and Neutron diffraction experiments were performed at the Helmholtz-Zentrum Berlin für Materialien und Energie. The data treatment was performed by F.O.X. (Free Objects for Xtallography) and FullProf Suite software.

The presentation will give an overview of the results of this structural study with a certain focus on the statistic disorder of the methylammonium molecule, and anion displacement. Moreover, structural trends will be discussed.

1. Bärnighausen, H., *Group-subgroup relations between space groups: A useful tool in crystal chemistry.* MATCH Communications in Mathematical and in Computer Chemistry 1980. **9**: p. 139-175.
2. Bock, O. and U. Müller, *Symmetrieverwandtschaften bei Varianten des ReO3-Typs.* Zeitschrift für anorganische und allgemeine Chemie, 2002. **628**(5): p. 987-992.
3. Glazer, A.M., *The classification of tilted octahedra in perovskites* Acta Crystallographica B, 1972. **28**: p. 3384-3392, .
4. Im, J.H., et al., *6.5% efficient perovskite quantum-dot-sensitized solar cell.* Nanoscale, 2011. **3**(10): p. 4088-93.

P027

Ammonothermal Synthesis and Crystal Growth of Indium Nitride

J. Hertrampf[1], N. Alt[2], E. Schlücker[2], R. Niewa[1]
[1]*University of Stuttgart, Institute for Inorganic Chemistry, Stuttgart, Germany*
[2]*University of Erlangen-Nuremberg, Institute of Process Machinery and Systems Engineering, Erlangen, Germany*

Due to their optical and electronic properties group-III nitrides are of high interest for optical devices (LEDs, lasers) and electronic applications [1]. All binary group-III nitrides are known, except for thallium(III) nitride [2].

There are two main challenges in InN synthesis: the limited thermal stability (decomposition above ~500 °C [3]) and the low heat of formation [−28.6 ± 9.2 kJ/mol [4]). In the past, InN was obtained as microstructures or thin films by various procedures like ammonolysis of In_2O_3 [5], molecular beam epitaxy [6] or metal-organic vapor deposition [7]. In 1995 Dwilinksi predicted InN synthesis via the ammonothermal route by stabilizing effects of the applied ammonia and nitrogen pressure [8].

We report the first growth of small InN crystals in the µm range from ammonothermal synthesis. A series of experiments with different molar ratios of $InCl_3$ and KNH_2 were carried out in customized autoclaves [9]. The most successful experiments were conducted with molar ratios of exact $n(KNH_2)/n(InCl_3) = 3$. Compared to ammothermal GaN synthesis, which typically proceeds either in ammono-basic or ammono-acidic milieu, these conditions may be suspected to represent an ammono-neutral medium. Depending on the temperature we observe different morphologies of the InN crystallites: Higher temperatures of 773 K lead to crystals with high aspect ratio and rod shape up to 2.5 µm in length (Fig.1), lower temperatures of 663 K lead to crystals with low aspect ratio and hexagonal platelet shape up to 2.0 µm in diameter (Fig. 2). Variations in pressure (200 to 250 MPa) have little influence on the morphology.

[1] M. R. Krames, O. B. Shchekin, R. Müller-Mach, G. O. Müller, L. Zhou, G. Harbers, M. G. Craford, *J. Disp. Technol.* **2007**, *3*, 160.
[2] T. M. M. Richter, R. Niewa, *Inorganics*, **2014**, *2*, 29.
[3] J. W. Trainor, K. Rose, *J. Electron. Mater.* **1974**, *3*, 821.
[4] M. R. Ranade, F. Tessier, A. Navarotsky, R. Marchand, *J. Mater. Res.* **2001**, *10*, 2824.
[5] B. Schwenzerl, L. Loeffler, R. Seshadri, S. Keller, F. F. Lange, S. P. DenBaars, U. K. Mishra, *J. Mater. Chem.* **2004**, *14*, 637.
[6] J. Wu, W. Walukiewicz, K. M. Yu, J. W. Ager III., E. E. Haller, H. Lu, W. J. Schaff, Y. Saito, Y. Nanishi, *Appl. Phys. Lett.* **2002**, *80*, 3967.
[7] T. Matsuoka, H. Okamoto, N. Nakao, H. Harima, E. Kurimoto, *Appl. Phys. Lett.* **2002**, *81*, 1246.
[8] R. Dwilinski, A. Wysmolek, J. Baranowski, M. Kaminska, R. Doradzinksi, J. Garczynski, L. Sierzputowski, H. Jacobs, *Acta Phys. Pol. A* **1995**, *88*, 833.
[9] N. S. A. Alt, E. Meissner, E. Schluecker, *J. Cryst. Growth* **2008**, *350*, 2.

Figure 1: InN rods from synthesis at 773 K.

Figure 2: InN platelets from ammonothermal synthesis at 663 K.

Figure 1

Figure 2

P028

The Elusive $LiBi_3S_5$: Synthesis, Characterization, and Topological Analysis

D. Wiedemann[1], S. Nakhal[1], B. Stanje[2], O. Dolotko[3], M. Wilkening[2], M. Lerch[1]

[1]*Technische Universität Berlin, Institut für Chemie, Berlin, Germany*
[2]*Technische Universität Graz, Institut für Chemische Technologie von Materialien, Graz, Austria*
[3]*Technische Universität München, Heinz Maier-Leibnitz Zentrum, Garching b. München, Germany*

Semiconducting chalcogenides are sought-after materials for thermoelectric refrigeration and portable power generation.[1] Because of their favorable properties, ternary and quaternary alkali-metal bismuth sulfides (*e.g.*, $RbBi_3S_5$, KBi_3S_5, and $Na_{0.66}K_{0.33}Bi_3S_5$) have been a focus of research. The lightest member of this class, $LiBi_3S_5$, has hitherto eluded directed synthesis and proper characterization, in spite of having been announced as early as in 1977.[2]

Inspired by this, we have synthesized $LiBi_3S_5$, characterized it structurally, and analyzed its topology with respect to possible pathways for lithium-ion migration. Two solid-state routes starting from $LiBiS_2$ and Bi_2S_3 led to success in synthesis. The crystal structure was then determined using powder neutron diffraction and NMR experiments, followed by inspection of the probable conduction pattern using Hirshfeld surface and Voronoi-Dirichlet partitioning (VDP) analyses of the void structure.

$LiBi_3S_5$ was found to be of the $AgBi_3S_5$ type with a strongly disordered cation lattice (three/two positions mainly occupied by bismuth/lithium, respectively; see Fig. 1). The topological analyses of genuine and derived structural models showed generation of Frenkel defects by displacement of lithium ions into tetrahedral voids to be possible. Subsequent 1D migration may only occur along [010] in channels including the positions Li5 (see Fig. 2). In reality, these will be blocked by bismuth ions, thus making the disordered material a poor long-range lithium-ion conductor.

This work is based upon experiments performed at the SPODI instrument operated by FRM II at the Heinz Maier-Leibnitz Zentrum (MLZ), Garching, Germany. Financial support by the Deutsche Forschungsgemeinschaft (FOR 1277: "Mobilität von Lithiumionen in Festkörpern [molife]") is gratefully acknowledged.

[1] D. M. Rowe, *Thermoelectrics Handbook: Macro to Nano*, CRC Press, Boca Raton (USA), **2006**.
[2] V. B. Lazarev, S. I. Berul', A. F. Trippel', *Russ. J. Inorg. Chem.* **1977**, *22*, 1218-1220.

Fig. 1. Crystal structure of $LiBi_3S_5$ (view along [010], ions with arbitrary radii, unit cell in black).

Fig. 2. Detail of ordered model of $LiBi_3S_5$ with migration channel (dark grey: cations, light grey: sulfide ions, black: elementary voids and channels; ions with arbitrary radii, unit cell in black).

Figure 1

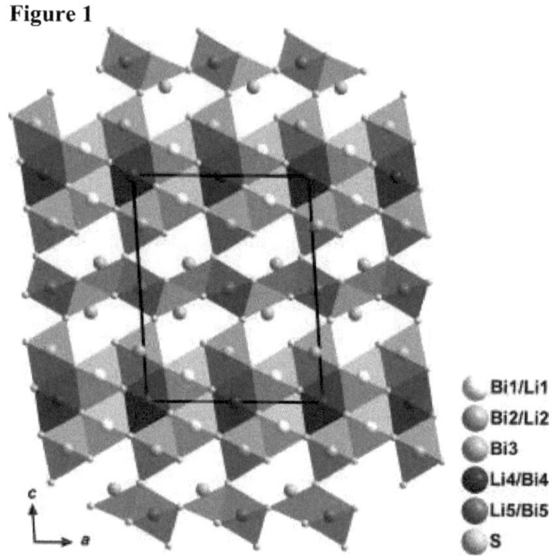

Figure 2

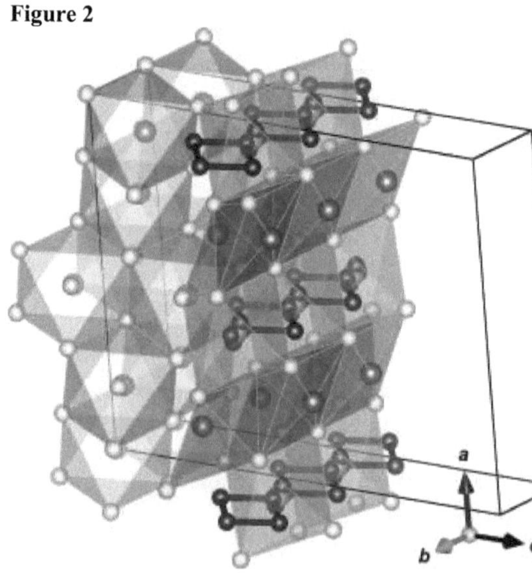

P029

Non-Hydrothermal Synthesis of $NaYF_2[CO_3]$ and $Na_2EuF_3[CO_3]$ and the Crystal Structure of $Na_2SmF_3[CO_3]$

H. Henning[1], T. Schleid[1]
[1]*University of Stuttgart, Institute for Inorganic Chemistry, Stuttgart, Germany*

For the synthesis of rare-earth metal(III) fluoride oxocarbonates(IV) *RE*F[CO_3] (*RE* = Sc, Y, La, Ce - Lu), a variation of the metathesis method first published by *Glaser* and *Meyer* [1] utilizing *RE*F_3 and $Li_2[CO_3]$ proved to be very successful, but resulted in the *in-situ* decomposition of any compound containing RE^{3+} cations with a smaller ionic radius than Nd^{3+}. This decomposition was suspected to be a result of the rising *Lewis*-acidity of the entire system.

In attempts to circumvent this issue, $Na_2[CO_3]$ was used to lower the acidity within the reaction mixture. Surprisingly, in the case of samarium, europium and yttrium quite unexpected products were obtained. Of these new products $NaYF_2[CO_3]$ and $Na_2EuF_3[CO_3]$ were already known to literature [2, 3], whereas these have been originally synthesized under hydrothermal conditions.

The new compound $Na_2SmF_3[CO_3]$ is isostructural to the already reported $Na_2EuF_3[CO_3]$ [3] and crystallizes in the orthorhombic space group *Pbca* with the lattice parameters a = 663.34(4) pm, b = 1080.92(7) pm and c = 1411.54(9) pm for Z = 8. The crystal structure contains a single crystallographically unique Sm^{3+} cation, two distinguishable Na^+ cations, three F^- anions and one triangular $[CO_3]^{2-}$ unit, consisting of carbon and three distinct oxygen atoms. Sm^{3+} is coordinated by six F^- anions (d(Sm–F) = 212 - 268 pm), which form two different triangular faces of a strongly distorted capped square antiprism, whereas the remaining corners are occupied by two oxocarbonate units connecting via common edges and vertices, respectively (d(Sm–O) = 243 - 270 pm, Figure 1). These $[CO_3]^{2-}$ anions (d(C–O) = 120 - 134 pm) connect *via* their corners and edges, respectively, to a single neighbouring Sm^{3+} cation, while the two aforementioned $(F^-)_3$ triangles coordinate also another Sm^{3+} cation each. This mode of connection creates hexagonal honeycomb-like layers (Figure 2), which stack offset from each other and are interpenetrated by the sixfold coordinated Na^+ cations within and in between these layers, occupying distorted octahedral voids (d(Na1–O) = 210 - 301 pm, d(Na1–F) = 224 - 226 pm; d(Na2–O) = 232 - 243 pm, d(Na2–F) = 226 - 273 pm).

Figure 1. Coordination environment of the Sm^{3+} cations in the crystal structure of $Na_2SmF_3[CO_3]$.

Figure 2. Honeycomb layer of the interconnected polyhedra around Sm^{3+} in the crystal structure of $Na_2SmF_3[CO_3]$ as viewed along [001].

[1] J. Glaser, H.-J. Meyer, *Z. Anorg. Allg. Chem.* **2010**, *636*, 2622-2625.
[2] A. B. Ali, S. Kodjikian, V. Maisonneuve, M. Leblanc, *Solid State Sci.* **2006**, *8*, 1322-1329.
[3] N. Mercier, M. Leblanc, *Acta Crystallogr.* **1994**, *C 50*, 1864-1865.

Figure 1

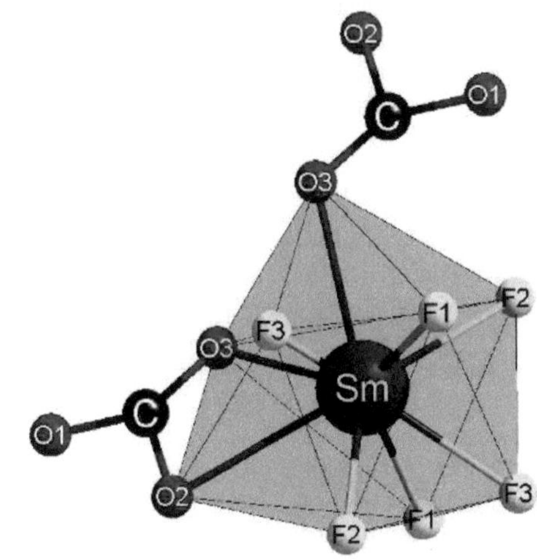

Figure 2

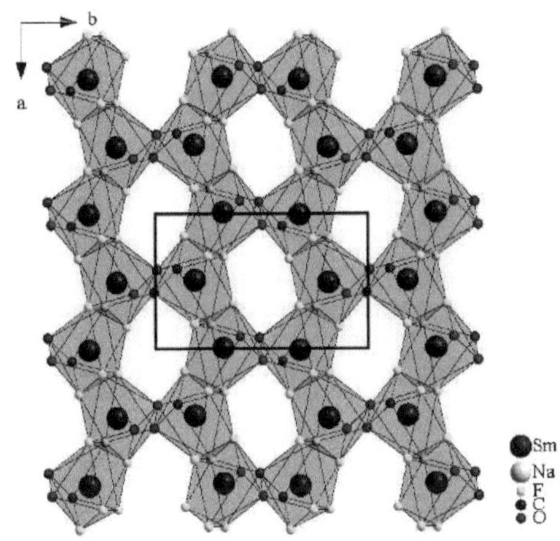

P030

The New Quaternary Anion-Rich Lanthanoid Compound $Pr_{10}N_6Te_5Cl_2$

M. Foltin[1], F. Lissner[1], T. Schleid[1]
[1]*University of Stuttgart, Institute for Inorganic Chemistry, Stuttgart, Germany*

Introduction

The ternary systems of lanthanoid(III) nitride tellurides exhibit three compositions so far, $Ln_4N_2Te_3$ (*Ln* = La - Nd, Gd - Dy) [1, 2], $Ln_{13}N_5Te_{12}$ (*Ln* = La - Nd) [3] and Ln_3NTe_3 (*Ln* = Gd - Ho) [4], which were examined intensively already. However, there are only few examples of quaternary compounds explored hitherto, consisting of the same three elements (lanthanoid, nitrogen and tellurium) as basic components plus an extra one. At first the crystal structure of the cesium-containing phases $Cs_2Ln_6N_2Te_7$ (*Ln* = Gd, Tb) [5] exhibits infinite chains of vertex-connected $[NLn_4]^{9+}$ tetrahedra as main structural feature and a coordination number of six for each Ln^{3+} cation. The oxide nitride tellurides Ln_3ONTe_2 (*Ln* = Gd, Tb) [6], on the other hand, show a tightly packed double strand built up by two different types of edge-connected tetrahedral units ($[NLn_4]^{9+}$ and $[OLn_4]^{10+}$) with an average surrounding of seven anions for each Ln^{3+} cation. Recent investigations revealed now a new quaternary lanthanoid nitride telluride with chlorine as fourth element: $Pr_{10}N_6Te_5Cl_2$.

Materials and Methods
$Pr_{10}N_6Te_5Cl_2$ was obtained by reactions of praseodymium metal with tellurium, sodium azide, praseodymium trichloride and sodium chloride as flux in evacuated silica tubes at 900 °C. The reactions typically ran for two weeks, including four days of heating and cooling, and resulted long, red needles of the crystalline title compound.

Results
The new praseodymium(III) nitride telluride chloride $Pr_{10}N_6Te_5Cl_2$ crystallizes orthorhombically in space group *Pbam* with the lattice parameters a = 1190.98(8), b = 1350.41(9), c = 1487.80(9) pm for Z = 4, and appears with a unique crystal structure in solid-state chemistry. The large unit cell is filled with seven crystallographically independent Pr^{3+} cations, which are surrounded by about eight anionic partners for each Pr^{3+} site, and seems to be tightly packed overall. But not only the Pr^{3+} cations show high coordination numbers. Almost perfect $(Pr_8)^{24+}$ cubes are centered by the $(Te1)^{2-}$ anions, a known, but unusual constellation for tellurium. Otherwise the smaller N^{3-} anions appear in the familiar tetrahedral cationic environment. These $[NPr_4]^{9+}$ units are connected via *trans*-oriented edges to 1D-$\{[NPr^{e}_{4/2}]^{3+}\}$ chains on the one hand and to more complex $[N_2Pr_3]^{3+}$ double strands by the fusion of two regular $[NPr^{e}_{4/2}]^{3+}$ chains of *trans*-oriented $[NPr_4]^{9+}$ tetrahedra. Both one-dimensionally infinite formations run along [100] at y/b = $^1/_4$ plus $^3/_4$ and z/c = 0 for the single $[NPr_{4/2}]^{3+}$ chains as well as y/b = 0 plus $^1/_2$ and z/c = $^1/_2$ for the $[N_2Pr_3]^{3+}$ double chains, which are charge-balanced by the remaining distinguishable Te^{2-} and Cl^- anions.

Figure 1. View at the crystal structure of $Pr_{10}N_6Te_5Cl_2$ along [100] (*top*) and its two different types of chains containing edge-connected $[NPr_4]^{9+}$ tetrahedra (*bottom*).

References

[1] F. Lissner, Th. Schleid, *Z. Anorg. Allg. Chem.* **2005**, *631*, 1119-1124.
[2] M. L. Foltin, F. Lissner, S. Strobel, Th. Schleid, *Z. Anorg. Allg. Chem.* **2014**, *640*, 1247-1253.
[3] F. Lissner, M. L. Foltin, Th. Schleid, *Eur. J. Inorg. Chem.* **2016**, in press.
[4] M. L. Foltin, Th. Schleid, *Z. Anorg. Allg. Chem.* **2015**, *641*, 292-297.
[5] F. Lissner, Th. Schleid, *J. Alloys Compds.* **2006**, *418*, 68-72; *Z. Anorg. Allg. Chem.* **2008**, *634*, 2061.
[6] M. L. Foltin, P. Talmon-Gros, Th. Schleid, *Z. Anorg. Allg. Chem.* **2014**, *640*, 2823-2829.

Figure 1

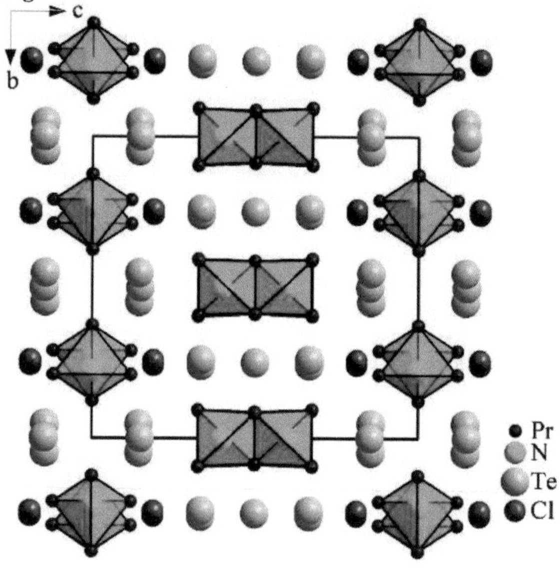

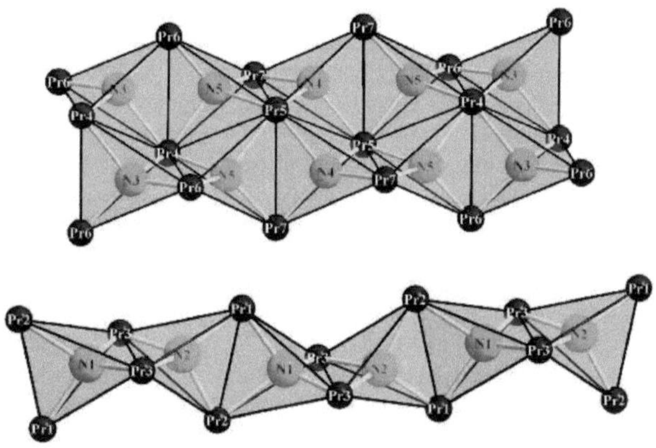

P031
Chloride Substituted Hybrid Perovskites $MAPbI_{3-x}Cl_x$ - a systematic, crystallographic study -

J. Steckhan[1,2], A. Franz[1], S. Schorr[3]
[1]*Helmholtz-Zentrum Berlin für Materialien und Energie, Abteilung Struktur und Dynamik von Energiematerialien, Berlin, Germany*
[2]*Universität Potsdam, Potsdam, Golm, Germany*
[3]*Freie Universität Berlin, Institut für Geologische Wissenschaften, Berlin, Germany*

Hybrid perovskites as $MAPbI_3$ and $MAPbCl_3$ are perovskites with an ABX_3-structure, where A is an organic molecule, B for example a lead(II)-cation and X a halide anion. Both structures consist of corner linked PbX_6-octahedra. Located in the space generated by these octahedra is a methyl ammonium (MA) cation. While $MAPbCl_3$ belongs to the cubic space group *Pm-3m*, $MAPbI_3$ belongs to the tetragonal space group *I4/mcm*. The symmetry is lowered by tilting and/or distortion of the PbX_6^--octahedra and by atom displacement.

Perovskites are widely known for their enormous possibility of element substitution, which leads to a huge variety of physical properties. The substitution of iodide by chloride is described as difficult respectively not possible [3]. In our study we examine systematically the influence of a decreasing chloride ratio in reaction solutions.

The chlorine substituted polycrystalline $MAPbI_{3-x}Cl_x$ was synthesized using the method described by Baikie et. al. [1] from an equimolar mixture of methylammonium halide and lead(II) halide in a mixture of γ-butyrolactone and N,N-dimethylformamide. The halide ratio was changed by mixing the respective halide compounds. Polycrystalline $MAPbCl_3$ was prepared following the synthesis of Leguy et. al. [4] by precipitation by lead(II) acetate from a methyl amine solution that was neutralized by an excess of hydrochloride acid.

Synchrotron X-ray , lab X-ray and neutron diffraction experiments were performed at the Helmholtz-Centre Berlin for Materials and Energy for structural investigations. The data were analyzed by Fullprof Suite Software.

References

[1] T. Baikie, N. S. Barrow, et. al., *J. Mater. Chem. A* (2013), 3, 9298-9307.
[2] M. T. Weller, O. J. Weber, et. al., *Chem. Commun.* (2015), 51, 4180-4183.
[3] S. Colella, E. Mosconi, et. al., *Chem. Mater* (2013), 25, 4613-4618.
[4] A. M. A. Leguy, Y. Hu, et. al., *Chem. Mater* (2015), 27, 3397-3407.

Figure 1

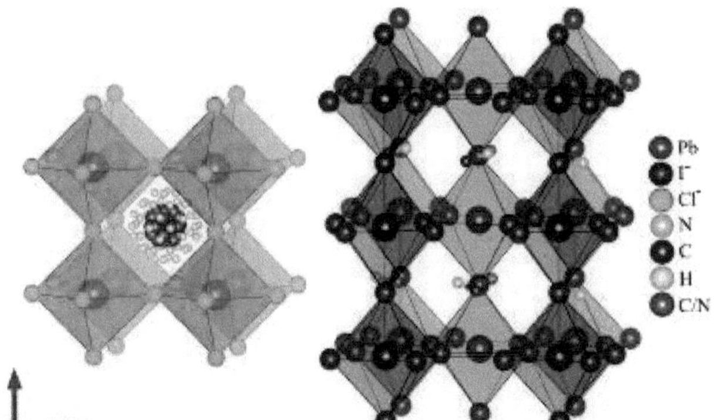

Figure 1: Crystal structure of $MAPbCl_3$ (left) [1] and $MAPbI_3$ (right; hydrogen not displayed) [2].

P032
Donor-Unsupported Lithium-Phosphanylmethanide $Li[CH_2P(tBu)_2]$ - Crystal Structure from X-Ray Powder Diffraction and DFT-D calculations

A. Bodach[1], L. Fink[1], K. Samigullin[1], E. Alig[1], M. Wagner[1], H.- W. Lerner[1]
[1] *Goethe University Frankfurt, Institute of Inorganic and Analytical Chemistry, Frankfurt am Main, Germany*

Organolithium compounds have gained prominence in organic and organometallic syntheses. Phosphanyl-substituted lithiumorganyls give access to new frustrated lewis pairs “FLPs”, such as $(3,5\text{-}(CF_3)_2C_6H_3)_2B\text{-}CH_2\text{-}PtBu_2$, synthesized *via* $Li[CH_2P(tBu)_2]$. $Li[CH_2P(tBu)_2]$ was prepared using a new synthesis protocol starting from $CH_3P(tBu)_2$ and *t*BuLi, without LiH formation during β-H elimination. The compound is self-igniting on air and cannot be recrystallized, while the solid-state structure is of general importance for the understanding of the reactivity [1]. To determine the crystal structure, high-resolution laboratory X-ray powder diffraction data were used. Two structure solutions with similar quality criteria were refined with Rietveld methods in space groups *I*2/*a* and *Iba*2, respectively. Both space groups are minimal non-isomorphic supergroups of *Ia*. The positions of the phosphorus atoms are similar in both solutions. To decide which solution is the correct one, we have carried out standard CASTEP DFT-D calculations [2]. A set of reference structures (CSD) was used for validation. By application of the RMSCD values [3], the calculations indicated that the solution with space group *Iba*2 is significantly wrong, while the crystal structure with space group *I*2/*a* is confirmed. To investigate the volume contractions of the unit cells, calculated with CASTEP, we have carried out thermal expansion studies with extrapolation of the volume to 0 K. During our CASTEP calculations thermal effects were not considered, therefore a possible dynamic disorder of *t*Bu groups could explain the volume differences.
For the final structure refinement, optimized restraints for the CH_3 groups and Li atoms were generated from CASTEP calculations (Figure 1 and 2). Finally, a low temperature data set was used for validation of the crystal structure with space group *I*2/*a*.

Figure 1: Molecular structure of $Li[CH_2P(tBu)_2]$

Figure 2: Rietveldplot; refinement of the crystal structure of $Li[CH_2P(tBu)_2]$

References:

[1] K. Samigullin, M. Bolte, H.-W. Lerner, M. Wagner, *Organometallics* **2014**, *33*, 3564-3569.
[2] S. J. Clark, M. D. Segall, C. J. Pickard, P. J. Hasnip, M. J. Probert, K. Refson, M. C. Payne, *Z. Krist.* **2005**, *220*, 567-570.
[3] J. van de Streek, M. A. Neumann, *Acta Cryst.* **2014**, *B70*, 1020-1032.

Presented results submitted to ZAAC

Figure 1

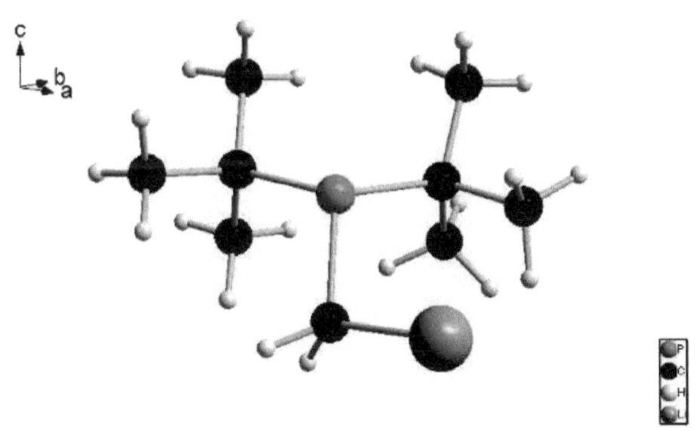

Figure 2

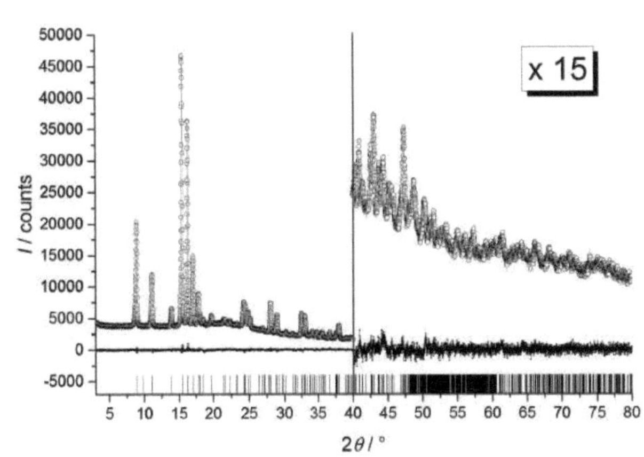

P033

$NaSc[SeO_3]_2$: A New Alkali-Metal Scandium Oxoselenate(IV)

S. Greiner[1], T. Schleid[1]
[1]*University of Stuttgart, Institute for Inorganic Chemistry, Stuttgart, Germany*

The new oxoselenate(IV) $NaSc[SeO_3]_2$ crystallizes orthorhombically in the centrosymmetric space group *Pnma* with the lattice parameters $a = 1279.28(9)$ pm, $b = 546.37(4)$ pm and $c = 814.02(6)$ pm for Z = 4 (CSD-430618). For the yttrium analogues $AY[SeO_3]_2$ (*A* = Na, K, Rb, Cs) [1, 2] several different structures and space groups are known. From these, $NaY[SeO_3]_2$ crystallizes also orthorhombically, but in the *non*-centrosymmetric space group $P2_1cn$, and the potassium and rubidium compounds exhibit the same structure as $NaSc[SeO_3]_2$. $CsY[SeO_3]_2$, however, favours the cubic space group *Pa* [2], whereas the lanthanum compound $NaLa[SeO_3]_2$ [1] is known to adopt the monoclinic space group $P2_1/n$. The Sc^{3+} cations in $NaSc[SeO_3]_2$ (see Table 1 for the atomic positions) are surrounded by six terminal $[SeO_3]^{2-}$ units in an octahedral coordination with Sc^{3+}–O^{2-} distances between 205 and 211 pm. These distances agree well with the six Sc^{3+}–O^{2-} distances in $Sc_2[SeO_3]_3$, which range from 202 to 220 pm [3]. The charge equalization is guaranteed through the Na^+ cations, which reside in eightfold oxygen coordination spheres provided by five pyramidal $[SeO_3]^{2-}$ units (Figure 1, *top*). The Na^+–O^{2-} distances cover an interval from 250 to 288 pm and the Se^{4+}–O^{2-} distances occur between 167 and 170 pm with O–Se–O angles from 98 up to 104°, which fit very well with all the values in the literature [1, 2]. The interconnection of the $[ScO_6]^{9-}$ octahedra to a three-dimensional network is ensured by the ψ^1-tetrahedral $[SeO_3]^{2-}$ anions via common vertices (Figure 1, *bottom*). Isostructural oxoselenates(IV) are already also known with iron, gallium and indium as trivalent metals for the examples of $AFe[SeO_3]_2$ (*A* = Na, K) [4, 5], $AGa[SeO_3]_2$ (*A* = Na, K) [6] and $AIn[SeO_3]_2$ (*A* = Na, K) [7].

References

[1] R. E. Morris, J. A. Hriljac, A. K. Cheetham, *Acta Crystallogr.* **1990**, *C 46*, 2013–2017.
[2] S. E. Bang, D. W. Lee, K. M. Ok, *Inorg. Chem.* **2014**, *53*, 4756–4762.
[3] J. Wontcheu, Th. Schleid, *Z. Anorg. Allg. Chem.* **2003**, *629*, 1463–1465.
[4] G. Giester, M. Wildner, *J. Alloys Compds.* **1996**, *239*, 99–102.
[5] G. Giester, *Z. Kristallogr.* **1993**, *207*, 1–7.
[6] D. W. Lee, K. M. Ok, *Inorg. Chem.* **2013**, *52*, 5176–5184.

[7] D. W. Lee, S. B. Kim, K. M. Ok, *Inorg. Chem.* **2012**, *51*, 8530–8537.

Figure 1. Atomic positions and U_{eq} values (in pm²) for $NaSc[SeO_3]_2$.

Figure 2. Coordination sphere of the Na^+ cations in $NaSc[SeO_3]_2$, which are surrounded by eight O^{2-} anions of five $[SeO_3]^{2-}$ units as bicapped trigonal prisms *(top)*. Crystal structure of $NaSc[SeO_3]_2$ with $[ScO_6]^{9-}$ octahedra connected to $[SeO_3]^{2-}$ pyramids via common vertices *(bottom)*.

Figure 1

Atom	Site	*x* / *a*	*y* / *b*	*z* / *c*	U_{eq}
Na	4*c*	0.3873(2)	1/4	0.4886(4)	334(7)
Sc	4*c*	0.09975(8)	1/4	0.29975(14)	139(3)
Se1	4*c*	0.06988(4)	1/4	0.71873(8)	145(2)
Se2	4*c*	0.29214(4)	1/4	0.00395(8)	165(2)
O1	4*c*	0.0265(3)	1/4	0.5226(5)	208(10)
O2	4*c*	0.1659(3)	1/4	0.0626(5)	282(11)
O3	8*d*	0.0030(2)	0.4887(6)	0.8003(4)	243(9)
O4	8*d*	0.2105(2)	0.5094(6)	0.3758(4)	353(10)

Figure 2

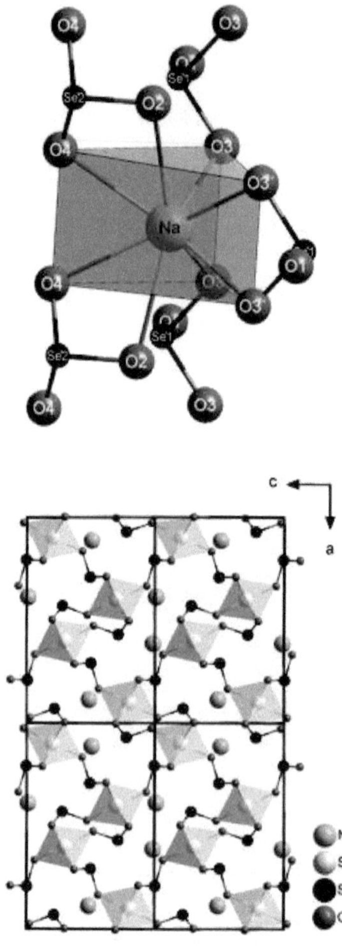

P034

$Rb_3[NO_3][B_{12}H_{12}]$: The Nitrate Dodecahydro-*closo*-Dodecaborate of Rubidium - A New Compound with a Known Crystal Structure

F. Kleeberg[1], J. Mehnert[1], T. Schleid[1]
[1]*University of Stuttgart, Institute for Inorganic Chemistry, Stuttgart, Germany*

Introduction

Compounds of mixed-anionic *closo*-dodecaborates like $A_3X[B_{12}H_{12}]$ (*A* = K - Cs, *X* = Cl - I) are known in literature for a decent time [1, 2], whereas reports about compositions with more complex anions, *e.g.* $[NO_3]^-$ or $[BF_4]^-$, are quite rare [3 - 5]. It took more than 30 years until each one of these compounds was fully characterized by single-crystal X-ray diffraction, all showing the same distorted perovskite-like structure mostly with trigonal *anti*-$LaAlO_3$-type arrangement [6, 7]. The here presented $Rb_3[NO_3][B_{12}H_{12}]$ was first described by *Kuznetsov et al.* [5] in

1980, however further characterization was still necessary, since it structurally differs slightly from the orthorhombic $Cs_3[NO_3][B_{12}H_{12}]$ [3].

Structure Description

The new *anti*-perovskite compound with the composition $Rb_3[NO_3][B_{12}H_{12}]$ crystallizes in the monoclinic space group $P2_1/c$ with a = 720.52(3) pm, b = 2146.15(9) pm, c = 1910.06(8) pm and β = 98.958(3)° for Z = 8. The crystal structure can be described best as a monoclinically distorted variation of the cubic *anti*-$SrTiO_3$- or orthorhombic *anti*-$GdFeO_3$-type that $Cs_3[NO_3][B_{12}H_{12}]$ adopts. The nitrogen atoms of the corresponding nitrate anions reside at the *Wyckoff* position 4*e*, just like the six crystallographically different Rb^+ cations. The centers of gravity of the three different *quasi*-icosahedral $[B_{12}H_{12}]^{2-}$ anions occupy the special *Wyckoff* positions 2*a* and 2*c* and the general *Wyckoff* position 4*e*, thus erecting a cubic closest package together with the Rb^+ cations. The nitrate anions reside in $^1/_4$ of the resulting octahedral voids built up by six metal cations, whereas Rb^+ itself is octahedrally coordinated by four $[B_{12}H_{12}]^{2-}$ and two $[NO_3]^-$ anions. Two $[B_{12}H_{12}]^{2-}$ icosahedra interacting via three hydrogen atoms of triangular faces (CN = 6) and two hydrogen atoms of two further of them each coordinating via edges (CN = 4) at the Rb^+ cations (d(Rb-H) = 289 - 386 pm, CN = 10; d(Rb-O) = 280 - 318 pm, CN = 2), provide for almost cuboctahedral overall coordination spheres. The cuboctahedral Rb^+-coordination of the $[B_{12}H_{12}]^{2-}$ anions and the $\{[NO_3]Rb_6\}^{5+}$ octahedra connected via all corners are characteristic for a cubic perovskite-type structure.

Figure 1. View at the *anti*-perovskite-like crystal structure of $Rb_3[NO_3][B_{12}H_{12}]$ along [100].

References

[1] E. L. Muetterties, J. H. Balthis, Y. T. Chia, W. H. Knoth, H. C. Miller, *Inorg. Chem.* **1964**, *3*, 444-451.
[2] N. T. Kuznetsov, G. S. Klimchuk, O. A. Kanaeva, K. A. Solntsev, *Russ. J. Inorg. Chem.* **1976**, *21*, 505-506.
[3] I. Tiritiris, *Doctoral Thesis*, University of Stuttgart, **2004**.
[4] Ch. Funk, *Bachelor Thesis*, University of Stuttgart, **2011**.
[5] O. A. Kanaeva, N. T. Kuznetsov, O. O. Sosnovskaya, L. V. Goeva, *Russ. J. Inorg. Chem.* **1980**, *25*, 1315-1321.
[6] I. Tiritiris, Th. Schleid, *Z. Anorg. Allg. Chem.* **2003**, *629*, 1390-1402.
[7] I. Tiritiris, J. Weidlein, Th. Schleid, *Z. Naturforsch.* **2005**, *60 b*, 627-639.

Figure 1

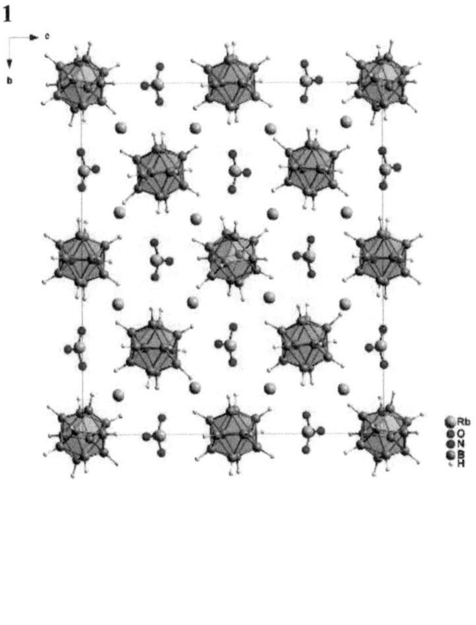

P035

Realstruktur als „Falle" bei der Strukturlösung auf Basis von Pulverbeugungsdaten

A. Leineweber[1]
[1]*Institut of Materials Science, Freiberg, Germany*

Kristalldefekte sind Abweichungen vom streng-periodischen Aufbau von Kristallen (wenn man Gitterschwingungen außen vor lässt) wie Punktdefekte, Versetzungen und sowie die Begrenztheit eines Kristalls (Oberfläche). In polykristallinen Ensembles kommen noch Inhomogenitäten der Zusammensetzung, der Kristallorientierung, der Defektdichte und der Kristallitgröße hinzu. All diese Effekte, in ihrer Gesamtheit als Realstruktur bezeichnet, beeinflussen das Ergebnis eines (Pulver-)Beugungsexperiments innerhalb der kinematischen Beugungstheorie dahingehend, dass die Bragg-Reflexe verbreitert werden bzw. Intensität von Bragg-Reflexen in diffuse Streuung umverteilt wird. Dieses erschwert in jedem Fall eine eindeutige und vollständige Interpretation der Daten auf Basis der meist allein sichtbaren Bragg-Reflexe. Vielfach kann das Problem der Realstruktur über eine Verfeinerung von Auslenkungsparametern und wenigen Besetzungsparametern sowie eine Berücksichtigung von oft lediglich phänomenologisch beschriebener Linienverbreiterung gelöst werden, wobei die Haupterkenntnisse der Strukturbestimmung weitgehend unverfälscht bleiben.

Der hier zu präsentierende Beitrag beschäftigt sich mit Fällen, bei denen diese einfache Vorgehensweise zu mehr oder weniger falschen Schüssen führen kann oder eine sinnvolle Interpretation der Beugungsdaten überhaupt scheitert. Insbesondere werden Fälle aus folgenden Gruppen diskutiert:

(1) Vorkommen starker Besetzungsfehlordnung, welche in Extremfällen zu mit den Beugungsdaten gleichwertig aber nicht äquivalenten Strukturbeschreibungen führen kann (verschiedene Raumgruppen), von denen aber oft nur eine erkannt und als „wahre" Struktur dargestellt wird.

(2) Vorkommen starker anisotroper Mikrodehnungsverbreiterung als Folge von (lokalen) Scherspannungen im Polykristall. Die *hkl*-abhängige Reflexverbreiterung kann dann einer Reflexaufspaltung zugeschrieben werden, welche durch weitere Linienverbreiterung verschmiert ist. So kann fälschlicherweise eine Symmetrie zugeordnet werden, welche niedriger ist als die des bei rein hydrostatischer Spannung vorliegenden Grundzustands.

(3) Vorkommen starker Stapelfehlordnung, welche zu einer starken Verschmierung von (bestimmten) Braggreflexen bis hin zu Streaks führen kann, so dass solche verschmierten Reflexe unter Pulverbeugungsbedingungen kaum mehr als solche erkennbar sind. So können im Extremfall lediglich die unverbreiterten Reflexe erkennbar bleiben. Diese mögen indizierbar sein, erlauben aber keine Bestimmung eines überzeugenden Strukturmodells, während die manchmal schwach sichtbaren „diffraction bands" (d. h. die Überbleibsel der *Streaks* von der Stapelfehlordnung) als solche unerkannt bleiben.

Ausgehend von Erfahrungen mit derartigen Fällen werden Empfehlungen präsentiert, mit Hilfe derer problematische Fälle erkannt werden können und wie entsprechende Beugungsdaten interpretiert werden können. Eine Charakterisierung der Realstruktur sollte auch helfen, ihre Ursachen zu erkennen. So kann dann über Beeinflussung der Synthesebedingungen Realstruktur zur Eigenschaftseinstellung gezielt modifiziert werden. Nicht zuletzt kann dann Realstrukturbeeinflussung auch eingesetzt werden, um detaillierte Beugungsuntersuchungen erschwerende Realstrukturphänomene zu reduzieren.

P036

New Mixed Ternary Strontium Trielides at the Border between Zintl and Laves Phases

C. Meyer[1], C. Röhr[1]
[1]*Albert-Ludwigs-Universität Freiburg, Institut für anorganische und analytische Chemie, Freiburg, Germany*

The binary alkaline-earth trielides A_xM_y (A = Ca, Sr, Ba; M = Al, Ga) exhibit a fascinating structural diversity ranging from electron precise Zintl compounds (e.g. $SrAl_2$ forming the KHg_2-type structure with a four-bonded Al^- network [1]) via electron inprecises Zintl compounds like Ca_3Ga_5 or Eu_5Ga_9 (both built up from structure elements of the KHg_2- and the CrB-type structure) [2,3] to Laves phases like h.p.-$SrAl_2$ [4] or Sr_5Al_9 [5].

By substituting the triel element in binary trielides against each other, the phase stabilities of mixed compounds $Sr_x(M1/M2)_y$ (fig. 1) have been systematically investigated by means of single crystal diffraction. The 'coloring' of the anionic network is discussed in relation to electronic and geometric aspects based on the results of FP-LAPW bandstructure calculations, i.e. Bader charges and volumes.
For $Sr_3Al_xGa_{5-x}$ a series of new ternary trielides could be obtained although both binary phases Sr_3Al_5 and Sr_3Ga_5 are unknown.

The phase width of these mixed trielides ranges from x = 2.4 to 3.6 (Ca_3Ga_5-type structure, SG *Cmcm*; for x = 2.4/3.6: a = 468.69(5)/471.30(4), b = 1133.3(2)/1135.28(7), c = 1572.1(2)/1577.0(1) pm, $R1$= 0.0442/0.0398). In accordance with the calculated Bader charges, the more electronegative element gallium prefers the occupation of the more negatively charged two-bonded M^- position $M(1)$ [fig. 2 (a)].

Similarly, a new mixed Eu_5Ga_9-type trielide $Sr_5Al_{6.8}Ga_{2.2}$ [SG *Cmcm*, a = 473.80(3), b = 1127.41(7), c = 2687.3(2) pm, $R1$= 0.0456, fig. 2 (b)] could be synthesized even though a binary binary gallide of this composition is not known and the binary aluminide is not isotypic (see below). Like for the Ga/In distribution in the Ca_3Ga_5-type structure, Ga prefers to take the two-bonded M^- position $M(1)$.

In contrast, the Laves-related phase Sr_5Al_9 consists of blocks of kagome nets (fig. 2 [c]).
In this structure, only a very small amount of Al could be substituted by Ga ($Sr_5Al_{8.5}Ga_{0.5}$, SG *R-3m*, a = 586.34(7), c = 3556.8(5), $R1$=0.0752. Due to the large difference of Bader charges, only one of three M positions is partly taken by Ga. As well as the larger triel element indium [6], gallium occupies the $M(1)$ position at the boundary of the kagome net blocks.

[1] W. Harms, M. Wendorff, C. Röhr, *Z. Naturforsch.* **62b**, 177-194 (2007).
[2] M. L. Fornasini, F. Merlo, *Z. Kristallogr.* **22**, 382-390 (2006).
[3] M. Rhode, M. Wendorff, C. Röhr, *Z. Anorg. Allg. Chem.*, **632**, 1195-1205 (2006).
[4] G. Cordier, E. Czech, H. Schäfer, *Z. Naturforsch.*, **B37**,1442-1445 (1982).
[5] N.B. Manyako, O.S. Zarechnyuk, T.I. Yanson, *Kristallografiya* **32**, 339-342 (1987).
[6] M. Wendorff, C. Röhr, *Z. Anorg. Allg. Chem.* **632**, 2164 (2006).

Fig. 1: Phase widths of mixed strontium aluminides/gallides

Fig. 2: Crystal structure types observed in ternary Sr-Al-Ga compounds: (a) Ca_3Ga_5, (b) Eu_5Ga_9, (c) Sr_5Al_9

Figure 1

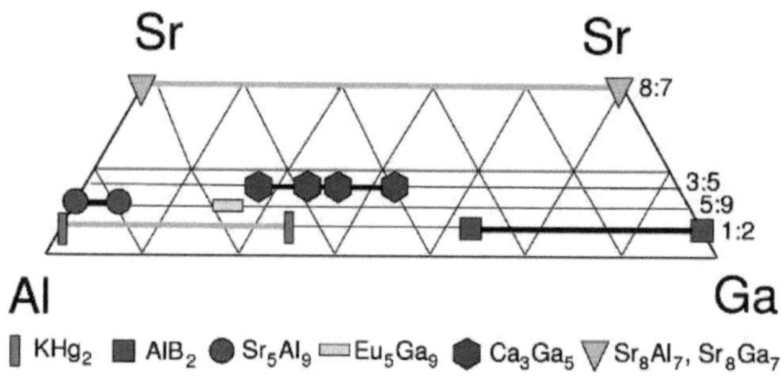

Figure 2

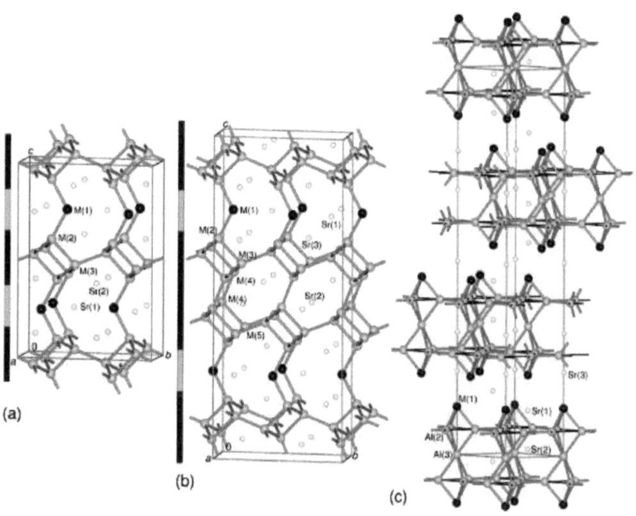

P037

Crystal and electronic structure of Rb_2In_3 and ternary gallium variants

M. Falk[1], C. Röhr[1]
[1]*Universität Freiburg, Institut für Anorganische und Analytische Chemie, Freiburg, Germany*

Alkali gallides and indides show a rich structure diversity, reaching from simple intermetallics [1] via Zintl phases to cluster compounds and boron-analogues [2]. The polyanions formed in the latter compounds can be explained by the Wade-Mingos or Wade-Jemmis rules. In systematic experimental and bond theoretical studies, the "coloring" of the polyanions in the Cs_2In_3-type structure has been examined, i. e. the influence of the different size and electronegativity of gallium and indium was investigated.

There are three known binary compounds crystallizing in the Cs_2In_3-type structure: Cs_2In_3 [3], Rb_2In_3 [4, 5] and K_2Ga_3 [6]. Based on Rb_2In_3, the amount of weighed gallium has been successively increased up to 50%. The elements were weighed under argon atmosphere in tantalum tubes and sealed. Due to the known incongruent melting behaviour of the binary phase Rb_2In_3 at 490 °C [7], an elaborate temperature program has been used. The samples were heated to 540 °C, cooled rapidly to 450 °C and slowly to 300 °C, then cooled to room temperature within 14 hours. The lattice parameters and the gallium/indium distribution of the crystals obtained were determined using single crystal X-ray diffraction.

Rb_2In_3 crystallizes in a tetragonal structure, containing two Rb and two In positions (fig. 1). Four In(1) and two In(2) atoms form distorted In_6 octhedra, which are connected via In(1)-In(1) *exo*-bonds to form layers perpendicular [001]. The 6 x 3 (In) + 4 (charge) = 22 electrons provided fit the electron count of the octahedral *closo*-cluster ($2n + 2 = 14$ v.e.), when the four *exo*-bonds at In(1) and the two lone pair electron pairs of In(2) are substracted. In agreement, the calculated tDOS of Rb_2In_3 shows a local minimum at the Fermi level (fig. 2). An AIM analysis results in Bader charges of -0.37 for In(1) and -0.61 for In(2). The Bader volumes amount to 36.4 for In(1) and 48.2 x 10^6 pm^3 for In(2). The successive replacement of In by Ga up to 36 % ($Rb_2In_{1.91}Ga_{1.09}$, *I*4/*mmm*, a = 647.33(6), c = 1607.5(2) pm, $R1$ = 0.0587) shows that Ga occupies the *M*(1) position exclusively. This position has a smaller Bader charge, but also a much smaller volume than In(2). The "coloring" of the polyanion is therefore not determined by electronegativity (Ga: 1.82, In: 1.49 after Allred-Rochow), but by geometrical factors (metallic radii: Ga: 126 pm, In: 144 pm). This result is in accordance with the lattice parameters of the obtained compounds, i. e. the decrease of the a axis and the volume (and thereby a decrease of the *exo*-bonds between the octahedra) and the small increase of the c axis. The results match the situation in $K_2Ga_{2.32}In_{0.68}$ found by Müller [8], where In occupies the *M*(2) position exclusively. An investigation of the system Cs-Ga-In is the purpose of current research.

[1] M. Wendorff, C. Röhr, *Z. Anorg. Allg. Chem.* **631**, 338-349 (2005).
[2] C. Meyer, M. Falk, C. Röhr, *Acta Crystallogr.* **A71**, s485 (2015).
[3] S. P. Yatsenko, K. A. Tschuntonow, A. N. Orlow, Y. P. Yarmolyuk, Y. N. Grin, *J. Less-Common Met.* **108**, 339-343 (1985).
[4] S. C. Sevov, J. D. Corbett, *Z. Anorg. Allg. Chem.* **619**, 128-132 (1993).
[5] G. Cordier, V. Müller, *Z. Kristallogr.* **203**, 150-151 (1993).
[6] R. W. Henning, J. D. Corbett, *Inorg. Chem.* **38**, 3883-3888, (1999).
[7] A. D. Pelton, S. Larose, *J. Phase Equilib.* **12**, 377-379 (1991).
[8] V. Müller, PhD thesis, TH Darmstadt (1995).

Fig. 1: Crystal structure of $Rb_2In_{3-x}Ga_x$.

Fig. 2: DOS of Rb_2In_3.

Figure 1

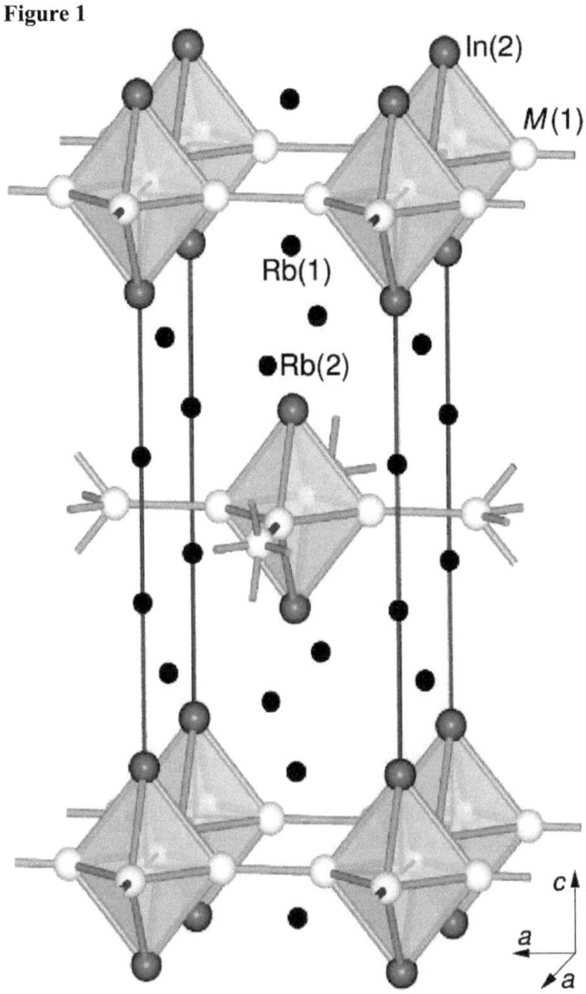

Figure 2

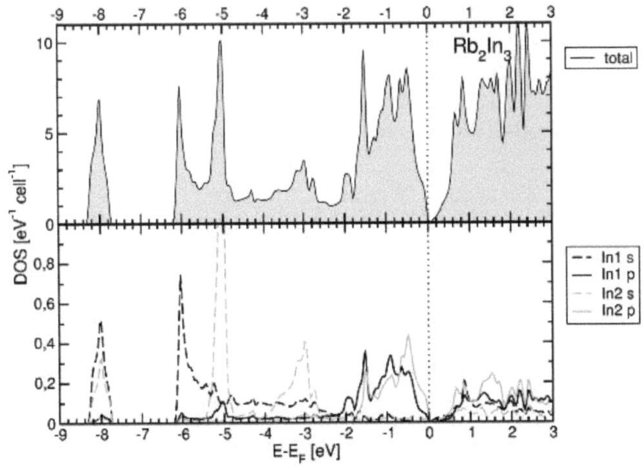

P038

Approaching Polysulfuric Acids: The New Hydrogentrisulfate Anion $[HS_3O_{10}]^-$ in the Crystal Structures of $Na[HS_3O_{10}]$, $K[HS_3O_{10}]$, and $Rb[HS_3O_{10}]$

L. V. Schindler[1], M. S. Wickleder[1]
[1]*Justus-Liebig-Universität Gießen, Institut für anorganische und analytische Chemie, Gießen, Germany*

Sulfuric acid is a strong acid whose acidity is reflected in the structural richness of its corresponding salts. However, while a great number of well characterized sulfates has been reported, the knowledge of acidic hydrogensulfates is more limited. Nevertheless it is obvious that here hydrogen bonds are the stamping feature in the respective crystal structures.[1] Moving one step forward in the homologous series of polysulfuric acids we arrive at disulfuric acid, $H_2S_2O_7$. For the salts of this acid, the deviation between the known structures of disulfates and hydrogendisulfates is even more prominent. In fact just six salts containing the latter anion have been reported[2] and comprehensive studies were completely missing until recently, when we presented a first systematic investigation.[3] These studies show that only under strongly acidic conditions the $[HS_2O_7]^-$ anion can be gained. This is in line with the often claimed but never proved estimation that the acidity of polysulfuric acids $H_2S_nO_{3n+1}$ increases with growing numbers of *n*. While trisulfates have been reported for only a few examples[2b, 4], their hydrogenated congeners, $[HS_3O_{10}]^-$, are not known at all - in line with the assumption of increasing acidity of long chained polysulfuric acids. We now present another finding that helps to shed some light on the properties of polysulfuric acids and their salts, namely the first hydrogentrisulfates exhibiting interesting hydrogen bonding systems (Figure 1). Colorless and extremely sensitive crystals of $Na[HS_3O_{10}]$ (monoclinic, $P2_1/n$ (No. 14), $Z = 4$, $a = 707.36(2)$ pm, $b = 1378.56(4)$ pm, $c = 848.10(3)$ pm, $\beta = 94.817(1)°$, $V = 824.09(4)\cdot 10^6$ pm^3) and $K[HS_3O_{10}]$ (orthorhombic, *Pccn* (No. 56), $Z = 4$, $a = 1057.16(3)$ pm, $b = 807.81(2)$ pm, $c = 897.57(2)$ pm, $V = 766.51(3)\cdot 10^6$ pm^3) were obtained from the reaction of the appropriate sulfate salts Na_2SO_4 and K_2SO_4 with fuming sulfuric acid, the reaction of Rb_2CO_3 with a remaining quantity of water in the starting material with neat SO_3 yielded colorless crystals of $Rb[HS_3O_{10}]$ (orthorhombic, *Pnma* (No. 62), $Z = 4$, $a = 891.43(3)$ pm, $b = 1095.34(4)$ pm, $c = 839.37(3)$ pm, $V = 819.58(5)\cdot 10^6$ pm^3). The hydrogentrisulfate salts are an important milestone in our efforts of exploring polysulfuric acids and their salts.

Figure 1. Chains and dimers in the crystal structures of $Na[HS_3O_{10}]$, $K[HS_3O_{10}]$, and $Rb[HS_3O_{10}]$.

[1] a) E. Kemnitz, C. Werner, S. Troyanov, *Eur. J. Solid State Inorg. Chem.* **1996**, *33*, 563-580; b) E. Kemnitz, C. Werner, S. Troyanov, *Eur. J. Solid State Inorg. Chem.* **1996**, *33*, 581-596.

[2] a) I. D. Brown, D. B. Crump, R. J. Gillespie, *Inorg. Chem.* **1971**, *10*, 2319-2323; b) J. Bruns, T. Klüner, M. S. Wickleder, *Chem. Asian J.* **2014**, *9*, 1594-1600; c) J. Bruns, O. Niehaus, R. Pöttgen, M. S. Wickleder, *Chem. Eur. J.* **2014**, *20*, 811-814; d) C. Logemann, H. Kleineberg, J. Ohlert, M. S. Wickleder, *Z. Anorg. Allg. Chem.* **2013**, *639*, 2796-2803; e) K. Rieß, Dissertation thesis, Carl von Ossietzky Universität Oldenburg (Oldenburg), **2012**; f) J. W. M. Steeman, C. H. MacGillavry, *Acta Crystallogr.* **1954**, *7*, 402-404.

[3] L. V. Schindler, M. Daub, M. Struckmann, A. Weiz, H. Hillebrecht, M. S. Wickleder, *Z. Anorg. Allg. Chem.* **2015**, *641*, 2604-2609.

[4] C. Logemann, Mathias S. Wickleder, *Z. Anorg. Allg. Chem.* **2010**, *636*, 2075-2075.

Figure 1

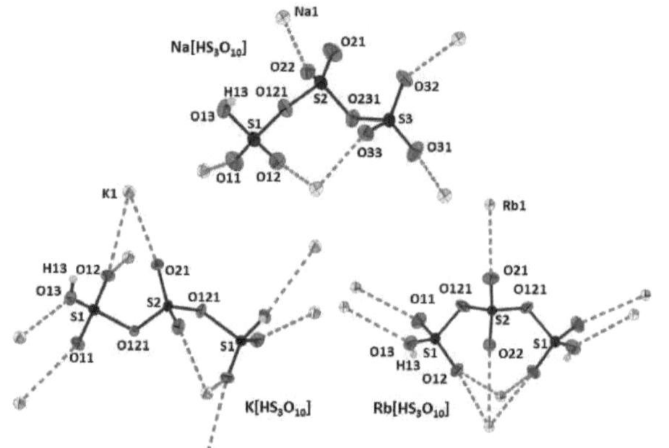

P039

A New Layered Europium(II) Oxotantalate(V): $Li_4Eu_3Ta_6O_{20}$

C. Funk[1], J. Köhler[2], T. Schleid[1]
[1]*University of Stuttgart, Institute for Inorganic Chemistry, Stuttgart, Germany*
[2]*Max Planck Institute for Solid State Research, Nanochemistry, Stuttgart, Germany*

Oxide based *perovskite* compounds have been studied for decades and $SrTiO_3$-type cubic $EuTiO_3$ for example was published in 1953 and exhibits interesting properties like its antiferromagnetic behaviour below 6 K [1, 2]. Out of these *perovskite* materials, the *Ruddlesden-Popper* phases were discovered nearly 60 years ago [3]. One example is $Li_2SrTa_2O_7$, which was investigated by *Floros et al.* in 1999 [4], and another one $Li_2EuTa_2O_7$ (Figure 1, *left*), first prepared in our group [5]. Now it was possible to synthesize a new compound with the composition of $Li_4Eu_3Ta_6O_{20}$ in squeeze-closed tantalum capsules. Eu_2O_3, Ta_2O_5, $Li_2[CO_3]$ and Ta powder were pressed into pellets and heated up to 1250 °C for 112 hours with several intermediate grindings. Just like $Li_2EuTa_2O_7$ before $Li_4Eu_3Ta_6O_{20}$ crystallizes tetragonally in the space group *I4/mmm* with $a = 394.29(2)$ pm, $c = 2609.41(19)$ pm and $c/a = 6.618$ for Z = 1. The tantalum atoms are octahedrally surrounded by six oxygen atoms and the vertex-connected $[TaO_6]$ octahedra are stacked to three layers (Figure 1, *right*). The oxygen octahedra, which coordinate lithium in the layer between, are tetrahedrally distorted and the tantalum atoms shift towards their oxygen ligands coordinating lithium as well (Figure 2d). The europium atoms are situated in cuboctahedra formed by oxygen (Figure 2b), but the europium(II) site is not fully occupied for charge balancing. A layer of lithium, coordinated tetrahedrally by oxygen atoms in first sphere, separates these bigger triple blocks from each other (Figure 2a). The longest distance between tantalum and oxygen is 215 pm and the shortest one 187 pm. The Eu-O distances range from 268 to 285 pm, while the Li-O distances cover the interval from 206 pm (tetrahedron) to 265 pm (elongated octahedron).

Figure 1. The crystal structures of $Li_2EuTa_2O_7$ (*left*) and $Li_4Eu_3Ta_6O_{20}$ (*right*), both shown as (010) projections.

Figure 2. The coordination polyhedra of $Li_4Eu_3Ta_6O_{20}$: a) $[LiO_4]$ tetrahedron, b) $[EuO_{12}]$ cuboctahedron, c) and d) $[TaO_6]$ octahedra.

[1] J. Brous, I. Fankuchen, E. Banks, *Acta Crystallogr.* **1953**, *6*, 67-70.

[2] Z. Guguchia, H. Keller, A. Bussmann-Holder, J. Köhler, R. K. Kremer, *Eur. Phys. J.* **2013**, *B 86*, 409-503.
[3] R. S. Ruddlesden, P. Popper, *Acta Crystallogr.* **1957**, *10*, 538-539.
[4] N. Floros, C. Michel, M. Hervieu, B. Raveau, *J. Mater. Chem.* **1999**, *9*, 3101-3106.
[5] A. C. Müller, *Diploma Thesis*, Univ. Stuttgart **2009**.

Figure 1

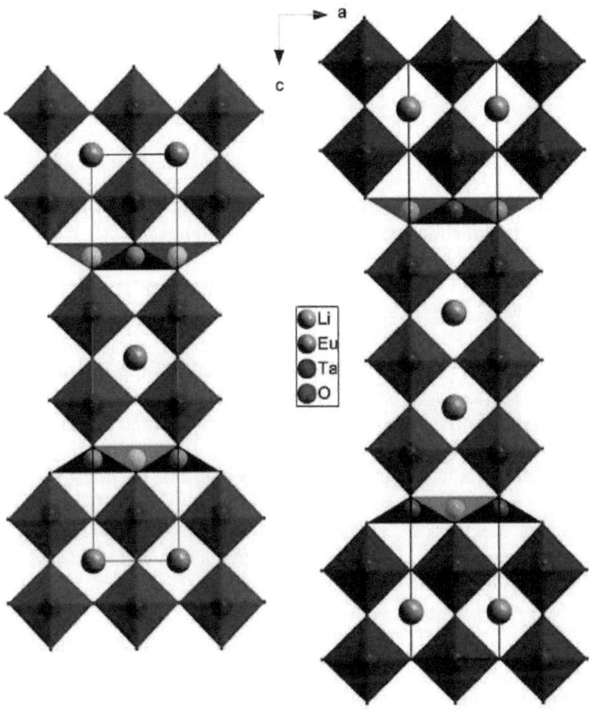

Figure 2

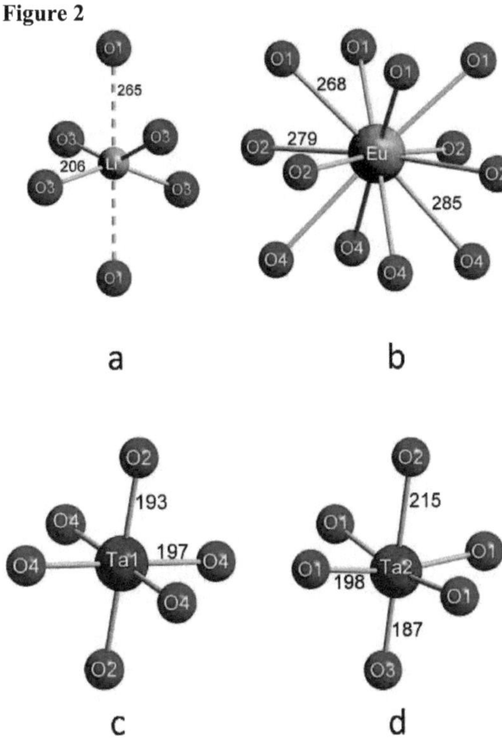

P040
Neutron powder diffraction study of the $Cu_xZn_{2-x}TiO_4$ cubic spinels at ambient conditions

M. He[1], B. Winkler[1], J. Ruiz-Fuertes[1], T. Bernert[2], N. Schrode[1], W. Morgenroth[1], E. Suard[3]
[1]*University of Frankfurt, Frankfurt am Main, Germany*
[2]*Max-Planck-Institut für Kohlenforschung, Mülheim/Ruhr, Germany*
[3]*Institut-Laue-Langevin, Grenoble cedex 9, France*

$ZnTi_2O_4$ is a cubic inverse spinel, which can be doped with transition metal cations to systematically change the physical properties, such as the band gap [1]. $Cu_xZn_{2-x}TiO_4$ cubic spinels with $0 \leq x < 1$ have been synthesized by the mechanochemical activation assisted solid-state synthesis and the structural and vibrational changes have been investigated [1]. However, neutron diffraction is superior to X-ray diffraction for the detection of the distribution of Cu^{2+} ($3d^9$) and Zn^{2+} ($3d^{10}$) in the $Cu_xZn_{2-x}TiO_4$ spinels because the neutron cross section of Cu^{2+} is sufficiently higher than that of Zn^{2+} [2] while the X-ray scattering factors are too similar to distinguish Cu^{2+} and Zn^{2+} in the same set of sites.

In this study, high-resolution neutron powder diffraction data [3] of the $Cu_xZn_{2-x}TiO_4$ spinels were collected on the D2B diffractometer at the Institut Laue-Langevin at ambient conditions in order to gain further insight into structural changes due to the incorporation of Cu^{2+}. The data collections were carried out on samples of around 5 g in the angular range 2θ = 0-160° with a step size of 0.05° and a monochromatic wavelength of 1.594 Å. Each individual scan had 25 steps and was repeated 10 times to increase statistics and get reliable neutron intensities. It took 4 h to obtain a complete diffraction pattern. Rietveld refinements [4-6] were used to analyse the data in the *d*-spacing range of 0.8-5.3 Å. A typical powder pattern is shown in figure 1.

The results showed less than 1% Cu^{2+} in the tetrahedral sites of the $Cu_xZn_{2-x}TiO_4$ spinels. With the substitution of Zn^{2+} by Cu^{2+} the lattice parameter a decreased moderately and linearly for $x \leq 0.6$ and contracted much more strongly at higher Cu^{2+} concentrations. Incorporation of Cu^{2+} into the $ZnTi_2O_4$ cubic spinel led to a decrease of the average octahedral bond length *d*(M-O) from 2.034(2) Å for the Zn_2TiO_4 end-member to 2.022(2) Å for x = 0.85(6). No significant change was observed for the average tetrahedral bond length, which was *d*(T-O) = 1.986(2) Å. The average bond lengths *d*(M-O) and *d*(T-O) are plotted against the Cu^{2+} concentration of the $Cu_xZn_{2-x}TiO_4$ spinels, as presented in figure 2.

Figure 1 Rietveld refinement results for the sample $Cu_xZn_{2-x}TiO_4$ with $x = 0.5$.

Figure 2 The average octahedral bond length *d*(M-O) and the average tetrahedral bond length *d*(T-O) versus the Cu^{2+} concentration of the $Cu_xZn_{2-x}TiO_4$ spinels. The lines are a guide to the eye.

M. He acknowledges grants from the University of Frankfurt and from the China Scholarship Council.

[1] J. Ruiz-Fuertes, T. Bernert, M. He, B. Winkler, V. L. Vinograd, V. Milman, *Appl. Phys. Lett.* **2014**, *105*, 071911.
[2] B. Grabcev, S. Todireanu, V. Cioca, *J. Appl. Crystallogr.* **1979**, *12(4)*, 399-400.
[3] ILL: Grenoble, **2015**, DOI: 10.5291/ILL-DATA.5-21-1089
[4] A. Larson and R. B. Von Dreele, GSAS, **2004**.
[5] B. H. Toby, *J. Appl. Cryst.* **2001**, *34*, 210-213.
[6] S. C. Vogel, *J. Appl. Cryst.* **2011**, *44*, 873-877.

Figure 1

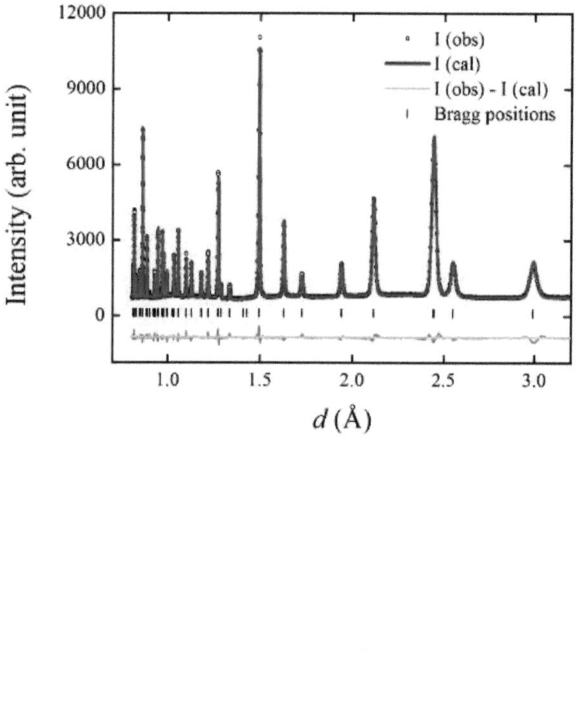

Figure 2

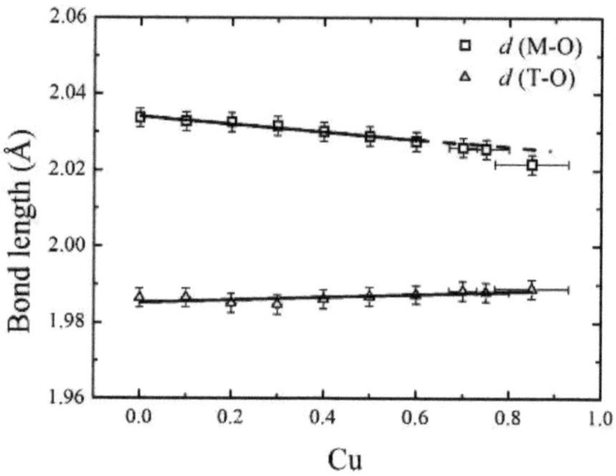

P041

Structure of the intermediate phase (INT) between sodalite (SOD) and cancrinite (CAN) - first approaches

H. Petersen[1], L. Robben[1], T. M. Gesing[1]
[1] *Universität Bremen, Institut für anorganische Chemie und Kristallographie, Bremen, Germany*

Hermeler et al. [1] synthesized single crystals of the intermediate phase (INT) between sodalite (SOD) and cancrinite (CAN). In single crystal diffraction they observed an ordered cancrinite structure in (hk0) and diffuse streaks in the (001) direction. The intensity maxima of these diffuse streaks could be correlated to the stacking sequence of SOD and CAN, implying a one dimensional disorder in the stacking sequence of the structure. However, in a new measurement of these crystals the diffraction pattern shows no difference to the results of Hermeler et al. [1].

A powder sample of the INT phase was synthesized hydrothermally. The reaction solution containing Na_2SiO_3 : $NaAlO_2$: Na_2CO_3 : NaOH in the molar ratio of 1 : 1 : 2.47 : 8 was heated for 48 h to temperatures ranging from 333 K to 473 K with temperature steps of 20 K. Rietveld analyses of the X-ray data obtained for the sample produced at 473 K show a disordered CAN structure only. The disorder of the CAN phases increases with decreasing synthesis-temperature. From the syntheses at 353 K and 333 K the INT phase could be observed. X-ray powder diffraction (XRPD) of the INT phase shows only the common reflections of the sodalite and cancrinite framework, which could be indexed using a hexagonal unit cell with lattice parameters a = 1268.9(6) pm and c = 1585(2) pm. The comparison of the powder sample with a powder diffractogram calculated from the single-crystal diffraction data shows equal intensity ratios, demonstrating the formation of the same phase.

Bonaccorsi et al. [2] correlated the numbers of stacked layers *N* in the ABC-6 family with the *c* lattice parameter and constrained them with the occurrence of actually observed cages (CAN (2); SOD (3), Losod (LOS) (4), Liottite (LIO) (6), Giuseppettite (GIU) (8)). Each possible cage contains a corresponding number of layers (number in brackets), resulting in the formula giving all possible combinations of cages in the structure. Through the possible amount of inserted templates in the cages the possible combination are reduced and the right sequence can be found. The combination of the cages under the assumption of a closed packed structure gives the stacking sequence of the phase. According to these calculations the hexagonal unit cell obtained in the PXRD experiment implies for the INT phase a stacking sequence of ABCACB, resulting in SOD and CAN cages.

Acknowledgment: The authors gratefully thank Prof. J.-C. Buhl for the single-crystal samples used in [1].

[1] G. Hermeler, J.-C. Buhl, W. Hoffmann, The Influence of carbonate on the synthesis of an Intermediate phase between sodalite and cancrinite, Catal. Today. 8 (1991) 415-426.
[2] E. Bonaccorsi, S. Nazzareni, Crystal chemical models for the cancrinite-sodalite supergroup: the structure of a new 18-layer phase, Zeitschrift Für Krist. - Cryst. Mater. 230 (2015) 345-351. doi:10.1515/zkri-2014-1819.

P042

Electrochemical formation of metal nitride surface layers in molten salt systems

T. Lehmann[1], R. Niewa[1]
[1] *University of Stuttgart, Institute for Inorganic Chemistry, Stuttgart, Germany*

Transition metal nitride films possess excellent properties like high hardness, wear and corrosion resistance, and high melting point. From the environmental point of view it is important to develop a nitriding process without the use of harmful gases, high temperatures and rather complicated and expensive apparatus. Molten salt electrochemical processes may be able to overcome these challenges [1].

Surface nitriding of transition metal nitrides was carried out in an electrochemical cell made of quartz glass. The potentiostatic electrolysis has been conducted according to the following principle: At 723 K a transition metal electrode is anodically polarized in a molten LiCl-KCl:Li_3N system. A small amount of Li_3N serves as a nitrogen ion source. Alternatively alkaline earth metal nitrides can be used. Additionally, a nitrogen gas electrode allows the successive supply of nitride ions via cathodic reduction of nitrogen gas to produce nitride ions [2].

A second possibility to form metal nitrides is the cathodic polarization of a transition metal electrode in a molten eutectic mixture of LiCl-KCl at 723 K with KNO_3 dissolved in the melt. The electrochemical surface nitridation may be rationalized as follows: nitrate ions are successively reduced via nitrite and nitrosyl ions to provide adsorbed nitrogen versatile for surface nitridation [3].

With these two methods an electrochemical surface nitriding of an iron electrode (α-iron) to form γ′-Fe_4N was achieved. The resulting iron nitride layer and electrolysis can be controlled by several factors such as thickness of the electrodes, duration of the electrolysis, voltage, choice of the electrode material and pretreatment of the electrodes. These influences will discussed.

[1] H. Tsujimura, T. Goto, Y. Ito, *Electrochimica Acta* **2002,** *47*, 2725-2731.

[2] T. Goto, Y. Ito, *Electrochimica Acta* **1998**, *43*, 3379-3384.

[3] T. Goto, M. Tada, Y. Ito, *Electrochimica Acta* **1994**, *39*, 1107-1113.

P043

Synthesis and Crystal Structure of Dy_2WO_6

K. Dorn[1], T. Schleid[1], I. Hartenbach[1]
[1] *University of Stuttgart, Institute for Inorganic Chemistry, Stuttgart, Germany*

Chloride derivatives of lanthanoid(III) oxidotungstates(VI) with the formula $Ln_3Cl_3[WO_6]$ represent an interesting class of compounds, due to their isolated trigonal prismatic $[WO_6]^{6-}$ units and their abilities to host active cations for luminescence applications. The attempt to extend the existence range of this series to the dysprosium compound by annealing equivalent amounts of $DyCl_3$, Dy_2O_3 and WO_3 with LiCl as fluxing agent at 750 °C in evacuated silica ampoules for two days resulted in colourless, needle-shaped single crystals of Dy_2WO_6 instead of the intended dysprosium(III) chloride oxidotungstate(VI) with the formula $Dy_3Cl_3[WO_6]$. The title compound crystallizes orthorhombically in the space group *Pbca* with the lattice constants a = 554.79(2) pm, b = 1580.97(6) pm and c = 1062.42(4) pm for Z = 8 (CSD-430462). The crystal structure comprises two crystallographically distinguishable Dy^{3+} cations (see Table 1 for the atomic positions), which are coordinated by seven O^{2-} anions each, forming monocapped trigonal prisms and is thus isotypic with the formula-analogous erbium(III) oxidotungstate(VI) Er_2WO_6 [1, 2]. The $[(Dy1)O_7]^{11-}$ polyhedra are connected to each other via common edges creating corrugated 2D-$\{[(Dy1)O_4]^{5-}\}$ layers spreading out parallel to (010), whereas the also undulated 2D-$\{[(Dy2)O_4]^{5-}\}$ layers perpendicular to [010] are built up by $[(Dy2)O_7]^{11-}$ polyhedra mutually linked by both edges and vertices. Both types of layers are interconnected by sharing common edges to form a 3D-$\{[Dy_2O_6]^{6-}\}$ framework apt to take up W^{6+} cations residing in distorted octahedral voids. The $[WO_6]^{6-}$ octahedra are vertex-fused to 1D-$\{[WO_{4/1}O_{2/2}]^{4-}\}$ strands along [100] creating zigzag chains with W···W···W angles of almost 90° (Figure 1, *top*). The distorted $[WO_6]^{6-}$ octahedra share common edges and vertices with the 2D-$\{[(Dy1)O_4]^{5-}\}$ layers, while with the 2D-$\{[(Dy2)O_4]^{5-}\}$ layers only mutual edges are found. The *non*-tungsten bonded $(O6)^{2-}$ anions represent the centers of $[ODy_4]^{10+}$ tetrahedra forming 1D-$\{[ODy_{4/2}]^{4+}\}$ chains perpendicular to (100) (Figure 1, *middle*) by sharing common trans-oriented edges. They finally generate a primitive rod-packing along [001] together with the 1D-$\{[WO_{4/1}O_{2/2}]^{4-}\}$ strands (Figure 1, *bottom*).

[1] A. V. Tyulin, V. A. Efremov, V. K. Trunov, *Kristallografiya* **1984**, *29*, 692-696.
[2] A. V. Tyulin, V. A. Efremov, V. K. Trunov, *Sov. Phys. Crystallogr.* **1984**, *29*, 410-412.

Table 1. Atomic positions and equivalent isotropic displacement parameters (U_{eq} in pm^2) for Dy_2WO_6.

Figure 1. View at the 1D-$\{[WO_{4/1}O_{2/2}]^{4-}\}$ chain (*top*) and the 1D-$\{[ODy_{4/2}]^{4+}\}$ strand (*middle*), both running parallel to [100], which are also shown in the complete crystal structure of Dy_2WO_6 as viewed along the *a* axis (*bottom*).

Figure 1

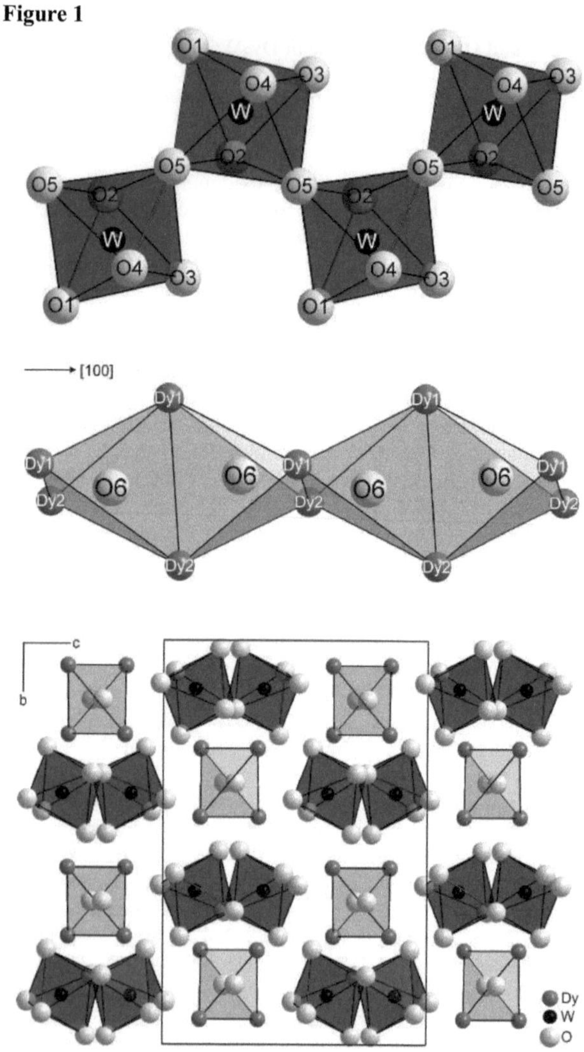

Figure 2

Atom	Site	x/ a	y/ b	z/ c	U_{eq}*
Dy1	8c	0.09647(5)	0.43705(2)	0.14078(3)	60(1)
Dy2	8c	0.05050(6)	0.26723(2)	0.35639(3)	64(1)
W	8c	0.10672(5)	0.11988(2)	0.12042(3)	54(1)
O1	8c	0.4062(9)	0.3942(3)	0.0113(5)	107(10)
O2	8c	0.0772(8)	0.0212(3)	0.2116(4)	101(9)
O3	8c	0.3833(8)	0.0764(3)	0.0449(5)	92(10)
O4	8c	0.1732(9)	0.2335(3)	0.0641(4)	99(10)
O5	8c	0.3382(8)	0.1701(3)	0.2692(4)	72(9)
O6	8c	0.3182(8)	0.3529(3)	0.2709(4)	62(9)

* $U_{eq} = \frac{1}{3}(U_{11} + U_{22} + U_{33})$.

Crystal Structure, Phase Transitions and Optical Properties of the Perovskite related layered hybrids ($(C(NH_2)_3)_2MI_4$; M = Pb, Sn

M. Daub[1], C. Haber[1], H. Hillebrecht[1]
[1]*Albert-Ludwigs-Universität Freiburg, Institut für Anorganische und Analytische Chemie, Freiburg, Germany*

Fig.1: Crystal structures of $(C(NH_2)_3)_2MI_4$; M = Pb, Sn measured at different temperatures.

Hybrid organic-inorganic lead halide perovskites $APbX_3$, in particular $(CH_3NH_3)PbI_3$, have emerged as excellent light harvesters in solid-state mesoscopic solar cells. [1] The very fast increasing power conversion efficiencies up to now over 20 % of such devices has induced an increase of research activities. Hybrid perovskites are nowadays discussed for a couple of different (optoelectronic) applications, such as photovoltaics, light emitting diodes, photodetectors and lasers. [2] Recently, it was shown that also more complex variants like A_2MX_4 with a layered structure can be used in solar cells. [3,4] Such perovskite-related layered hybrids can be classified into different types according to the crystallographic direction along the sheets formally where cut out from a perovskite framework. Therefore (($C(NH_2)_3)_2MI_4$;M = Pb, Sn can be classified as (110) oriented. [5]

In accordance with the literature [6] we observed two phase transitions for $(C(NH_2)_3)_2PbI_4$. The HT-phase (($C(NH_2)_3)_2PbI_4$-I, *Pnma*, Z =4, a = 13.037(5)Å, b = 13.612(5)Å, c = 9.364(4)Å) is stable above 356 K. Between 356 K and 307 K a twofold superstructure is formed caused by a partial reorganization of the cations (($C(NH_2)_3)_2PbI_4$-II, *Pnnm*, Z = 8, a = 27,033(5)Å, b = 9.310(2)Å, c = 12.918(3)Å). Below 307 K the structure becomes monoclinic (($C(NH_2)_3)_2PbI_4$-III, $P2_1/n$, Z = 8, a = 9.228(2)Å, b = 27.184(7)Å, c = 12.715(3)Å, β = 90.465(14)) caused by a small rotation of the PbI_6-octahedra and the complete ordering of the cations. To the best of our knowledge, $(C(NH_2)_3)_2SnI_4$ has so far not been mentioned in the literature. In analogy to the Pb compound, also the Sn compound shows two phase transitions. While the HT-phase as well as the room temperature phase are isotypically to the corresponding lead compound (($C(NH_2)_3)_2SnI_4$-I: *Pnma*, Z =4, a = 13.038(7)Å, b = 13.574(7)Å, c = 9.454(5)Å and $(C(NH_2)_3)_2SnI_4$-II: *Pnnm*, Z = 8, a = 26.891(5)Å, b = 9.303(2)Å, c = 12.943(3)Å), the LT-phase realizes a triclinic variant (P , Z = 4, a = 9.2660(5)Å, b = 12.6698(6)Å, c = 14.4401(7)Å, α = 66.9768(24)°, β = 85.8839(25)°, γ = 88.2602(26)°).

The room temperature vibrational spectra of the title compound show mainly the expected modes for the guanidinium cation. In the Raman spectra additionally a strong band of the n(M-I) mode is observed. From UV-VIS and PL spectroscopy the band edges are around 2.0 eV (M = Sn) / 2.3 eV (M = Pb) and are therefore in agreement with the observed colors, that are red (M = Sn) and yellow/orange, (M = Pb),

[1] M. M. Lee, J. Teuscher, T. Miyasaka, T. N. Murakami, H. J. Snaith, *Science* **2012** *338*, 643-647.
[2] Q. Chen, N. De Marco, Y. Yang, T.-B. Song, Ch.-Ch. Chen, H. Zhao, Z. Hong, H. Zhou, Y, Yang *Nano Today* **2015**, *10*, 355-396.
[3] M. Daub, H. Hillebrecht, *Angew. Chem.* **2015**, *127,* 11168-11169; *Angew. Chem. Int Ed.*, **2015**, *54*, 11016-11017.
[4] D. H. Cao, C. C. Stoumpos, O. K. Farha, J. T. Hupp, M. G. Kanatzidis *J. Am. Chem. Soc.* **2015**, *137*, 7843-7850.
[5] Z. Cheng, J. Lin, *CrystEngComm*, **2010**, *12*, 2646-2662.
[6] M. Szafranski, A. Katrusiak *Phys. Rev. B* **2000**, *61*, 1026-1035.

P044

Figure 1

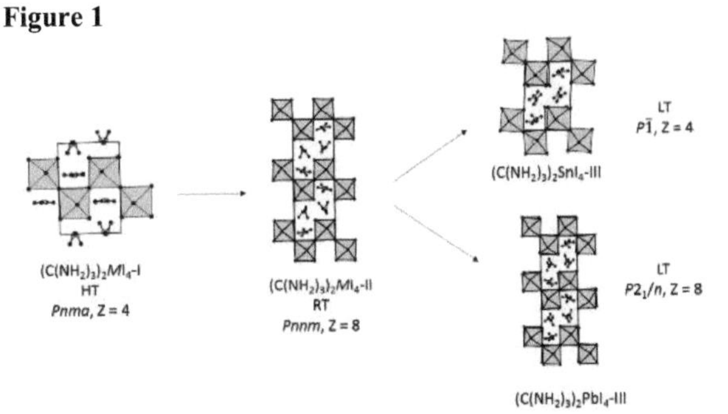

P045

Systematic X-Ray Study of Barium Long-Chain Carboxylates

J. Volk[1], W. Frank[1]
[1]*Heinrich-Heine-Universität Düsseldorf, Institut für Anorganische Chemie und Strukturchemie II, Düsseldorf, Germany*

The alkaline earth metal soaps are materials with a wide range of industrial applications. Therefore the structural chemistry of these compounds is of general and practical interest. However, reports concerning the crystalline state of these amphiphilic materials are rare,[1-2] and particularly barium salts of aliphatic carboxylic acids have not been systematically studied. Here we report a X-Ray study of anhydrous barium carboxylates ($Ba(C_nH_{2n-1}O_2)_2$ in the range n 8 to 22).
Very fragile platelets of barium soaps were crystallized from a H_2O/BuOH solution obtained by the reaction of ammonium carboxylate with barium chloride. Any kind of mechanical stress has to be avoided to get crystals of sufficient quality for single crystal X-ray structure determinations. Polycrystalline powders are more easily to obtain by the reaction of carboxylic acids with aqueous ammonia solution and addition of barium chloride solution. X-Ray powder patterns of different barium long chain carboxylates (with 8-22 carbon atoms) have been indexed and the symmetry of the unit cells has been determined with TOPAS[3]. The structures of barium pelargonate[1], laurate, myristate and pentadecanoate could be solved and refined with single crystal data, those of barium undecanoate, palmitate, stearate and nonadecanoate were determined from X-ray powder diffraction data by Rietveld refinement using TOPAS[3]. All compounds crystallize triclinic (space group type *P*-1). For all studied barium soaps the unit cell parameters *a* are close to 4.4 Å, the parameters *b* close to 4.6 Å and parameter *c* increases with the length of the according aliphatic chain.
The homotypic solids of the barium soaps have a bilayered structure and exhibit a clear separation of hydrophilic inorganic regions including distorted tetragonal prismatic coordinated barium centres and hydrophobic organic regions including the alkyl chains (Figure 1). The inorganic basis of the lamellar structure is dominated by Coulomb interactions. The carboxylates act as chelating and bridging ligands linking the metal atoms along the *a* and *b* axes. The layers are linked by *van der Waals* interactions between the terminal $-CH_3$ groups of the aliphatic chains, only. Figure 2 shows the linear dependence of unit cell parameter *c* and the number of carbon atoms in the chain and documents that *c* grows about 2.5 Å with every additional carbon atom in the aliphatic chain. The unit cell parameters *a* and *b* remain almost constant. The reported structure of barium valerate[4] can also be classified as homotypic with the solids described before and the corresponding length of the unit cell has also been included in Figure 2.
In conclusion, crystal data of an almost complete series of barium salts of aliphatic carboxylic acids (barium soaps) with 8 to 22 carbon atoms and selected structures derived from single crystals or crystal powders show the substances to be homotypic in the complete range of chain lengths.

[1] J. Volk, W. Frank, *Z. Krist. Suppl.* **2014**, *34*, 157.
[2] J. Volk, W. Frank, *Z. Anorg. Allg. Chem.* **2014**, *640*, 2368.
[3] A. A. Coelho, *TOPAS-Academic V5*, Brisbane, Australien, **2012**.
[4] P. Torres, P. Norby, J. C. Grivel, *J. Anal. Appl. Pyrolysis* **2015**, *116*, 120-128.

Figure 1

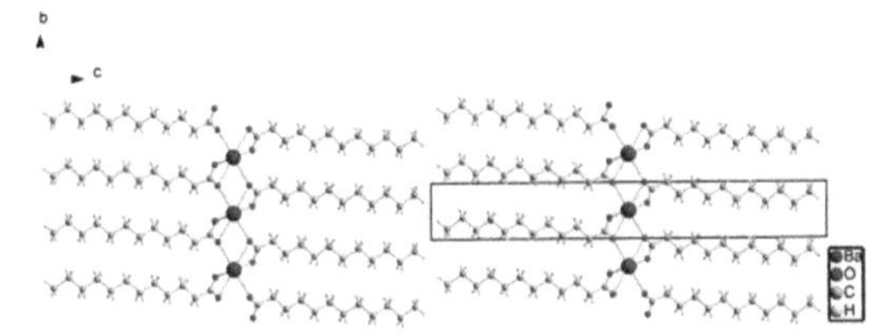

Fig. 1: Packing diagram of barium laurate.

Figure 2

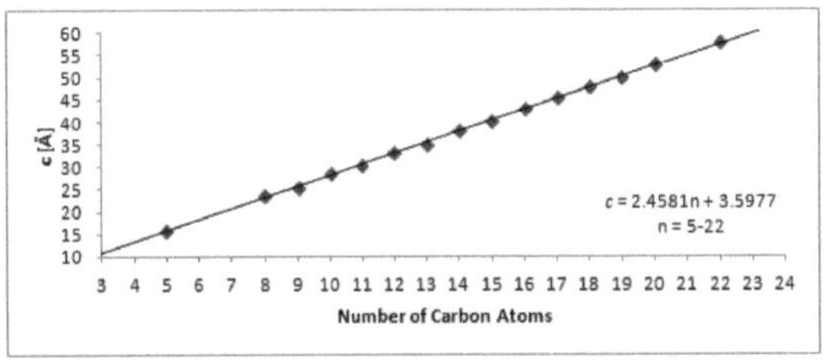

Fig. 2: Dependence of unit cell parameter *c* and the number of carbon atoms in the aliphatic chain.

P046

Rietveld Refinement of Crystal Structures of Group 14 Element Tetrachloridogallates

S. Schloots[1], W. Frank[1]
[1]*Heinrich-Heine-Universität Düsseldorf, Institut für Anorganische Chemie und Strukturchemie II, Düsseldorf, Germany*

In the past we reported the results of crystal structure determinations of tin(II)- and lead(II)-tetrachloridoaluminates.[1] Because of the similar ionic radii of aluminium and gallium and the chemical similarity of compounds of these elements in general, we assumed that group 14 element tetrachloridogallates can also be prepared.[2-3] The combined results of thermal analyses and X-ray powder diffraction experiments revealed the existence of two crystalline intermediate compounds in each of the two systems $SnCl_2$-$GaCl_3$ and $PbCl_2$-$GaCl_3$. Here we report the syntheses and the crystal structures of $Sn(GaCl_4)_2$ (**1**) and $SnCl(GaCl_4)$ (**2**) just as $Pb(GaCl_4)_2$ (**3**) and $PbCl(GaCl_4)$ (**4**). Polycrystalline powders of **1**-**3** were obtained by heating the corresponding stoichiometric quantities of MCl_2 and $GaCl_3$ (M=Sn, Pb) above the melting point in evacuated glass ampoules and fast cooling to room temperature. We did not succeed to crystallize **4** from the melt. However, crystalline powder of **4** was obtained in glass ampoules by heating stoichiometric quantities $PbCl_2$ and $GaCl_3$ above the melting point (to 350 °C) and subsequent fast cooling in a first step. Then the polycrystalline sample obtained by this procedure (containing crystallites of **3** and $PbCl_2$[4]) was hold for about 2 h at 180 °C giving almost pure **4**. Crystallographic data of **1-4** received from powder diffraction via Rietveld refinement using TOPAS[5] can be found in the table below.
1 as well as **3** crystallize isotypic with $M(AlCl_4)_2$ (M = Sn, Pb)

which we described before[1]. The solids of **1** and **3** contain two crystallographically independent $[GaCl_4]^-$ tetrahedra and one M^{2+} cation distorted triple-capped trigonal prismatic surrounded by nine Cl atoms belonging to a total of five chloridogallate anions. The 3D network of **2** and **4**, which are isotypic to $SnCl(AlCl_4)$[1], is composed of the anionic $[GaCl_4]^-$ moieties and a cationic $[MCl]^+_n$ partial structure in form of a "waved ladder" structure along the crystallographic *c*-axis.
In conclusion we present a complete set of intermediate compounds obtained in the quasi-binary phase systems $SnCl_2$-$GaCl_3$ and $PbCl_2$-$GaCl_3$.

Empirical formula	$Sn(GaCl_4)_2$ **(1)**	$SnCl(GaCl_4)$ **(2)**	$Pb(GaCl_4)_2$ **(3)**	$PbCl(GaCl_4)$ **(4)**
Crystal system	Orthorhombic	Monoclinic	Orthorhombic	Monoclinic
Space group; *Z*	*Pbca;* 8	*Cc;* 4	*Pbca;* 8	*Cc;* 4
Unit cell dimensions	*a* = 12.21684(12) Å, *b* = 10.30269(11) Å, *c* = 20.2968(2) Å	*a* = 9.51233(13) Å, *b* = 16.2778(2) Å, *c* = 5.79565(6) Å, *β* = 119.2798(9)°	*a* = 12.2538(2) Å, *b* = 10.2768(2) Å, *c* = 20.3616(3) Å	*a* = 9.54700(6) Å, *b* = 16.31252(11) Å, *c* = 5.70444(4) Å, *β* = 118.5889(5)°
R_p; R_{wp}	0.0287; 0.0369	0.0318; 0.0393	0.0285; 0.0365	0.0343; 0.0436

[1] S. Schloots, W. Frank, *Z. Krist. Suppl.* **2014**, *34*, 136.
[2] L. Pauling, *J. Am. Chem. Soc.* **1927**, *49*, 765-790.
[3] A. E. van Arkel, P. Aßmann, G. Borelius, G. Chaudron, E. Daniels, R. Gadeau, W. Geibel, W. Graßmann, C. Hayward, G. Jantsch, *Reine Metalle: Herstellung· Eigenschaften· Verwendung*, Springer-Verlag, **2013**.
[4] K. Sahl, J. Zemann, *Naturwissenschaften* **1961**, *48*, 641-642.
[5] A. A. Coelho, *TOPAS-Academic V5*, Brisbane, Australien, **2012**.

P047

$Eu_2Se[SiO_4]$: A Selenide *Ortho*-Oxosilicate with Trivalent Europium

D. Zimmermann[1], T. Schleid[1]
[1]*University of Stuttgart, Institute for Inorganic Chemistry, Stuttgart, Germany*

A most common oxophosphate structure with additional foreign anions can be found as apatite, $Ca_5(F,Cl,OH)[PO_4]_3$ [1], in the lithosphere of planet earth, or written as rare-earth metal oxosilicate with chalcogenide anions and cation defects with the formula $M_{4.667}Ch[SiO_4]_3$ (M = Sc, Y, La - Lu) [2-4]. But nature is creative and so we have also access to other compositions, such as $M_4Ch_3[Si_2O_7]$ [5] or $M_2Ch[SiO_4]$ (Ch = O - Te) with different distinctions like A-, B- or C-type [6], which are highly depending on the chalcogenide anions that are not bound directly to the oxosilicate units. Now $Eu_2Se[SiO_4]$ was obtained in attempts to synthesize the still hypothetical compound $Eu_{13}F_{30}Se$ as analogon to $Eu_{13}F_{30}S$ [7]. The title compound was prepared from europium metal (Eu), europium trifluoride (EuF_3), elemental selenium (Se) and sodium chloride (NaCl) as flux in sealed silica capsules supplemented with secondary fused silica ampoules for protection at 850 °C for ten days. After the reaction and yielded as a by-product, it was possible to isolate plate-shaped crystals with orange colour. $Eu_2Se[SiO_4]$ crystallizes orthorhombically in the space group *Pbcm* with lattice parameters like a = 611.65(5) pm, b = 708.46(6) pm, c = 1093.08(9) pm for Z = 4. The structure exhibits two crystallographically different europium positions. $(Eu1)^{3+}$ carries six O^{2-} (d(Eu-O): 243 - 255 pm) and two Se^{2-} anions (d(Eu-Se): 294 - 308 pm) as well as one additional Se^{2-} with a longer distance of 352 pm as ligands. The second cation $(Eu2)^{3+}$ has also six O^{2-} anions (d(Eu-O): 243 - 246 pm) and two Se^{2-} anions (d(Eu-Se): 312 pm) as next neighbours, but now shows two additional selenide ligands with even larger distances of 376 pm as compared to $(Eu1)^{3+}$. Both coordination environments are strongly distorted, in the case of $(Eu1)^{3+}$ attempting to form a trigonal prism with three caps, whilst the coordination sphere should result in a square prism with two caps, one atop and one underneath the squares, for $(Eu2)^{3+}$ (Figure 1). The whole structure is creating a framework using isolated oxosilicate tetrahedra $[SiO_4]^{4-}$ linked via the short europium(III)-oxygen contacts forming channels apt to take up the Se^{2-} anions (Figure 2). Since Eu_2Se_3 is unknown, there are only a few europium(III) compounds containing dianionic selenium.

[1] J. M. Hughes, M. Cameron, K. D. Croxley, *Am. Mineral.* **1989**, *14*, 870-876.
[2] C. Wickleder, I. Hartenbach, P. Lauxmann, Th. Schleid, *Z. Anorg. Allg. Chem.* **2002**, *628*, 1602-1606.
[3] I. Hartenbach, Th. Schleid, *Z. Anorg. Allg. Chem.* **2004**, *630*, 1727-1727.
[4] I. Hartenbach, Th. Schleid, *Z. Kristallogr.* **2005**, *220*, 206-210.
[5] M. Grupe, F. Lissner, Th. Schleid, W. Urland, *Z. Anorg. Allg. Chem.* **1992**, *616*, 53-60.
[6] I. Hartenbach, S. F. Meier, J. Wontcheu, Th. Schleid, *Z. Anorg.*

Allg. Chem. **2002**, *628*, 2907-2913.
[7] D. D. Zimmermann, Th. Schleid, *Z. Anorg. Allg. Chem.* **2012**, *636*, 1633-1633.

Figure 1. Coordination environments for both Eu^{3+} cations in the crystal structure of $Eu_2Se[SiO_4]$ with rather elongated contacts to the Se^{2-} anions.

Figure 2. Crystal structure of $Eu_2Se[SiO_4]$ with the Se^{2-} anions residing in channels along [010] of the 3D-$\{(Eu_2[SiO_4])^{2+}\}$ framework.

Figure 1

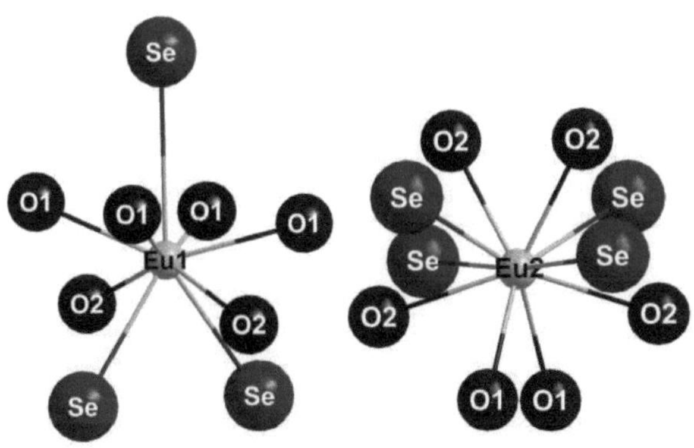

Figure 2

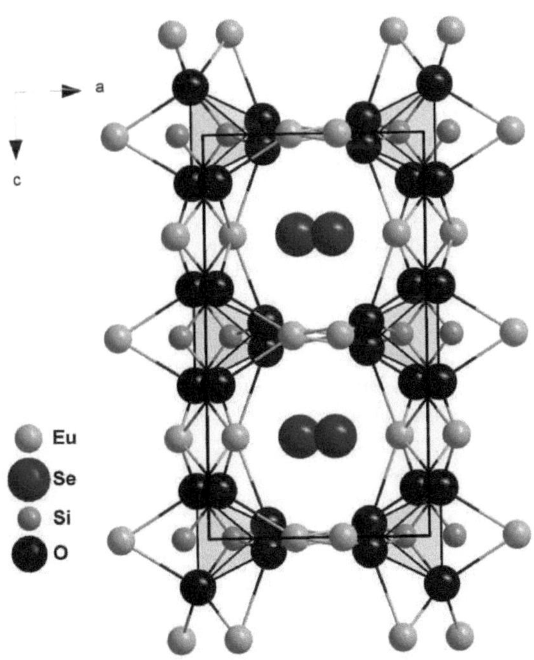

P048

Formation, structures and thermal behavior of Ga-rich alloys Mo/Ga/In, Mo/Ga/Sn and Mo/Ga/In/Sn

J. Platzek[1], H. Hillebrecht[1]
[1]*Universität Freiburg, Freiburg, Germany*

The structure of gallium-rich molybdenum intermetallics $(MoGa_5)_nGa$ (n = 6,8) and $(MoGa_5)_nX$ (X = In, Sn) (n = 4,6,8) are structurally closely related. The common feature are $MoGa_{10}$ polyhedra ("centaur"-polyhedra) which are connected to octameric units with a cube-like arrangement. With n = 8 there is a rhombohedral unit cell (V_8Ga_{41} type [1]) with isolated units, with n = 6 there are dimers in Mo_6Ga_{31} and a monoclinic structure [2] and with n = 4 there is a tetragonal structure with linear chains ($Mo_4Ga_{20}S$ type [3]).

Now we have investigated the reactivity of Mo with Ga-rich alloys Ga/Sn, Ga/In, and Ga/Sn/In according to the following temperature program:

RT —60°C/h→ 900°C (24h) —4°C/h→ 400°C —100°C/h→ RT

X-ray measurements show, that with the binary systems rhombohedral structures are formed. In agreement to the results of the refinement and EDX measurements the approximate compositions are $Mo_8Ga_{40}Sn$ and $Mo_8Ga_{40}In$, respectively. With the ternary alloy we observed the formation of the tetragonal structure with increased In/Sn content: $Mo_4Ga_{16}(In,Sn)_{\sim4}$. EDX measurements showed a In/Sn ratio of 2:1. At lower temperatures (500°C) tetragonal structures are formed in all cases.

The hardness of single crystals of the rhombohedral representatives could be determined at 477-573 Hv for Mo_8Ga_{41}. 317-378 Hv for $Mo_8Ga_{41-x}In_x$ and 485-540 Hv for $Mo_8Ga_{41-x}Sn_x$

DTA measurements reveal for all rhombohedral phases a sharp melting point around 833°C on heating. The subsequent cooling curve showed two peaks with a first onset temperature at 814°C and a second one around 770°C. After the heating cycle the XRD patterns showed the monoclinic structure. In contrast to this the tetragonal structures showed a broad melting peak with an onset around 670°C. The onset of the cooling curve was at 735°C with an unchanged XRD pattern of the sample after heating and cooling.

[1] K. Yvon, *Acta Crystallogr.* B, **1975**, *31*, 117-120
[2] K. Yvon, *Acta Crystallogr. B*, **1974**, *30*, 853-861.
[3] M. Wolcyrz, R. Andruszkiewicz, K. Lukaszewicz, *Acta Crystallogr. C*, **1989**, *45*, 991-993.

P049

Hydrogenation Properties of the Binary Palladium Rich Compounds $MgPd_2$ and MPd_3 (M: Pb, Bi)

A. Götze[1], A. Peretzki[1], H. Kohlmann[1]
[1]*Institut für Anorganische Chemie, Leipzig, Germany*

The hydrides MPd_3H_x (M: Mg, Mn, In, Tl; filled $AuCu_3$ type structure) form by hydrogen uptake from the binary intermetallics.[1-4] The hydrogenation could be observed by *in situ* neutron powder diffraction and a mechanism was proposed.[3]

$MgPd_2$[5], $PbPd_3$[6] and $BiPd_3$[7] as further candidates for hydrogenation were synthesized from the elements with iodine as mineralizing agent. The hydrogenation properties of these binary compounds were investigated by *in situ* thermal analysis (DSC) and *in situ* temperature resolved X-ray as well as neutron powder diffraction at hydrogen pressures up to 50 bar.

The DSC of $MgPd_2$ shows three irreversible thermal effects. X-ray powder diffraction of $MgPd_2$ before and after the first thermal effect at about 200°C shows a cell volume increase up to 5.6 % (*Pnma*, a = 5.5901(5) Å; b = 5.0490(3) Å; c = 8.4697(7) Å). In the hydrogenation process the lattice parameters a and c increase and b decreases thus reducing the distortion of octahedral sites. The last two thermal effects show the decomposition of this intermediate and the formation of $MgPd_3H_x$. $PbPd_3$ forms $PbPd_3D_{0.13(1)}$ (filled $AuCu_3$ type) under 50 bar deuterium pressure. $BiPd_3$ forms $BiPd_3D_{0.19(1)}$ (filled $ZrAl_3$ type) under 70 bar deuterium pressure and its hydrogenation was observed by *in situ* neutron powder diffraction.

[1] H. Kohlmann, N. Kurtzemann, R. Weihrich, T. Hansen, *Z. Anorg. Allg. Chem.* **2009**, *635*, 2399-2405.
[2] P. Oennerud, Y. Andersson, R. Tellgren, P. Nordblad, F. Bourée, G. André, *Solid State Comm.* **1997**, *101*, 433-437.
[3] H. Kohlmann, *J. Solid State Chem.* **2010**, *183*, 367-372.
[4] N. Kurtzemann, H. Kohlmann, *Z. Anorg. Allg. Chem.* **2010**, *636*, 1032-1037.
[5] C. Wannek, B. Habrecht, *J. Solid State Chem.* **2001**, *159*, 113-120.
[6] M. Ellner, *J. Less-Comm. Met.* **1981**, *78*, 21-32.
[7] K. Schubert, S. Bhan, T. K. Biswas, K. Frank, P. K. Panday, *Naturwissenschaften* **1968**, *55*, 542-543.

P050

Nitrogen Ordering in Inverse Perovskite Nitrides

D. Stoiber[1], P. Höhn[2], R. Niewa[1]
[1]University of Stuttgart, Institute for Inorganic Chemistry, Stuttgart, Germany
[2]Max Planck Institute for Chemical Physics of Solids, Dresden, Germany

Large single crystals of up to 4 mm edge size of (Sr_3N)Sb, (Sr_3N_x)Sn and (Sr_3N_x)Pb were grown from various alkali metal melts [1]. Crystal structures were elucidated by single crystal X-ray and powder synchrotron diffraction. The antimony containing compound crystallizes as an ideal cubic inverse perovskite in space group *Pm-3m* as earlier deduced from microcrystalline powder samples [2]. Thus, it shows full nitrogen occupation of the octahedral voids formed exclusively by the strontium framework. In comparison, the compounds (Sr_3N_x)Sn and (Sr_3N_x)Pb, formerly described with disordered nitrogen occupation with the cubic perovskite cell [3], exhibit nitrogen deficiency in a 2 x 2 x 2 enlarged supercell with space group *Fm-3m*. The *F*-centering is a consequence of only every second nitrogen position being fully occupied ((Sr_3N_x)Sn: a = 1045.97(1) pm and (Sr_3N_x)Pb: a = 1050.22(1) pm). The once equal Sr-octahedra can now be distinguished in larger partially occupied and smaller fully nitrogen-centered ones. These results are in accordance to the ionic character of the compounds, since they lead to overall formulae of ($Sr_3N_{2/3}$)Sn and ($Sr_3N_{2/3}$)Pb retaining the necessary charge balance.

[1] P. Höhn, H. Borrmann, U. Burkhardt, A. Ovchinnikov, W. Schnelle, K. Zechel, ECSSC15, Vienna 2015, PTue012, Book of Abstracts p. 189.
[2] F. Gäbler, M Kirchner, W. Schnelle, U. Schwarz, M. Schmitt, H. Rosner, R. Niewa, *Z. Anorg. Allg. Chem.* **2004**, *630*, 2292.
[3] F. Gäbler, M Kirchner, W. Schnelle, M. Schmitt, H. Rosner, R. Niewa, *Z. Anorg. Allg. Chem.* **2005**, *631*, 397.

P051

Formation and stability of mullite-type aluminumborates

K. Hoffmann[1,2,3], T. J. Hooper[4], J. V. Hanna[4], M. M. Murshed[2,3], H. Zhao[5], U. Kolb[5], M. Fischer[1], O. Dolotko[6], P. Kudějová[6], Z. Révay[6], H. Schneider[1], T. M. Gesing[2,3], R. X. Fischer[1]
[1]Universität Bremen, Kristallographie, Bremen, Germany
[2]Universität Bremen, Chemische Kristallographie fester Stoffe, Institut für Anorganische Chemie und Kristallographie, Bremen, Germany
[3]Universität Bremen, MAPEX Center for Materials and Processes, Bremen, Germany
[4]University of Warwick, Department of Physics, Coventry, Great Britain
[5]Johannes Gutenberg-Universität Mainz, Institut für Physikalische Chemie, Mainz, Germany
[6]Technische Universität München, Heinz Maier-Leibnitz Zentrum (MLZ), Garching, Germany

Aluminum borates belong to the family of mullite-type materials with the characteristic chain of edge-sharing AlO_6 octahedra [1-2]. According to the phase diagram [3] $Al_4B_2O_9$ (A_2B) and $Al_{18}B_4O_{33}$ (A_9B_2) are reported to be the thermodynamically stable phases in the Al_2O_3-B_2O_3 system. Additionally, Mazza et al. [4] published a series of compounds with a composition given as $Al_{6-x}B_xO_9$ ($1 \leq x \leq 3$), and an average structure close to mullite pointing out, that further heat treatment yields the well-known stable equilibrium phases.

We found the compounds described by Mazza et al. [4] to crystallize at 1223 K in the compositional range between 13.17 - 25.45 wt.% B_2O_3 in the precursor phase. The a and b lattice parameters can be well resolved showing a continuous change indicating a solid solution between the endmembers $x = 1$ and $x = 2$. ^{11}B NMR data show boron to be 3- and 4-fold coordinated with an increase of BO_4 with increasing initial boron content. Due to enlarged B-O distances, neutron diffraction and DLS modeling [5] were performed to calculate probable split positions for oxygen. Linking two octahedral chains by BO_3 and varying the weighting factor for B-O from 1 - 2.5, an increasing distortion is observed for the octahedra while the B-O distances approach the ideal distance of 137 pm in a nearly planar 3-fold coordination. PGAA was performed to determine the Al/B ratio.

Series of A_9B_2 ($\geq$ 1373 K) were synthesized as a function of the initial boron content. Most of the analytical methods are not sensitive for boron, therefore it is a continuous discussion to differentiate between $Al_{18}B_4O_{33}$ and $Al_{20}B_4O_{36}$ (Al_5BO_9). From PGAA experiments a decreasing Al/B ratio with increasing initial boron is observed. An increase of BO_4 relatively to BO_3 is observed from ^{11}B NMR data with increasing initial boron content. Both NMR and PGAA data indicate a solid solution between Al and B.

Series of samples with the A_2B structure were synthesized at 1223 K. A trend for an increase of BO_4 with increasing initial boron content is observed from ^{11}B NMR data. Electron-diffraction experiments were performed using samples differing in their initial composition. Disorder and superstructure effects could be observed for crystals synthesized with an excess of boron. An *ab-intio* structure solution was carried out with direct method for crystals without distortion. In contrast to our previous studies [6] the channel oxygen atom O5 is not observed independent from the initial boron content. This is confirmed by DFT calculations.

1. Fischer, R.X. and H. Schneider, *Crystal chemistry of borates and borosilicates with mullite-type structures: a review.* European Journal of Mineralogy, 2008. **20**: p. 917-933.
2. Fischer, R.X. and H. Schneider, *The mullite-type family of crystal structures*, in *In: Mullite*, H. Schneider and S. Komarneni, Editors. 2005, Wiley - VCH, Weinheim. p. 1-46, 128-140.
3. Gielisse, P.J.M. and W.R. Foster, *The system Al_2O_3-B_2O_3.* Nature, 1962. **195**: p. 69-70.
4. Mazza, D., M. Vallino, and G. Busca, *Mullite-type structures in the system Al_2O_3-Me_2O_3 (Me = Na,K) and Al_2O_3-B_2O_3.* Journal of the American Ceramic Society, 1992. **75**(7): p. 1929-1934.
5. Baerlocher, C., A. Hepp, and M. W.M., *DLS-76.* 1977.
6. Fischer, R.X., et al., *Crystal structure of synthetic $Al_4B_2O_9$: A member of the mullite family closely related to boralsilite.* American Mineralogist, 2008. **93**: p. 918-927.

P052

Subsequent condensation in [$Zn_2(NH_3)_2(NH_2)_3$]Cl and [$Zn(NH_3)_2(NH_2)$]Br: towards ammonothermal crystal growth of Zn_3N_2

T. M. M. Richter[1], N. S. A. Alt[2], E. Schlücker[2], R. Niewa[1]
[1]University of Stuttgart, Institute of Inorganic Chemistry, Stuttgart, Germany

[2]Friedrich-Alexander-University Erlangen-Nürnberg, Institute of Process Machinery and System Engineering, Erlangen, Germany

We present two possible intermediates in a conceivable ammonothermal synthesis of Zn_3N_2, namely $[Zn_2(NH_3)_2(NH_2)_3]Cl$ ($P2_12_12$, a = 577.15(4) pm, b = 1023.59(6) pm, c = 654.56(4) pm, V = 386.69(5) 10^6 pm^3, Z = 2) and $[Zn(NH_3)_2(NH_2)]Br$ ($P2_1/n$, a = 760.55(4) pm, b = 597.72(4) pm, c = 1257.22(8) pm, β = 93.475(4)°, V = 570.48(6) 10^6 pm^3, Z = 4). Both compounds were obtained under ammonothermal conditions ($[Zn_2(NH_3)_2(NH_2)_3]Cl$: 97 MPa, 873 K; $[Zn(NH_3)_2(NH_2)]Br$: 230 MPa, 773 K) from Zn and NH_4Cl or NH_4Br respectively. For both halides, colourless plate shaped crystals were collected from the cold temperature zone of the autoclave, after chemical material transport from the hot dissolution zone. We expect the promising semiconductor Zn_3N_2 with a narrow band gap of 0.9 - 1.2 eV [1] to form under similar conditions at slightly higher temperatures. Most of the reported ammines of zinc halides such as $[Zn(NH_3)_4]Br_2$ [2] are synthesized at ambient temperature and the highest synthesis temperature reported for an amidozincate is 773 K for $Li_4[Zn(NH_2)_4](NH_2)_2$ [3]. Consequently, rising the synthesis temperature seems to correlate with an increasing condensation of the tetrahedrally coordinated Zn^{2+} units. This indicates a possible ammonothermal synthesis of Zn_3N_2, which consists of a 3-D-connection of $[ZnN_{4/6}]$-tetrahedra, may be accessible by further increasing the synthesis temperature.

The simultaneous occurrence of ammine and amide ligands in one complex is rarely observed (e.g., $[Cr_2(NH_2)_3(NH_3)_6]I_2$ [4] and $InF_2(NH_2)(NH_3)$ [5]). While coordination by four NH_2^- or four NH_3 ligands such as in $Li_4[Zn(NH_2)_4](NH_2)_2$[3] and $Zn(NH_3)_3F_2$[6] are well known for Zn^{2+}, the two title compounds manifest tetrahedral coordination of Zn via two ammine and two amide ligands ($[Zn(NH_3)_2(NH_2)]Br$) and one ammine and three amide ligands ($[Zn_2(NH_3)_2(NH_2)_3]Cl$), respectively. Whereas most of the known zinc ammines and amidozincates contain isolated $[Zn(NH_2)_4]^{2-}$ or $[Zn(NH_3)_4]^{2+}$ tetrahedra, the here presented $[Zn(NH_3)_2(NH_2)]Br$ and $[Zn_2(NH_3)_2(NH_2)_3]Cl$ contain ${}^1_\infty[Zn(NH_3)_2(NH_2)_{2/2}]$ 1-D chains and ${}^2_\infty[Zn(NH_3)(NH_2)_{3/2}]^+$ 2-D layers (see figure 1). In both compounds, the edge-sharing of the tetrahedra is realized via the electron lonepair at NH_2^-.

Figure 1: Section of the crystal structure of ($[Zn_2(NH_3)_2(NH_2)_3]Cl$, showing ${}^2_\infty[Zn(NH_3)(NH_2)_{3/2}]^+$ 2-D layers.

[1] S.-H. Yoo, A. Walsh, D. O. Scanlon, A. Soon, *RSC Adv.* **2014**, *4*, 3306.
[2] R. Eßmann, *J. Mol. Struct.* **1995**, *356*, 201.
[3] T. M. M. Richter, N. S. A. Alt, E. Schlücker, R. Niewa, *Z. Anorg. Allg. Chem.* **2015**, *641*, 1016.
[4] U. Zachwieja, H. Jacobs, *Z. Kristallogr.* **1993**, *206*, 247.
[5] D. Ketchum, G. Schimek, W. Pennington, J. Kolis, *Inorg. Chim. Act.* **1999**, *294*, 200.
[6] T. M. M. Richter, N. S. A. Alt, E. Schlücker, R. Niewa, *Inorg. Chem.* **2016**, submitted.

Figure 1

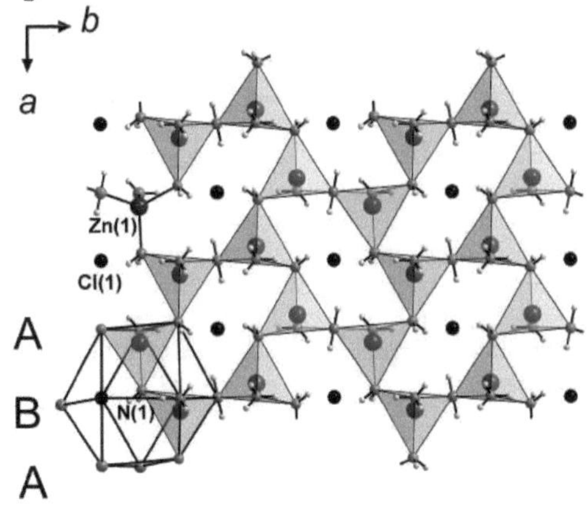

P053
Elimination of wurtzitic stacking faults and indium-doping in sphalerite ZnS by melt-freeze @ HP/HT conditions

J. Heinrich[1], M. Schwarz[2], G. Heide[1]
[1]*Tu Bergakademie Freiberg, Institut für Mineralogie, Freiberg, Germany*
[2]*TU Bergakademie Freiberg, Institut für Anorganische Chemie, Freiberg, Germany*

Introduction & Objectives
Different ways to obtain phase-pure sphalerite ZnS without any wurtzitic stacking faults or inclusions were investigated. The objective was to develop and establish a simple method to synthesize required quantities of doped and undoped sphalerite with an appropriate effort and time.

The material is to be used as a reference for the bacterial leaching of indium-bearing natural sphalerite (*sp*-ZnS:In), for which contents of up to 1560 ppm have been reported of the Freiberg district[1]. So far, leaching experiments on natural *sp*-ZnS:In ore were done using sulfuric acid where ferric ions were added and bacteria in cultivation medium containing ferrous ions. To determine and compare the yield regarding to different leaching methods, synthetic sphalerite with a known indium amount is required.

Materials & Methods
Precipitated ZnS nanopowder with high chemical purity was provided by a commercial supplier. X-ray diffraction of the pristine powder revealed a high density of wurtzitic stacking faults (Fig. 1), which could be completely removed by high-pressure-high-temperature (HPHT) treatment at ~8 GPa ~1600°C using a large volume toroid-type press[2].

Sphalerite samples with 100, 250, 500 and 1000 ppm doping level of indium were prepared by the same procedure, using mechanically homogenized ZnS nanopowder with corresponding amounts of In_2S_3 added. The microstructure of the retained sp-ZnS:In is indicative of a complete melting[3] of the mixture during HPHT treatment, so that a homogeneous dissolution of the In2S3 phase can be assumed, while phase separation was prevented due to rapid temperature quenching and freezing.

Results
Fig. 1 shows X-ray pattern of the original and the HPHT-treated material, respectively. In the latter case, practically no intensity of any wurtzite-type stacking faults in sp-Zns or any crystalline phase other than sphalerite can be observed. X-ray pure sphalerite was synthesized.

Conclusion & outlook
The synthesis of undoped and doped sp-ZnS containg 100, 250, 500 and 1000 ppm indium was successfully realized via a high-pressure high-temperature melt-freeze procedure using a large volume toroid-type high pressure apparatus. The material is currently being used as a reference in comparison to natural *sp-*ZnS:In in bacterial leaching experiments, using bacteria cultivated out of the environment of the mine "Forschungs- und Lehrbergwerk Reiche Zeche" of the TU Bergakademie Freiberg. Further doping experiments dealing with an iron doping and microstructural investigations to determine the distribution and incorporation of the In^{3+} and $Fe^{2+/3+}$ Ions in the crystal lattice of ***sphalerite are planned.***

Acknowledgement
This work is financially supported by the Dr.-Erich-Krüger-Stiftung of the TU Bergakademie Freiberg in the framework of the 2nd Krüger Research School: Biohydrometallurgical Center for Strategic Elements ("BHMZ").

[1]Seifert, T; Sandmann, D. (2006): Mineralogy and geochemistry of indium-bearing polymetallic vein-type deposits: Implications for host minerals from the Freiberg district, Eastern Erzgebirge, Germany. In: Ore Geology Reviews 28 (1), S. 1–31. DOI: 10.1016/j.oregeorev.2005.04.005.

[2]Khvostantsev, L.; Slesarev, V.N.; Brazhkin V.V (2004): Toroid Type High Pressure Device: History and Prospects High Pressure Research, 371-383.

[3] Addamiano, Arrigo; Dell, Paul A. (1957): The Melting Point of Zinc Sulfide. In: General Electric Co.

Fig.1. XRD of synthetic sphalerite, vertical displacement of 23,000 counts to separate the original material (dotted line) from the synthesized material (solid line). Sphalerite reflexes are marked by an asterisk, phase determination were used by reference pattern from PDF #01-071-4763. Diffraction features assigned to wurtzite-2H type stacking faults are marked by a rectangle.

Figure 1

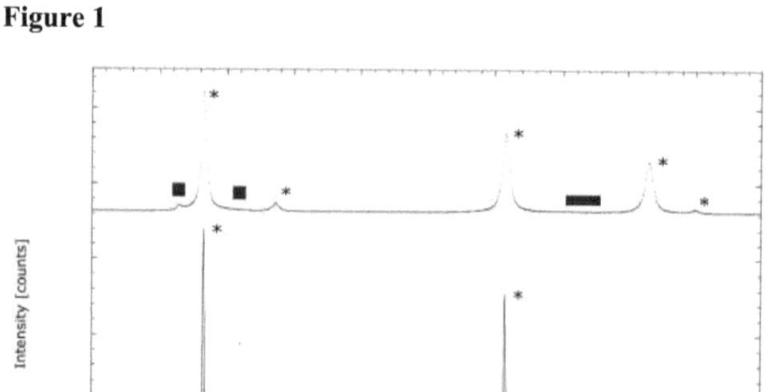

P054
$LiCu_{0.5}Mn_{1.5}O_4$ and $LiNi_{0.25}Cu_{0.25}Mn_{1.5}O_4$ Spinel Electrodes for Li-ion Batteries
Ö Budak[1]
[1]*Karlsruhe Institute of Techonology, IAM-ESS, Eggenstein-Leopoldshafen, Germany*

$LiCu_{0.5}Mn_{1.5}O_4$ and $LiNi_{0.25}Cu_{0.25}Mn_{1.5}O_4$ Spinel Electrodes for Li-ion Batteries

Öznil Budak*, oznil.budak@partner.kit.edu, Natalia Bramnik, natalia.bramnik@kit.edu, Aiswarya Bhaskar, aiswarya.bhaskar@kit.edu, Nilüfer Kiziltas-Yavuz, niluefer.yavuz@kit.edu, Helmut Ehrenberg, helmut.ehrenberg@kit.edu, Michael Knapp, michael.knapp@kit.edu, Institute for Applied Materials (IAM), Karlsruhe Institute of Technology (KIT)*

Introduction
The substituted manganese spinels $LiM_xMn_{2-x}O_4$ (M = Cr, Fe, Co, Ni, Cu) are attractive candidates for the development of advanced positive electrodes for lithium-ion batteries (LIBs). Among them, $LiNi_{0.5}Mn_{1.5}O_4$ (LNMO) received most attention due to the high operation voltage and high theoretical capacity. In the spinel structure of LNMO (S.G.) the lithium ions are located on the tetrahedral site (8a) and the electrochemically active Ni^{2+} ions occupy the octahedral site (16d). In contrast to LNMO, $LiCu_{0.5}Mn_{1.5}O_4$ is prone to cation disorder. The presence of transition metals on the tetrahedral site can have a strong impact onto the lithium transport, and, therefore, on the electrochemical performance. The elucidation of the interplay between the cation distribution and the electrochemical performance is, therefore, very important for the development of Cu-based spinel electrode materials. In this work, a comparative study of $LiCu_{0.5}Mn_{1.5}O_4$ and $LiNi_{0.25}Cu_{0.25}Mn_{1.5}O_4$ spinels as cathode materials for LIBs is reported with the focus on the structural characterization and correlation of the structural parameters with cycling behaviour.

Objectives
The structural characterization of $LiCu_{0.5}Mn_{1.5}O_4$ and $LiNi_{0.25}Cu_{0.25}Mn_{1.5}O_4$ spinels in the pristine state are performed for the samples prepared at different synthesis conditions to reveal the influence of the annealing temperature, annealing time and atmosphere on the phase purity and cation distribution. The evolution of the crystal structure during lithiation-delithiation is studied by *in-situ* synchrotron diffraction.

Materials and Methods
$LiCu_{0.5}Mn_{1.5}O_4$ and $LiNi_{0.25}Cu_{0.25}Mn_{1.5}O_4$ spinels are synthesized by a citric acid-assisted sol-gel method. The final calcination is performed at different temperatures: 600 °C, 750 °C and 900 °C. The obtained powder are characterized by Scanning Electron Microscopy (SEM) and X-Ray Diffraction (XRD). In this study, the *in-situ* diffraction data are obtained at the Powder Diffraction beamline (MSPD) at ALBA in Barcelona.

Results
Figure 1 shows XRD patterns for the $LiCu_{0.5}Mn_{1.5}O_4$ samples annealed at 600 °C, 750 °C and 900 °C for 4h in air. In addition to the main phase with spinel structure space group), the samples contain different ratios of CuO (S.G. C12/c1), respectively, fractions of CuO phase are 7.5, 5.9 and 5.8 % obtained at 600 °C, 750 °C and 900 °C. However, the materials annealed at 750 °C and 900 °C contain Li_2MnO_3 (S.G. C12/m1) admixtures (respectively, 3.9 and 13.7 %). The integrated intensity ratio of the 400 / 311 reflections of the spinel phase decreases with increasing temperature. This decrease indicates the increased amount of

transition metal ions on the tetrahedral site (8a). The Rietveld analysis was performed with a structural model based on a partial occupation on the 8a site by copper ions, which increases from ~2% to ~18% for the samples annealed at 600°C to 900°C, respectively.

Figure 1 XRD patterns of $LiCu_{0.5}Mn_{1.5}O_4$ annealed for 4h at 600 °C, 750 °C and 900 °C, respectively.

Conclusion
The annealing temperature has a strong effect on the cation distribution and phase ratios. The correlation between electrochemical performance and the structural characteristics will be discussed.

Figure 1

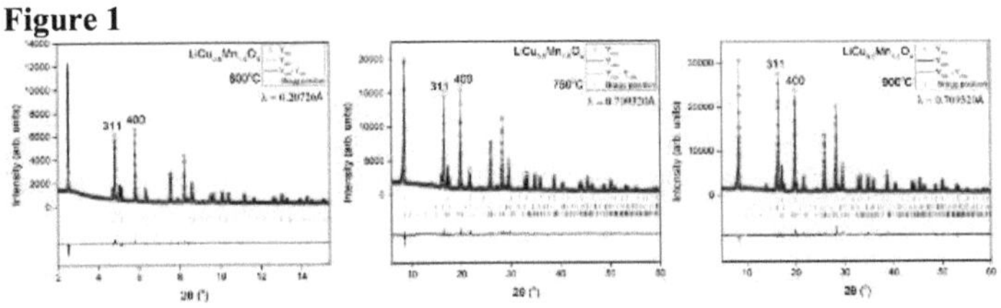

P055
Studies on Hydrogen Bonds and Local Disorder in Silicates and Phosphates using Neutron and X-ray Diffraction
M. Meven[1], D. G. Gatta[2]
[1]*RWTH Aachen, Institut für Kristallographie, Aachen, Germany*
[2]*Università degli Studi di Milano, Dipartimento di Scienze della Terra, Milano, Italy, Italy*

Various minerals of the silicate and phosphate families have been subject to numerous studies because of their importance to many industrial processes and because of their frequent occurence (about 90%) in the earth's crust. Those members with complex framework structures including hydrogen bonds and/or site or other types of disorder require the combination of several techniques, e.g. of neutron and x-ray diffraction together with other methods, to reveal all structural details.Within the last few years we have been able to resolve successfully structural details of different minerals (e.g. epidote, zoisite, elbaite, brazilianite, paravauxite and most recently also hureaulite and axinite) and some of them will be presented at this conference.

[1] F. Cámara, G.D. Gatta, M. Meven and D. Pasqual; *Thermal expansion and high temperature structure evolution of zoisite by single-crystal X-ray and neutron diffraction*; Physics and Chemistry of Minerals, 39(1):27-45 (2011).
[2] G.D. Gatta, R.M. Danisi, I. Adamo, M. Meven and V. Diella; *A single-crystal neutron and X-ray diffraction study of elbaite*; Physics and Chemistry of Minerals, 39(7):577-588 (2012).
[3]G.D. Gatta, I. Adamo, M. Meven and E. Lambruschi; *A single-crystal neutron and X-ray diffraction study of pezzottaite, $Cs(Be_2Li)Al_2Si_6O_{18}$*; Physics and Chemistry of Minerals, (39):829-840 (2012).
[4] G.D. Gatta,P. Vignola,M. Meven and R. Rinaldi; *Neutron diffraction in gemology: Single-crystal diffraction study of brazilianite, $NaAl_3(PO_4)_2(OH)_4$*;American Mineralogist,98(8-9):1624-1630 (2013).
[5]. G.D. Gatta, P. Vignola and M. Meven; *On the complex H-bonding network in paravauxite, $Fe^{2+}Al_2(PO_4)_2(OH)_2{\cdot}8H_2O$: A single-crystal neutron diffraction study*; Mineralogical Magazine, Vol 78(4): 841-850 (2014)

Figure 1

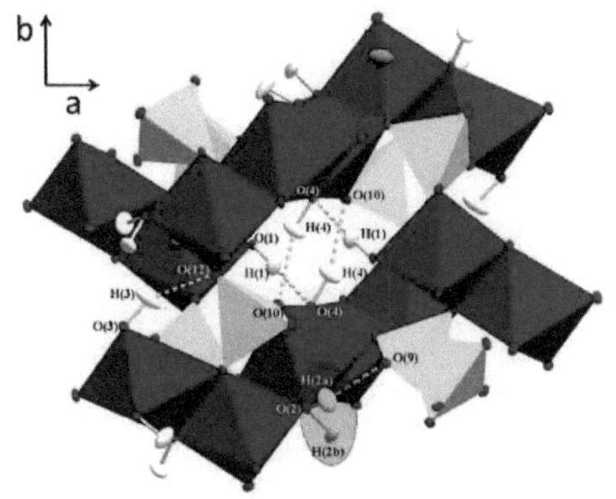

site disorder of H(2) in Brazilianite at RT [4]

P056
The first structurally characterized mixed diorganochlorofluorostannanes: Me_2SnClF and iPr_2SnClF
H. Reuter[1], K. Neue[1]
[1]*University of Osnabrück, Institute of Chemistry of New Materials, Osnabrück, Germany*

The general procedure for the synthesis of diorganodifluorostannanes, R_2SnF_2, involves the reaction of the corresponding dichloride or dibromide dissolved in ethanol with an aqueous solution of ammonium fluoride, NH_4F, or alkaline fluoride, MF, in a molar ratio of 1 : 2. Upon mixing both solutions the insoluble difluoride precipitates immediately as microcrystalline powder that can be filtered off and purified by washing with water.

By use of a small experimental variation, tetrahydrofuran instead of ethanol, we were able to isolate - besides the desired difluorides - the first single crystals suitable for single crystal X-ray diffraction of the mixed diorganochlorofluorostannanes, R_2SnClF. Up to now, we achieved structural information of this class of compounds in case of R = methyl (Me) and R = isopropyl (iPr).

While Me_2SnClF crystallizes as inversion twin in the chiral orthorhombic space group $P2_12_12_1$ with one formula unit in the asymmetric unit, iPr_2SnClF crystallizes in the centrosymmetric monoclinic space group $P2_1/n$ with two formula units in the asymmetric unit (Fig. 1). Despite these crystallographic differences both structures follow the same building principles: trigonal-bipyramidally coordinated tin atoms are bridged by μ_2-fluorine atoms in axial positions while the chlorine atoms adopt a non-bridging position within the equatorial plane of the tin atoms. All in all, polymeric one-dimensional chain structures result which are more or less bent at the bridging fluorine atoms and almost linear at the tin atoms.

The poster will give more information on the structure refinements and will also explain the structural differences and similarities of both compounds in more detail.

Figure 1

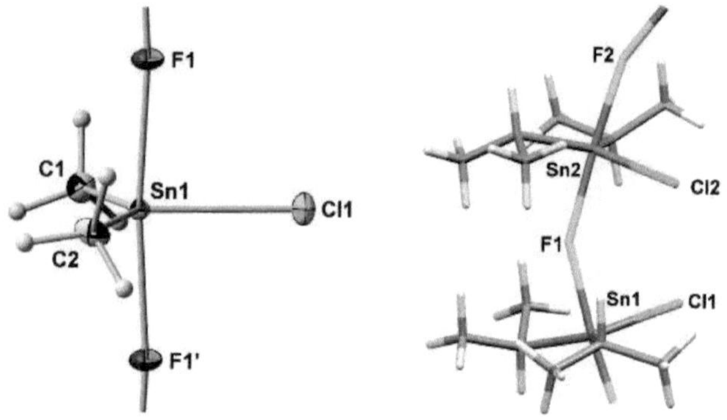

[1] M. S. Saidov, A. Yusupov, R. S. Umerov, *J. Cry. Growth* **1981**, *52*, 514-518.
[2] T. A. Campbell, M. Schweizer, P. Dold, A. Cröll, K. W. Benz, *J. Cry. Growth* **2001**, *226*, 231-239.
[3] I. G. Atabaev, N. A. Matchanov, M. U. Khazhiev, S. A. Yusupova, *Tech. Phys. Lett.* **2010**, *36*, 115-118.
[4] G. K. Azhdarov, Z. M. Zeynalov, Z. A. Agamaliyev, A. I. Kyazimova, *Crystallogr. Rep.* **2010**, *55*, 716-719.
[5] N. Usami, M. Kitamura, K. Obara, Y. Nose, T. Shishido, K. Nakajima, *J. Cry. Growth* **2005**, *284*, 57-64.
[6] W. G. Pfann, *Zone Melting*, 2. Aufl., Wiley, New York, **1966**.

P058

Crystal structures and thermoelectric properties of long-periodically ordered germanium arsenic tellurides

M. Nentwig[1], F. Fahrnbauer[1], M. Kasprick[1], P. Schultz[1], O. Oeckler[1]
[1]*University of Leipzig, IMKM, Leipzig, Germany*

Amorphous germanium arsenic tellurides (GAT materials) have been discussed as optical materials due to their low absorption coefficients for infrared radiation.[1] The corresponding crystalline phases form layered structures and thus exhibit pronounced anisotropy. In line with a broad range of layered chalcogenides, there has been considerable effort to examine the crystal chemistry of GAT materials as well as their electronic structures.[2-5] Phases with the general formula $(GeTe)_nAs_2Te_3$ form a homologous series of layered, long-periodically ordered compounds. For example, samples with the composition $(GeTe)_{10}As_2Te_3$ are obtained by slowly heating stoichiometric mixtures to 950 °C (silica glass ampoule, Ar atmosphere), quenching in water and subsequent annealing at 430 °C for 7-10 d, again followed by quenching in water.

Single-crystal X-ray diffraction revealed that $As_2Ge_7Te_{10}$ = $(GeTe)_7As_2Te_3$ exhibits a long-range ordered structure (space group *R*`3*m*, a = 4.135 Å, c = 102.3 Å) that consists of rocksalt-type building blocks separated by van der Waals gaps. They are comparable to the ones found for germanium antimony tellurides and related compounds. The rhombohedral 57*R* stacking sequence results in an extraordinarily large *c* lattice parameter. When such structures are formed, the extremely long diffusion pathways required to obtain completely ordered compounds often lead to samples that contain intergrown rocksalt-type slabs with varying thicknesses. As a result, the system $(GeTe)_nAs_2Te_3$ is prone to form both disordered and inhomogeneous samples, especially for $n > 9$. Yet, powder diffraction data reveal that such samples contain 57*R*-$As_2Ge_7Te_{10}$ as a main component. These inhomogeneous materials exhibit promising thermoelectric properties: the Seebeck coefficient as well as the lattice part of the thermal conductivity are in the same range as germanium antimony tellurides with high figures of merit ZT.[6] However, the high electrical conductivity and consequently the pronounced thermal transport still impede such ZT values for GAT materials.

$(GeTe)_nAs_2Te_3$ with n = 10 decomposes into phases with n = 7 and 12, which is a consequence of the very long diffusion pathways that are required to obtain long-periodically ordered compounds and of complicated features of the phase diagram associated with superior stabilities of certain compositions. Crystal structures of many other possibly metastable compounds can be derived from the lattice parameters as obtained e. g. by electron diffraction by a simple algorithm. Thus, they can be included in multi-phase Rietveld refinements.

[1] A. Feltz, *Amorphous Inorganic Materials and Glasses*, VCH, Weinheim, **1993**.
[2] S. Kuypers, G. van Tendeloo, J. van Landuyt, S. Amelinckx, H. W. Shu, S. Jaulmes, J. Flahaut, P. Laruelle, *J. Solid State Chem.*

P057

SiGe single crystal growth by vertical Float Zone (FZ) melting starting from Spark Plasma sintered feed rods

A. Wagner[1], A. Cröll[2], H. Hillebrecht[1]
[1]*Institut für Anorganische Chemie, Freiburg, Germany*
[2]*Institute for Geoscience, Cystallography, Freiburg, Germany*

The growth of SiGe single crystals is still an issue of research especially concerning crystal quality, growth phenomena on earth [1][2] as well as in space, properties [3] and homogeneity. Latter is mainly affected by the concentration dependence of the distribution coefficients of silicon and germanium in another and their deviation from unity which results in segregation independent of growth method [4]. One way to regulate the Ge distribution along the growth direction in the crystal is the modification of the Ge concentration in the feed material and the use of a vertical zone leveling process [5] starting with a seed to obtain single crystals.

In the following, the preparation of silicon-rich feed rods of the compositions $Si_{1-x}Ge_x$ with x= 0.11 and 0.20 by Spark Plasma Sintering (SPS) and the results of the growth experiments in a doubleellipsoide mirror furnace are presented. The starting material itself was produced by mechanical alloying of silicon and germanium powders and as Spark Plasma Sintering is known to be a current and pressure assisted sintering procedure with influence on diffusion and material properties [2], high-density compacts have been prepared in short times.

The as grown crystals have been characterized by means of Ge distribution by SEM and EDX along growth direction and by differential interference contrast (DIC) microscopy after etching. All grown crystals showed single crystalline areas, restricted by grain boundaries. The Ge incorporation along growth direction starting form uniform feed rods matched well with the expected, theoretical Pfann graphs for zone melting [6] but never resulted in a homogeneous concentration due to the long transient, whereas a segmented feed rod achieved a complete compensation of segregation after zone leveling. For latter crystal quality has been confirmed by WBSXT.

The aim of this work was the production of homogeneous SiGe single crystals, which can be used as *ZT*-reference material, which is known to be the figure of merit of a thermoelectric material [4]. It is needed for the calibration of the measuring equipment. Furthermore applications like single crystalline SiGe solar cells or substrate material for epitaxial applications could be possible, if dislocation densities are to be low in the grown crystals.

1988, 73, 192.
[3] G. van Tendeloo, D. van Dyck, S. Kuypers, H. W. Zandbergen, S. Amelinckx, *Phys. Status Solidi A* **1987**, *102*, 597.
[4] P. E. Lippens, E. Brousse, J. C. Jumas, *J. Phys. Chem. Solids* **1999**, *60*, 1663.
[5] P. E. Lippens, E. Brousse, J. Olivier-Fourcade, A. Gheorghiu de la Rocque, C. Sénémaud, *J. Phys. Chem. Solids* **2001**, *62*, 467.
[6] T. Rosenthal, M. N. Schneider, C. Stiewe, M. Döblinger, O. Oeckler, *Chem. Mater.* **2011**, *23*, 4349.

P059

Hydrogenation Properties of Laves-Phases $LnMg_2$ (Ln = La, Ce, Nd, Sm, Eu, Gd, Tb, Ho, Er, Tm, Yb)

A. Werwein[1], H. Kohlmann[1]
[1]*University of Leipzig / Faculty for Chemistry and Mineralogy, Institute for Inorganic Chemistry, Leipzig, Germany*

Laves-Phase are promising intermetallic compounds with the general composition AB_2. They are suitable for hydrogen storage due to their excellent ability to form hydrides.[1] Especially Laves-Phases containing lanthanides are very interesting, since many combination of elements are possible.[2] With the usage of magnesium as the B-atom a light weight metal with good hydrogen absorption is added. The ternary hydrides $LnMg_2H_7$ for Ln = La, Ce, Sm and $EuMg_2H_6$ are already known.[3-5] The title compounds were prepared by the elements and studied during the hydrogenation by *in- situ* thermal analysis (DSC) at starting hydrogen pressures of 5 MPa and temperatures up to 450°C. Except for the compounds $EuMg_2$ and $YbMg_2$ all form binary hydrides. The synthesis of the ternary hydrides were done by hydrogenation under 140 bar H_2 pressure and 140 °C. The compound $NdMg_2$ forms the ternary hydride $NdMg_2H_7$ (a = 6.2920(9) Å, c = 9.4218(16) Å, $P4_12_12$) which is isostructural to $LaMgH_7$. The hydride is thermally very stable up to 408 °C, which was observed by DSC experiments with one bar hydrogen pressure. The Laves-Phase EuMg2 forms a new hydride $EuMg_2H_x$ of black. color, whose X-ray powder diffraction pattern could be indexed to an orthorhombic unit cell (a = 6.64897(9) Å, b = 11.37107(16) Å, c = 10.70122(16) Å). The crystal structure was refined in space group $C222_1$ and represents an orthorhombically distorted hexagonal Laves-Phase type. Structural relationships between intermetallic compound and its hydride were proven by crystallographic group-subgroup relationships.

[1] P. Chen, M. Zhu *Materials Today* **2008**, 11, 36-43.
[2] K. A. Gschneidner , Jr. , V. K. Pecharsky, *Zeitschrift für Kristallographie*, **2006**, 221, 375-381.
[3] F. Gingl, K. Yvon, T. Vogt, A. Hewat, *Journal of Alloys and Compounds*, **1997**, 253-254, 313-317.
[4] H. Kohlmann, F. Werner, K. Yvon, G. Hilscher, M. Reissner, G. J. Cuello, *Chemistry - A European Journal*, **2007**, 13, 4178-4186.
[5] H. Kohlmann, F. Gingl, T. Hansen, K. Yvon, *Angewandte Chemie International Edition*, **1999**, 38, 1856-2070.

Figure 1

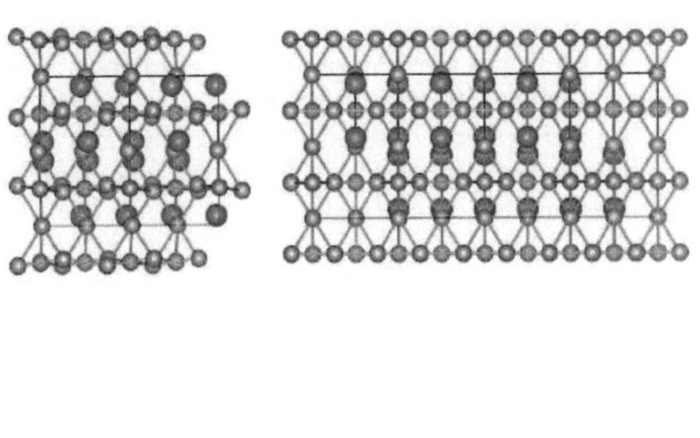

P060

Disorder in the crystal structure of dimethyldifluorostannane, Me_2SnF_2

H. Reuter[1], K. Neue[1], T. Gieschen[1]
[1]*University of Osnabrück, Institute of Chemistry of New Materials, Osnabrück, Germany*

Most diorganodifluorostannanes, R_2SnF_2, are white powders insoluble in water and non-coordinating organic solvents like *n*-hexane, toluene and others. As a result of this insolubility, crystal growth of diorganodifluorostannanes is a challenge up to now not solved satisfactorily. Single crystals suitable for X-ray diffraction experiments are only available from dimethyldifluorostannane, Me_2SnF_2, because of its low solubility in hot water, but have been studied only by Schlemper and Hamilton in 1966 [1]. Its structure, therefore, is often used as prototype for the structures of polymeric, insoluble diorganodifluorostannanes.

Schlemper and Hamilton found a tetragonal unit cell of space group I4/mmm with Sn, F and C in special positions resulting in a layer-structure perpendicular to the crystallographic c-axis consisting of octahedrally coordinated tin atoms linearly bridged by four equidistant μ_2-fluorine atoms while the methyl groups adopt a trans-position below and above the tin-fluorine plane. In this structure model the tin-fluorine distance corresponds to a/2.

In redetermining this structure by use of high quality data measured at 100 K with a Kappa APEX II diffractometer equipped with an area detector and low-temperature device we found a similar tetragonal unit cell as Schlemper and Hamilton at room temperature. Moreover, systematic absences confirmed the previously chosen space group I4/mmm so that we used the old structure model as starting point for our own structure refinements. Improvements, however, were limited, mainly because of large anisotropic displacement ellipsoids of the fluorine atoms. This problem could be solved by use of two split positions symmetrically to the crystallographic mirror plane perpendicular to the ab-plane.

The poster will bring face to face both structure models (Fig. 1) and will discuss the tin-fluorine distances and angles now derived from intensity data.

[1] Schlemper, E.O. and W.C. Hamilton, *The Crystal Structure of Dimethyltin Difluoride. An Example of Octahedral Coordination of Tin.* Inorg. Chem., **5** (1966) 995-998.

Figure 1

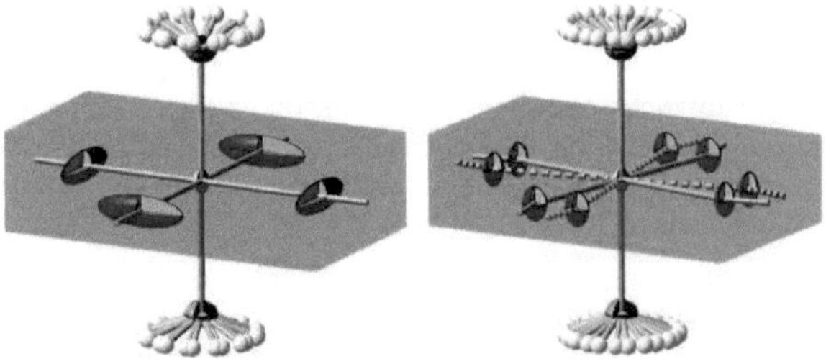

P061

Structural Studies of M_2(*m*-dobdc) Metal–Organic Frameworks and Their Interactions with CO_2, H_2, D_2, Xe and Kr gases

T. Runčevski[1], M. T. Kapelewski[1], J. Lee[1], Z. Smith[1], H. J. Park[1], C. M. Brown[2], J. R. Long[1]
[1]*UC Berkeley, Berkeley, United States*
[2]*National Institute of Standards and Technology, Gaithersburg, United States*

Metal-organic frameworks are a class of porous materials with promising applications in gas sorption, capture, storage and purification. One important family of these materials are the M_2(dobdc) (M = Mg, Mn, Fe, Co, Ni, Cu, Zn; dobdc^{4-} = 2,5-dioxido-1,4-benzenedicarboxylate) or M-MOF-74 frameworks. Many interesting features of M_2(dobdc) are due to the exposed coordination metal sites in the activated framework. While the applications of M_2(dobdc) are exciting, being able to tune these materials for more favorable properties is of great interest. Towards this end, a structural isomer of the M_2(dobdc) framework, M_2(*m*-dobdc) (*m*-dobdc4- = 4,6-dioxido-1,3-benzenedicarboxylate) was recently realized. Here, the carboxylate and oxido groups are situated at meta positions on the benzene ring, in contrast to the dobdc^{4-} linker, in which they are in para positions. Our studies on M_2(*m*-dobdc) indicate that, although it exhibits a similar channel structure to that of M_2(dobdc), it shows significant differences in adsorption behavior within the channels and, to some extent, on the exposed metal sites. Structural studies were performed by neutron and X-ray powder diffraction using H_2, D_2, CO_2, Xe, and Kr as probe gas molecules. It was shown that the gas molecules have different preferred sites of adsorption in M_2(*m*-dobdc) as compared to M_2(dobdc). The meta analogues were prone to disorder of the adsorbed gas molecules (especially for CO_2). Furthermore, Ni_2(*m*-dobdc) was found to have one of the largest heats of adsorption for hydrogen ever reported, a key characteristic for assessing and designing novel adsorbents for hydrogen storage applications. Detailed crystallographic studies were performed to explain these observations.

P062

Same Same but Different

C. J. Müller[1]
[1]*Lund University, Lund, Sweden*

The intermetallic compounds Ni_2In and HT-Ni_3Sn_2 form a solid solution. This is not surprising as In and Sn atoms are similar in size, and both intermetallic phases adopt the Ni_2In structure type ($P6_3/mmc$, Z=2). Within the alloy structures, In and Sn atoms randomly occupy the hexagonal lattice sites (2*c*) whereas Ni atoms are found in all octahedral interstices (2*a*) and in the trigonal bipyramidal interstices (2*d*). Notably, the random occupation of this lattice site increases from half to full occupation when going from Ni_3Sn_2 to Ni_2In. In contrast to this solubility range, the Ni_2In type superstructures of $Ni_{13}In_9$ and Ni_7In_3 (i.e. LT-Ni_2In) dissolve almost no Sn atoms on their pseudohexagonal lattice sites. And unsurprisingly, the solid solution is observed at temperatures above 400 °C only. [1]

At 300 °C, HT-modifications Ni_2In and Ni_3Sn_2 are no longer stabilized upon alloying with a third element. A miscibility gap has formed, and no alloys can be observed between the binary Ni_2In type superstructures LT-Ni_3Sn_2, $Ni_{13}In_9$ and Ni_7In_3. Interestingly, this gap is not completely empty as two novel ternary intermetallic compounds Ni_5InSn_2 and Ni_5In_2Sn appeared. [2]

Both compounds crystallize in novel, but different Ni_2In type superstructures. In both cases, In and Sn atoms do not occupy the pseudohexagonal lattice sites randomly any longer, and therefore the symmetry is reduced. But while Ni_5InSn_2 crystallizes in the rhombohedral space group *R*32, Ni_5In_2Sn was found to be orthohexagonal in *Cmcm*. Interestingly, they have same nominal amount of Ni atoms on the interstitial sites. And as will be presented and discussed, they both contain the same structural motif for the network atoms: honeycomb layers of In atoms in whose centers sole Sn atom reside.

References

[1] C. Schmetterer, A. Zemanova, H. Flandorfer, A. Kroupa, H. Ipser, *Intermetallics*, **2013**, *35*, 90-97
[2] C.J. Müller, *On Evil Twins and Their Absent Friends*, Dissertation, Lund University, 2015

P063

Biomimetic crystal growth of carbonate phases in gels: an in situ X-ray diffraction study

B. Maier[1], S. Grott[2], W. Schmahl[1]
[1]*LMU München, Department für Geo- und Umweltwissenschaften, München, Germany*
[2]*LMU München, Department für Chemie, München, Germany*

Carbonate biomineral crystals are composites consisting of organic and mineral components and typically exhibit a mesocrystalline microstructure, i.e. mutually co-aligned nanocrystals embedded in the organic matrix. That results in extraordinary mechanical properties and allows for intriguing crystal shapes. The biomineral mesocrystals are the basis for the complex hierarchical morphology of biomaterials.

Hydrogel matrices are popular crystallization substrates for biomimetic crystallization since they can be fine-tuned to adequately reproduce a wide range of biomineralization conditions. Hydrogels have successfully been used to obtain synthetic mesocrystals of a variety of phases that are relevant to biological hard materials.

The aim of this study is to observe the nucleation and growth of carbonate phases in hydrogels in situ depending on various parameters: type of hydrogel, solid content of hydrogel, solution concentration, addition of Mg to the hydrogel, and temperature. For that purpose a counter-diffusion system was scaled such that it could be fit on a X-ray diffractometer. Here we show the successful experimental setup as well as first results using gelantin as growth medium.

P064

The phase transition of $Pb_8F_{14}I_2$

M. Weil[1]
[1]*TU Wien, CTA-SC, Wien, Austria*

The crystal structure of the lead fluoroiodide $Pb_8F_{14}I_2$ has been determined on the basis of room temperature Weissenberg data in the orthorhombic space group *Bmmb* [1]. The nearly equal lengths of two axes (a, $b \approx 6.05$ Å, $c \approx 25$ Å) and an *I*-centred pseudo-tetragonal subcell make it appear likely that a phase transition into a tetragonal high-temperature structure occurs. Although the author of the original study has derived an idealized structure in space group *I*4/*mmm* with the cell relations $a_{\text{ideal.}} \approx a_{\text{room-temperature}}/\sqrt{2}$, $c_{\text{ideal.}} \approx c_{\text{room-temperature}}$, neither experimental details regarding a possible phase transition nor structure data of the high-temperature phase were reported at that time or afterwards.

In the current study, details of the first order phase transition (transition temperature monitored by temperature-dependent single crystal X-ray and DSC data) and the experimentally determined high-temperature structure are reported.

[1] B. Aurivillius, *Chem. Scripta.* 15, 153-164 (1980).

P065

Tilting Structures in *Inverse*-perovskites, M_3TtO (M = Ca, Sr, Ba, Eu; Tt = Si, Ge, Sn, Pb)

J. Nuss[1], A. W. Rost[1], C. Mühle[1], K. Hayama[1], V. Abdolazimi[1], H. Takagi[1]
[1]*Max-Planck-Institut für Festkörperforschung, Stuttgart, Germany*

Single crystal X-ray diffraction experiments were performed for the whole series of *inverse*-perovskites, M_3TtO (M = Ca, Sr, Ba, Eu; Tt = Si, Ge, Sn, Pb) in the temperature range of 50-500 K. Cubic M_3TtO (Tt = Sn, Pb) representatives show distinct anisotropic displacement parameters of the M atoms at 295 K, which become isotropic upon cooling for M = Ca, Sr, Eu. Their structures can be regarded as 'ideal' cubic perovskites at 50 K. In contrast, Ba compounds undergo a phase transition below ~150 K from cubic (*Pm*3*m*, *cP*5) to orthorhombic (*Ibmm*, *oI*20), with slightly tilted OBa_6 octahedra, Đ(O-Ba-O) ~174° (100 K). For M_3TtO (Tt = Si, Ge) representatives two cases can be distinguished. For bigger M atoms, like e.g. europium, they crystallize in *anti*-$GdFeO_3$ type of structure (*Pbnm*, *oP*20), with pronounced tilting of OEu_6 octahedra, Đ(O—Eu—O) ~160° (295 K). For the smaller sized Ca, orthorhombic symmetry (*Ibmm*) is found at room temperature. While Ca_3GeO changes into cubic at ~400 K, Ca_3SiO is the only representative in the M_3TtO family where all three polymorphs can be found within the temperature range 50-500 K: *Pbnm* — *Ibmm* — *Pm*3*m*. They show tiny differences in the tilting of the OCa_6 octahedra, expressed by the O—Ca—O bonding angles of ~171° (100 K), ~174° (295 K), and 180° (500 K), see figure 1. The diverse phase transitions are all accompanied by multiple twinning, in terms of 'pseudo merohedry' or 'reticular pseudo merohedry', respectively [1].

Figure 1: Structure field map (temperature versus Goldschmidt tolerance factor *t*) showing the stability regions of different distortion variants of the *inverse*-perovskites M_3TtO. The inserts show the octahedral representation of the 'ideal' cubic *inverse*-perovskites (right) and the distorted ones (left).

[1] Nuss, J., Mühle, C., Hayama, K., Abdolazimi, V. & Takagi, H. (2015) *Acta Cryst.* **B71**, 300-312.

Figure 1

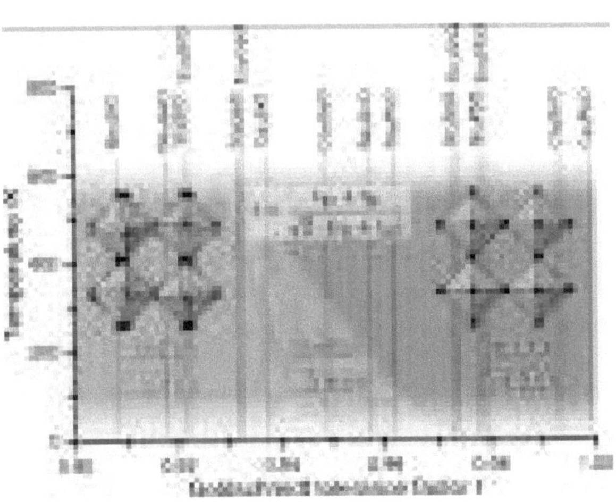

Crystal growth, crystal-chemical and optical characterization of Ba_2TiO_4

S. Lenz[1], M. Burianek[1], R. X. Fischer[1]
[1]*Universität Bremen, FB05-Geowissenschaften, Bremen, Germany*

Introduction

Barium orthotitanate (Ba_2TiO_4) is the only stable phase on the Ba-rich side of the binary system BaO-TiO_2. Many phases of this system have attracted the interest of research because of their high dielectric constants. Ba_2TiO_4 has two polymorphs: the monoclinic low-temperature β-Ba_2TiO_4 and the orthorhombic high-temperature α'-Ba_2TiO_4. The monoclinic polymorph is usually twinned either on (100) or (001) and crystallizes in space group $P2_1/n$. The orthorhombic polymorph crystallizes in space group $P2_1nb$. The crystal structures of beta low and alpha high-temperature phases are related to each other with α'-Ba_2TiO_4 exhibiting a tripled b-axis with respect to β-Ba_2TiO_4 [1]. Between both polymorphs a martensitic phase transformation occurs at 100-220°C [2]. Ba_2TiO_4 is a compound with the rare case of Ti^{4+} tetrahedrally coordinated by oxygen atoms.

Objectives

The overall aim of this work is to determine the electronic polarizability α_e of $^{[IV]}Ti^{4+}$ calculated from the refractive indices (RI) of Ba_2TiO_4. Because of the twinning in the low-temperature phase, optical data have to be measured on the high-temperature polymorph α'-Ba_2TiO_4 stabilized at room temperature.

Methods

Very accurate optical measurements can be carried out by the prism method, but this requires large single crystals. Due to the high melting point of approximately 1860 °C experiments to synthesize large boules of single-crystal α'-Ba_2TiO_4 had to be carried out by the Czochralski method with Ir-equipment under N_2-atmosphere.

Results-Conclusion

Orthorhombic α'-Ba_2TiO_4 cannot be grown from pure Ba_2TiO_4 melts. First growth products consisted of monoclinic β-Ba_2TiO_4 (Fig. 1). By doping small amounts of Nd_2O_3 (0.1-0.25 mole %) the orthorhombic polymorph α'-Ba_2TiO_4 could be stabilized down to room temperature. In this case the unfavorable phenomenon of spiral growth occured (Fig.2). However, small homogeneous crystals could be separated showing that this route is basically suitable to produce crystals for optical investigations. Preliminary results from spindle-stage measurements show that the RI of Ba_2TiO_4 are higher than 1.85. Further work is in progress to produce large crystals for the prism method using the crystals from the first experiments as seed crystals.

References:

[1] Günter, J. R. and Jameson, G. B. (1984), Acta Crystallographica Section C Crystal Structure Communications 40 (2), 207-210
[2] Chen, B., Liao, F. H., Jiao, H. and Jing, X. P. (2013), Phase Transitions 86 (4), 380-390.

Figure Captions:

Fig. 1: Undoped monoclinic Ba_2TiO_4, T_m=1860°C.
Fig. 2: Nd_2O_3 stabilized orthorhombic Ba_2TiO_4, spiral growth, T_m=1860°C.

P066

Figure 1

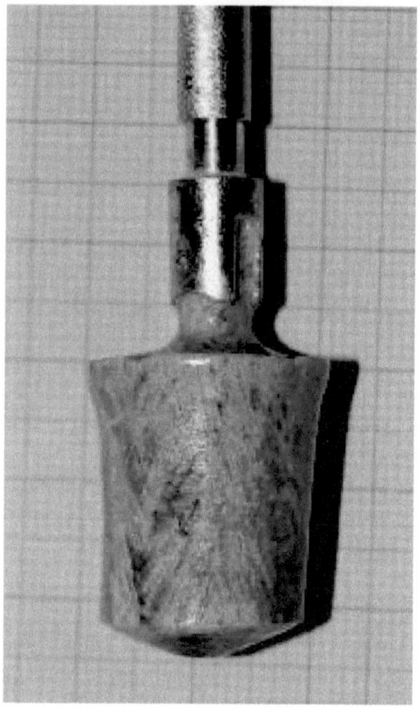

Figure 2

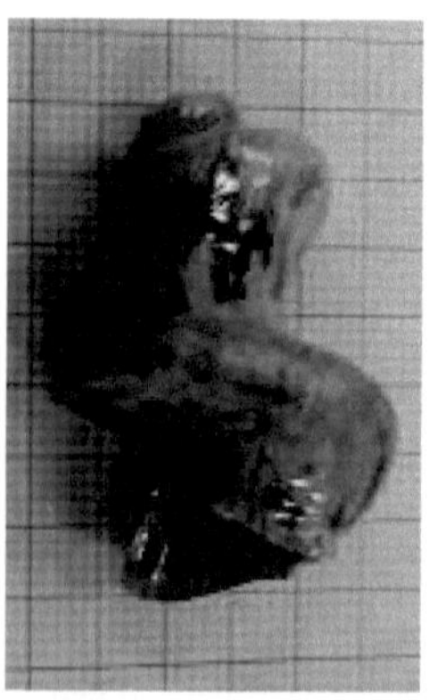

Lightning talks of young crystallographers

P067

1.6 Å Structure of PilBac1: Insights into Long-Range Electron Transfer in Bacteria and a Sulfur-SAD Success Story

M. Gorgel[1,2], A. Boeggild[2], J. Ulstrup[2], U. Mueller[3], M. Weiss[3], P. Nissen[2], T. Boesen[2]
[1]*Ludwig-Maximilians University, Gene Center, Munich, Germany*
[2]*Aarhus University, Department for Molecular Biology and Genetics, Aarhus, Denmark*
[3]*Helmholtz Zentrum Berlin für Materialien und Energie, Macromolecular Crystallography (HZB-MX), Berlin, Germany*

Bacterial nanowires are extracellular filaments that can transfer electrons from one cell to another and from a cell to an electron acceptor. This allows the metal reducing bacterium *Shewanella oneidensis* to grow on insoluble and toxic metals such as Fe(III), Mn(IV) and U(VI). These networks have already shown application in environmental sciences and nanotechnology, nevertheless, their optimal implementation requires an understanding of their underlying structure and mechanism as revealed by structural biology and nanoscience.

We therefore determined the 1.6 Å crystal structure of the putative nanowire subunit PilBac1. We phased the structure by sulfur-SAD despite several obstacles such as a low anomalous signal (a theoretical Bijvoet ratio of 0.9% at a wavelength of 1.8 Å), radiation damage-induced reduction of the cysteines and a multiplicity of only 5.5 [1]. The anomalous signal was improved by merging three data sets from different volumes of a single crystal, yielding a multiplicity of 17.5, and a sodium ion was added to the substructure of anomalous scatterers. The overall structure of PilBac1 reveals the common fold of type IV pili, but in a model of the assembled pilus we can clearly define individual clusters of closely spaced aromatic residues (3-5 Å). These short distances might allow for metallic-like electron transfer between the aromatic rings; however, the distances between the clusters will be too high for continuous electron transfer along the pilus and additional factors such as cytochromes will consequently be necessary.

In summary, our structure provides insight into the assembly of (non-conductive) pili in *Shewanella* and at the same time represents an exceptional case for sulfur-SAD phasing despite high radiation damage and a low anomalous signal.

References: 1: Gorgel *et al.* (2015). Acta Cryst. D. 71, 1095-1101.

P068

Temperature-induced phase transition in tetragonal copper sulfide at low temperatures

D. Zimmer[1], L. Bayarjargal[1], J. Zhang[2], E. Alig[3], L. Fink[3], E. Haussühl[1], C. Jin[2], B. Winkler[1]
[1]*Goethe University, Crystallography, Frankfurt am Main, Germany*
[2]*Institute of Physics Chinese Academy of Sciences, Beijing, Germany*
[3]*Goethe University, Chemistry, Frankfurt am Main, Germany*

Cuprous sulfides ($Cu_{2-x}S$; $0 \leq x \leq 0.25$) are nontoxic p-type semiconductors with a suitable band gap for the application as thin film solar cell absorbers [1]. However, their crystal structures and corresponding phase relations are complex due to the high mobility of copper atoms and vacancies within the sulfur sublattice and have therefore been in the focus of many studies for several decades. Since their electrical transport properties strongly depend on the stoichiometry and the crystal structure [2], the experimental and theoretical investigation of this structure-property relation is of great interest.

The aim of this study is to investigate the influence of structural changes in the tetragonal high pressure high temperature polymorph of Cu_2S [3, 4] on its electrical transport properties at low temperatures.

We synthesized polycrystalline samples of tetragonal Cu_2S in a large volume press at 5 GPa and 973 K. The samples were characterized by EDX and powder XRD. The evolution of the crystal lattice as a function of temperature was investigated in situ by powder XRD measurements between 295 and 87 K. Resistance measurements were performed by voltage-driven 4-wire measurements between 300 and 2 K. The heat capacity was determined experimentally by thermal-relaxation calorimetry between 300 and 2 K and theoretically investigated by DFT-based calculations.

The heat capacity and powder XRD measurements confirm previous findings [5], showing a low temperature phase transition. According to our data the transition takes place at ~205 K with a hysteresis of ±15 K (Fig. 1).

The resistance is strongly correlated with the structural transition and changes by several orders of magnitude across the transition, also showing a hysteresis on cooling and heating (Fig. 2).

The authors gratefully acknowledge financial support by the DFG (Projects Wi 1232 & BA 4020/2-1).

Figure 1: Heat capacity of tetragonal Cu_2S

Figure 2: Resistance of a tetragonal Cu_2S sample

[1] R.B. Hall & J.D. Meakin (1979). *Thin Solid Films* **63**, 203.
[2] K. Okamoto & S. Kawai (1972). *Jap. J. Appl. Phy.* **12**, 8, 1130.
[3] A. Janosi (1964). *Acta Cryst.* **17**, 311.
[4] B.J. Skinner (1970). *Econ. Geol.* **65**, 724.
[5] S. Stølen, F. Grønvold & E. F. Westrum Jr. (1990). *J. Chem. Thermodynamics* **22**, 1035.

Figure 1

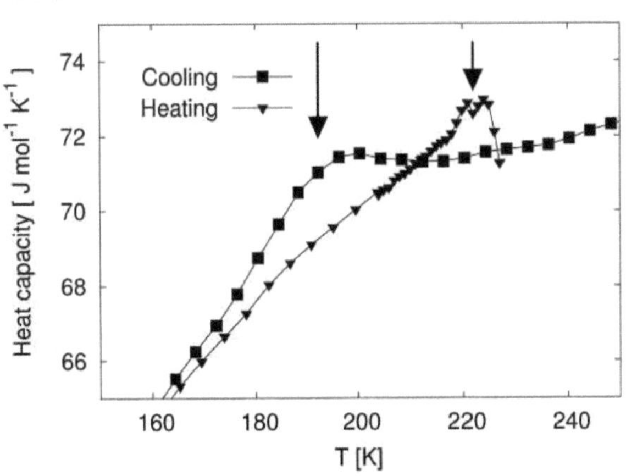

Figure 2

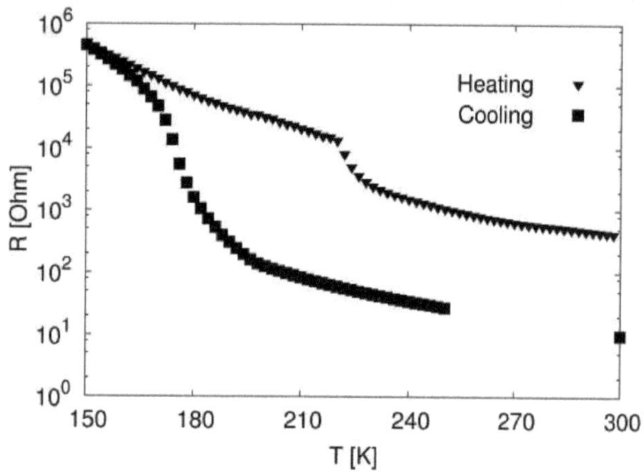

P069

Structural complexity of simple Fe_2O_3 oxide at high pressures and temperatures

E. Bykova[1], M. Bykov[1], C. McCammon[1], S. Ovsyannikov[1], H.- P. Liermann[2], I. Kupenko[3], A. Chumakov[3], R. Rüffer[3], M. Hanfland[3], V. Prakapenka[4], N. Dubrovinskaia[1], L. Dubrovinsky[1]

[1]*University of Bayreuth, Bayerisches Geoinstitut, Bayreuth, Germany*
[2]*Deutsches Elektronen-Synchrotron, P02.2 Extreme Conditions Beamline, Hamburg, Germany*
[3]*European Synchrotron Radiation Facility, Grenoble, France*
[4]*Argonne National Laboratory, GSECARS, Argonne, United States*

The Fe-O binary system is widely studied both in solid state and mineral physics. Oxides of iron are present in nature in the form of FeO (known at normal conditions as wüstite, rock-salt structure), Fe_2O_3 (hematite, corundum structure), and Fe_3O_4 (magnetite, inverse spinel structure). High-pressure behaviour of natural iron oxides was a subject of studies since early 60s but still many controversies, in particular, regarding crystal structures of high-pressure high-temperature (HPHT) polymorphs, remained. For instance, many structures have been proposed for a high-pressure polymorph of Fe_3O_4, among them monoclinic one with octahedrally coordinated iron atoms [1], $CaMn_2O_4$-type [2], and $CaTi_2O_4$-type (space group *Bbmm*, No. 63) structures [3,4]. Orthorhombic $GdFeO_3$ perovskite [5] and Rh_2O_3-II [6] structural types have been proposed as candidates for a high-pressure polymorph of Fe_2O_3 which exists above 45 GPa.

We report here the crystal structures of novel iron oxides, synthesized at HPHT conditions. HPHT polymorphs of Fe_2O_3 and Fe_3O_4, and iron oxides with completely new stoichiometry, as e.g. Fe_5O_7 and $Fe_{25}O_{32}$, were observed in our experiments in laser-heated diamond anvil cells. They were characterized using single-crystal X-ray diffraction and synchrotron Mössbauer source spectroscopy. Our crystallographic data allowed to explain electronic and magnetic changes in Fe_2O_3 and Fe_3O_4 induced by HPHT.

We found similarities in the crystal structures of Fe_5O_7, some high-pressure polymorphs of Fe_2O_3 and Fe_3O_4, and the recently discovered Fe_4O_5 [7] and Fe_5O_6 [8]. Based on the study of Guignard *et al.* [9] we suggest a homological series of iron oxides with a common formula mFeO·$n$$Fe_2O_3$ ($m \geq 1$, $n \geq 0$). The crystal structures of the iron oxides belonging to the homological series are composed of FeO6 trigonal prisms and octahedra. So far we experimentally observed six representatives of the series. Our results suggest that unusual trigonal prismatic oxygen coordination of iron may be common for iron-oxygen compounds at conditions of the Earth's lower mantle.

References

[1] H.-K. Mao, T. Takahashi, W. A. Bassett, G. L. Kinsland, L. Merrill, *J. Geophys. Res.* **79**, 1165 (1974).
[2] Y. Fei, D. J. Frost, H.-K. Mao, C. T. Prewitt, D. Haeusermann, *Am. Mineral.* **84**, 203 (1999).
[3] C. Haavik, S. Stølen, H. Fjellvåg, *Am. Mineral.* **85**, 514 (2000).
[4] L. S. Dubrovinsky, N. A. Dubrovinskaia, C. McCammon, G. K. Rozenberg, R. Ahuja, J. M. Osorio-Guillen, V. Dmitriev, H.-P. Weber, T. Le Bihan, B. Johansson, *J. Phys. Condens. Matter* **15**, 7697 (2003).
[5] J. S. Olsen, C. S. G. Cousins, L. Gerward, H. Jhans, B. J. Sheldon, *Phys. Scr.* **43**, 327 (1991).
[6] G. Rozenberg, L. Dubrovinsky, M. Pasternak, O. Naaman, T. Le Bihan, R. Ahuja, *Phys. Rev. B* **65**, 064112 (2002).
[7] B. Lavina, Y. Meng, R. T. Downs, P. F. Weck, S. R. Sutton, Y. Zhao, *Proc. Natl. Acad. Sci. USA* **108**, 17281 (2011).
[8] B. Lavina, Y. Meng, *Sci. Adv.* **1**, e1400260 (2015).
[9] J. Guignard, W. A. Crichton, *Mineral. Mag.* **78**, 361 (2014).

P070

High-pressure phase transition and single-crystal elasticity of $SrCO_3$ by Raman spectroscopy and Brillouin spectroscopy

N. Biedermann[1], S. Speziale[2], H. J. Reichmann[2], M. Koch-Müller[2], G. Heide[3]

[1]*European XFEL, HED Group, Hamburg, Germany*
[2]*Deutsches GeoForschungsZentrum GFZ, Potsdam, Germany*
[3]*Technische Universität Bergakademie Freiberg, Institut für Mineralogie, Freiberg, Germany*

Carbonates are key carriers of Carbon into the deep mantle and they were found to be stable at conditions of the lower mantle due to their presence as inclusions in transition-zone diamonds (Brenker et al., 2007). Strontianite ($SrCO_3$), which is a component in all natural carbonates, is isostructural to aragonite, a major high-pressure polymorph of $CaCO_3$, and therefore a model material to investigate the stability and physical properties of carbonates at high pressure and high temperature. Strontianite was synthesized at 4 GPa and 1273 K for 24h in a multi-anvil apparatus. We compressed single-crystals or powder samples in the diamond anvil cell and measured Raman scattering up to 78 GPa. A transition at 26 ± 1 GPa was observed during compression to a new high-pressure structure of $SrCO_3$. This post-aragonite phase was preserved until the highest experimental pressure and in decompression still present down to 18 GPa. An increase in the coordination number for Sr^{2+} produces a clear change in the vibrational spectrum of $SrCO_3$ and is in agreement with previously published high-pressure structures with space group *Pmmn* for aragonite-type carbonates (e.g. Ono et al., 2005; Oganov et al., 2006; Ono, 2007). In addition to our study of the high-pressure polymorphism of $SrCO_3$, we have used Brillouin spectroscopy to determine the elastic properties of synthetic $SrCO_3$ single-crystals at ambient conditions. The present study provides the first experimental data for the individual stiffness tensor c_{ij} of $SrCO_3$. Measuring Brillouin scattering along 69 different crystallographic directions in a set of three platelets we were able to determine c_{11}, c_{12}, c_{13}, c_{22}, c_{23}, c_{33}, c_{44}, c_{55} and c_{66} with 151.3(7), 63.2(9), 30(2), 106.4(7), 41(1), 75(1), 39.8(7), 28.0(3) and 31.5(4) GPa, respectively. Results for the adiabatic bulk modulus are at 59(1) GPa and in comparison with the elastic moduli of other aragonite-type carbonates it confirms a systematic linear relationship between cation radius and bulk modulus. The shear modulus for $SrCO_3$ presents a value of 32.0(5) GPa which is about 10 % smaller in comparison to that of aragonite (Liu et al., 2005). Although pure strontianite is not a relevant phase of the Earth's mantle, measurements of the acoustic velocities as well as studying the high-pressure phase behavior provide necessary reference information for the elasticity and stability of aragonite-type carbonates that might be present in the Earth's interior.

References:

Brenker et al. (2007) Earth and Planetary Science Letter 260, 1.

Ono et al. (2005) Phys Chem Minerals 32, 8.
Oganov et al. (2006) Earth and Planetary Science Letter 241, 95.
Ono (2007) Phys Chem Minerals 34, 215.
Liu et al. (2005) Phys Chem Minerals 32, 97.

P071

On the mechanism and accompanying effects of two-dimensional polymerization in a novel mono-layered monomer crystal

G. Hofer[1], T. Weber[2], A. D. Schlüter[1]
[1]*ETH Zürich, Polymer Chemistry, Zürich, Switzerland*
[2]*ETH Zürich, Laboratory of Crystallography, Zürich, Switzerland*

A breakthrough in the synthesis of sheet-like polymers was recently achieved based on the polymerization of trifunctional monomers [1]. By crystallizing these novel monomers prior to their connection, genuine two-dimensional polymerization can be achieved resulting in a two-dimensional polymer (2DP).

The chiral anthracene-based monomer molecules of the here presented 2DP crystallize in the polar space group *R*3 in a 2-cyanopyridine (*cpy*) solution. The crystal structure allows the anthracene moieties from the nearest neighbouring monomers to face each other close enough to achieve a controlled long-ranged ordered polymerization in the *ab*-plane of the crystal by continuous UV irradiation. This turns the monomer single crystal into a polymer single crystal, a process that can thermally be reversed. During either direction of this phase transition, the space group *R*3 is preserved. Both, polymerization and depolymerization can be frozen at any time by simply removing the crystal from the triggering source thus creating a disordered and meta-stable monomer/polymer single crystal. The polar *cpy* is incorporated into the structure in a partially disordered manner next to the anthracene moieties of monomers within the same layer, Figure 1.

This contribution will address all important issues of this unusual polymerization based on an analysis of the Bragg and diffuse X-ray [2] scattering of various states of the polymerization/depolymerization with the goal to provide a complete understanding of the process and its implications for structure formation and properties. This last aspect concerns the polar *cpy* molecules, which during polymerization undergo gradual positional changes that result in strong effects on the overall dipole moment.

Figure 1: Network structure of the two-dimensional polymer synthesized by [1].

[1] Kory, M. J., Wörle, M., Weber, T., Payamyar, P., van de Poll, S. W., Dshemuchadse, J., Trapp, N., Schlüter, A. D. (2014) *Nat. Chem.*, *6*, 779-784.
[2] Weber, T., Simonov, A. (2012) *Z. Kristallogr.*, *227*, 238-247.

Figure 1

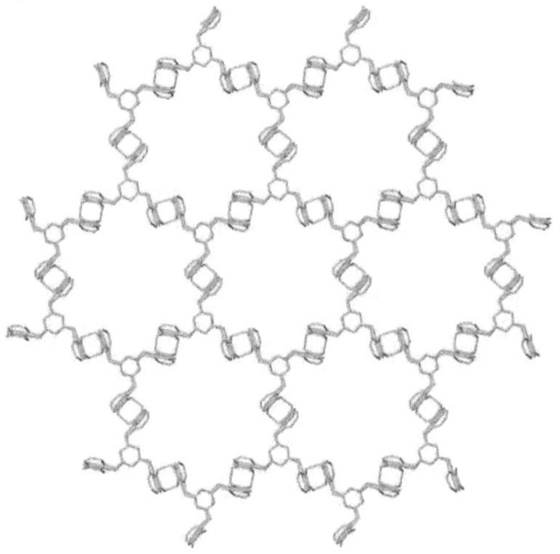

P072

Synthesis, Characterization, and investigation of electronic and optical properties of double perovskite oxide $BaSrMgTeO_6$

M. Ait Haddouch[1], A. abbassi [2], Y. Aharbil[1], H. Labrim[3], Y. Tamraoui[4], F.-E. Mirinioui[4], B. Manoun [4], H. Ez-Zahraouy[2], A. Benyoussef[2,5], L. Laânab[6], S. Benmokhtar[1]
[1]*université hassan II Casablanca, Laboratoire de Chimie Physique des Matériaux LCPM, casablanca, Morocco*
[2]*Université Mohammed V, Laboratoire de Magnétisme et Physique des Hautes Energies L.M.P.H.E. URAC 12, rabat, Morocco*
[3]*Centre National de l'Energie, des Sciences et des Techniques Nucléaires, rabat, Morocco*
[4]*Univ. Hassan 1er, laboratoire des Sciences des Matériaux, des Milieux et de la modélisation (LS3M), Khouribga, Morocco*
[5]*Institute of Nano-materials and Nanotechnologies, MAScIR, rabat, Morocco*
[6]*Université Mohammed V, Laboratoire Conception et Systèmes (LCS), rabat, Morocco*

A double perovskite $BaSrMgTeO_6$ has been synthesized and characterized by physical techniques: X-ray diffraction, Raman diffusion, and diffuse reflectance spectroscopy.The crystal structure was resolved at room temperature by the Rietveld refinement method, and revealed that $BaSrMgTeO_6$ crystallizes in the cubic system, space group Fm-3m with lattice parameter (a=8.018Å), and shows an almost perfect ordering between Mg^{2+} and Te^{6+} cations at the B substructure. The Raman spectra of this crystalline perovskite oxide is interpreted by means of factor group analysis in terms of space group Fm-3m. Assignments of the Mg/Te-O vibrational stretching and bending modes have been made. Based on density functional theory (DFT), and using full potential-linearized augmented plane wave (FP-LAPW) method with the generalized gradient approximation (GGA), implemented in the Wien2k package, we have investigated the electronic and the optical proprieties of such material. The optical band gap obtained with GGA approximation is equal to 2.8 eV, which is in good agreement with our experimental results. We found that the transmittance T is stable and reach the average of 90% in both experiment and theoretical studies.

P073

Controlled Stepwise Synthesis of a Cu MOF From Magnesium Precursor to Mesoporous Material

Q. Guo[1], U. Englert[1]
[1]*Institute of Inorganic Chemistry, Aachen, Germany*

A three-dimensional copper *metal-organic framework* (MOF) has been synthesized from a magnesium precursor and copper sulfate crystalized in space group *I*2₁3. The effective free volume calculated by PLATON analysis after removing all the guest molecules amount approximately 54.8 percent of the crystal volume (8773 Å^3 of 16007 Å^3 unit per cell volume). Topological analysis was carried out by *GTECS3D* to obtain an insight into the structure [*Z. Kristallogr. Suppl.* **2012**, *32*, 117.]. The structure can be viewed as a 4-connected topology with a Vertex symbol $3.3.10_2.10_2.10_3.10_3$, which corresponds to a **lcv** net. The 3D intersecting channel system has a cross section of approximately 9 Å. In additional, the oxygen atoms of sulfate ions and methyl groups from the ligand are exposed toward the channel, which facilitate the trapping of polar guests.

Figure 1

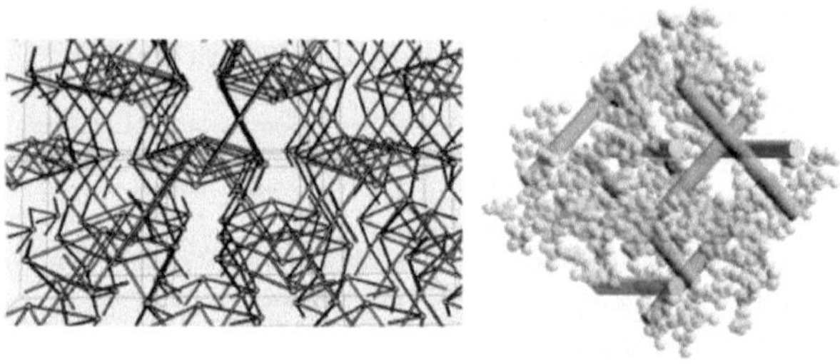

lcv network structure of the Cu MOF (*left*); 3D channel net running along the *a*, *b* and *c* axes (*right*).

P074

Anion recognition by dinuclear Ni(II) complexes with macrocyclic NHC ligands

P. J. Altmann[1], A. Pöthig[1]
[1]*TU München, Catalysis Research Center, Garching, Germany*

The tailored design of ligands to optimise the electronic and steric properties of molecular complexes is a common principle in homogeneous catalysis. In the last two decades, more sophisticated supramolecular concepts have been developed. We recently introduced the family of calix[4]imidazolium[2]pyrazole cyclophanes, which can act as a ligand precursor for bowl-shaped dinuclear transition metal complexes with a defined spatial arrangement of the two centres.[1] A new member of this family is presented, namely a binuclear Ni-NHC complex, which is able to selectively recognise and self-assemble with guests via tennis-ball like encapsulation. NMR titration studies showed, that addition of halides leads to formation capsules with cavities of defined size, in which only chloride and bromide can be incorporated. The solid state structures derived by SC-XRD studies suggest a stabilisation of the capsules through a combination of multiple hydrogen bonding and attractive Coulomb interactions between the guest and the metal centres.

[1] Philipp J. Altmann, Christian Jandl, Alexander Pöthig* *Dalton Trans.*, **2015**, *44*, 11278-11281.

P075

Structure Analysis of Human Prolidase Mutations gives insight into the Prolidase Deficiency disease mechanisms

P. Wilk[1,2], M. Ühlein[1], H. Dobbek[2], M. S. Weiss[1], U. Mueller[1]
[1]*Helmholtz-Zentrum Berlin, BESSY II, Macromolecular Crystallography (HZB-MX), Berlin, Germany*
[2]*Humboldt-Universität zu Berlin, Institute of Biology, Berlin, Germany*

Prolidase is the only enzyme in humans capable of cleaving peptide bonds preceding proline or hydroxyproline residue in dipeptides. It is involved in last steps of collagen catabolism and likely in regulation of peptidic hormones. The diminished or absent prolidase activity was related to a rare autosomal disease, manifested by variety of clinical symptoms, known as Prolidase Deficiency (PD). To date there is no definitive cure to PD, partially due to a limited understanding of the wild-type (wt) enzyme in terms of detailed substrate binding mode and as a consequence catalyzed reaction mechanism, as well as due to lack of any structural information of the mutants. Here we describe the high resolution crystal structures of wt human prolidase in the ligand-free form and in the substrate-bound state providing the missing structural information. We analyze the mode of substrate binding and a mechanism of hydrolysis utilized by prolidase pointing to position of a catalytic bridging molecule. We have also selected and characterized structurally eight clinically relevant mutations introducing single amino acid changes in the prolidase chain. Based on high resolution crystal structures we have identified different mechanisms of enzymes inactivation. Together this information shall ease the understanding of pathologies related to this enzyme.

Figure 1: The overall structure of human prolidase is shown in cartoon representation with subunit A colored according to secondary structure. Mn2+ ions in the active site are shown as magenta spheres and GlyPro substrate as stick representation in orange. The positions of mutated residues in chain A are indicated by cyan spheres. The inlets show two examples of enzymes inactivation mechanism: by losing of Mn2+ ion (Asp276Asn, upper panel) or by structure destabilization (Gly448Arg, lower panel). Structures of mutants are superimposed on wt protein, 2Fo-Fc maps contoured at 2 σ are shown in blue and anomalous difference maps (AnoDe) contoured at 10σ shows positions of manganese ions.

Figure 1

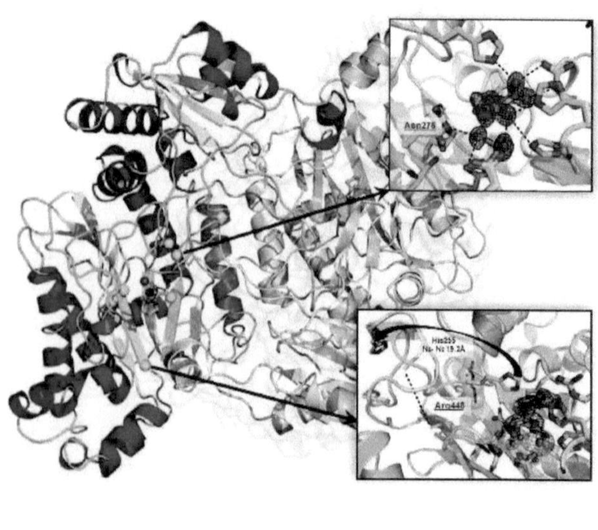

P076

Influence of Modulators for MOF Formation

J. Bergmann[1], H. Krautscheid[1]
[1]Universität Leipzig, Institut für Anorganische Chemie, Leipzig, Germany

Introduction

In the last two decades Metal-Organic Frameworks (MOFs) have emerged as a new class of porous materials with great potential for a wide range of applications, including gas storage, separation, transport, heterogeneous catalysis, sensors and drug delivery systems.[1,2]

Objectives

A well-known problem in the synthesis of MOFs is their sensitivity towards extraneous influences such as reaction temperature and time, solvent, pH-value or the presence of modulators. Aim of this work is the systematic investigation of different modulators on the formation of a Cd^{2+} based MOF with a substituted 1,2,4-triazolylisophthalate linker[3] (Fig. 1).

Materials & Methods

To obtain single crystals suitable for X-ray diffraction, solvothermal syntheses were performed in steel autoclaves with water/acetonitrile (1:1; *v:v*) as solvent.

Single crystal structure analyses[4] were performed on a STOE IPDS 1 or 2T diffractometer with Mo-K_{α} radiation (λ = 71.073 pm). To verify phase purity, powder X-ray diffractograms were measured on a STOE STADI-P (Cu-$K_{\alpha1}$; λ = 154.059 pm). Thermal behaviour was investigated by temperature dependent PXRD and TG-DTA (STA F1 Jupiter; NETZSCH).

Results

Depending on the modulator, three closely related modifications of a MOF with the general formula ${}_{\infty}^{3}$[Cd(Me-prz-trz-ia)] {Me-prz-trz-ia$^-$ = 5-(3-methyl-5-(pyrazine-2-yl)-4*H*-1,2,4-triazol-4-yl)isophthalate} could be synthesized and characterized. These three dimensional frameworks with **rtl** topology exhibit microporous channels with calculated pore diameters in the range of 300-550 pm[5] and calculated porosities up to 50 %[6] which make these compounds interesting for gas storage and separation.

Without any modulator a monoclinic phase ($P2_1/n$; $\mathbf{1^{mP}}$; z = 4) is formed.[7] Using modulators with carboxylic and/or amino groups a tetragonal phase ($P4_2/n$; $\mathbf{1^{tP1}}$; z = 8) is formed. At 55 °C a phase transition towards the monoclinic modification $\mathbf{1^{mP}}$ is observed. Synthesis in presence of isophthalic acid or aminoisophthalic acid leads to another tetragonal modification ($P4_2/m$; $\mathbf{1^{tP2}}$; z = 4).

Conclusion

The formation of MOF crystals is very sensitive regarding the presence of modulators. Especially modulators with one or more donor groups are able to coordinate to the metal ions and pre-organize them after getting replaced by the actual linker, leading to different modifications of related structures.

References:

[1] S. Kitagawa, R. Kitaura, S. Noro, *Angew. Chem.*, **2004**, *43*, 2334.
[2] A. U. Cajka, N. Trukhan, U. Müller, *Chem. Soc. Rev.*, **2009**, *38*, 1284.
[3] D. Lässig, J. Lincke, H. Krautscheid, *Tet. Lett.*, **2010**, *51*, 653-656.
[4] G. M. Sheldrick, *Acta Cryst.* **2008**, *A64*, 112-122.
[5] S. Bhattacharya, K.E. Gubbins, *Langmuir,* **2006**, *22*, 7726.
[6] *PLATON*, A. L. Spek, *J. Appl. Cryst.* **2003**, *36*, 7-13.
[7] J. Bergmann, K. Stein, M. Kobalz, M. Handke, M. Lange, J. Möllmer, F. Heinke, O. Oeckler, R. Gläser, R. Staudt, H. Krautscheid, *Microporous and Mesoporous Materials* **2015**, *216*, 56-63.

Fig. 1: Tested modulators (top) and reaction scheme of MOF synthesis (bottom).

Figure 1

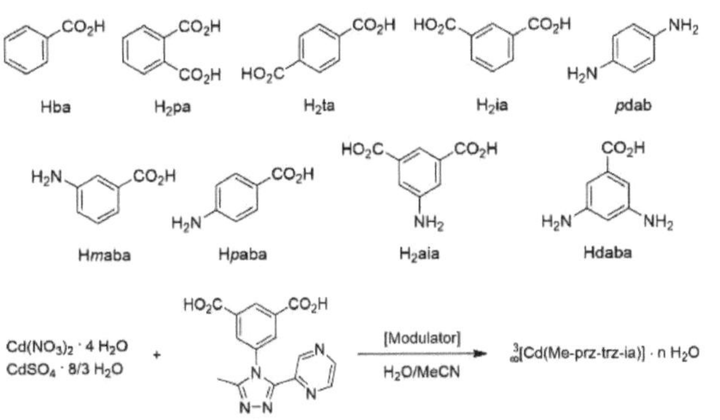

P077

Riddles of a novel Mn-polymer with a tetraazaadamantane-like ligand

D. Premužić[1], M. Hołyńska[1]
[1]Philipps-Universität Marburg, Fachbereich Chemie, Marburg, Germany

Symmetry plays an important role in the field of single-molecule magnets (SMMs), where axial symmetry of the system enhances the magnetic anisotropy.[1] Tripodal ligands can be used to impose a three-fold symmetry in their metal complexes. Tris(1-propan-2-onyl oxime)amine (Ox_3H_3) is an example of a tripodal oxime ligand with different known metal complexes, mainly containing Ni.[2] This ligand undergoes an unusual intramolecular cyclization leading to a 4,6,10-trihydroxy-1,4,6,10-tetraazaadamantane product.[3] In combination with Mn, Ox_3H_3 transforms *in situ* and yields different Mn^{IV}-dimers depending on the used additives. In order to improve the magnetic properties and obtain Mn^{III}-complexes with much higher intrinsic magnetic anisotropy, these complexes were reduced with KC_8. In the case of the azide-brigded complex a polymeric compound was synthesized (Fig. **1**). Herein we present the results obtained by various methods including NEXAFS, SQUID, and HF-EPR. The resulting data indicate a successful reduction, but the oxidation state of the Mn is still under discussion.

Fig. **1**: Molecular structure of the title polymeric compound (hydrogen omitted for clarity).

[1] T. Glaser, V. Hoeke, K. Gieb, J. Schnack, C. Schröder, P. Müller, *Coord. Chem. Rev.*, **2015**, *289-290*, 261-278. [2] A. N. Semakin, A. Yu. Sukhorukov, A. V. Lexiv, S. L. Ioffe, K. A. Lyssenko, Y. V. Nelyubina, V. A. Tartakovsky, *Org. Lett.*, **2009**, *11*, 4072-4075. [3] M. J. Goldcamp, S. E. Robison, J. A. Krause Bauer, M. J. Baldwin, *Inorg. Chem.*, **2002**, *41*, 2307-2309.

Figure 1

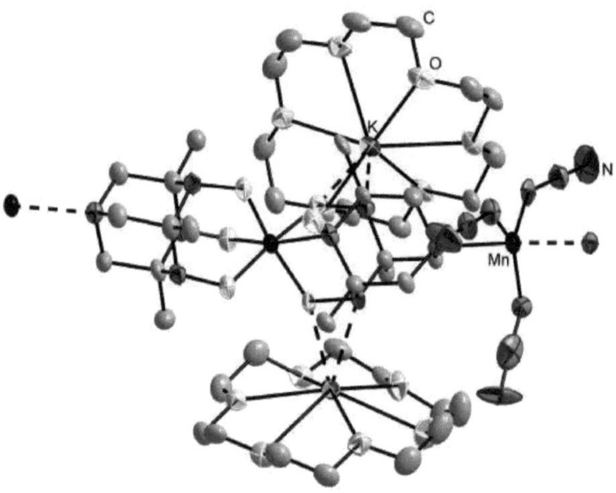

P078

Computational adsorption experiments at the (100)-pyrite-water interface: The influence of surface defects to the H_2O-adsorption model.

S. Meis[1], H. Gies[1], U. Magdans[2]
[1]*Ruhr-University Bochum, Bochum, Germany*
[2]*University of Potsdam, Potsdam, Germany*

Chemical reactions at the (100)-pyrite (FeS_2) surface play an important role in many environmental, geological and chemical processes, e.g. acid mine drainage, bacterial pyrite oxidation, oxidation in aquifers and coal liquefaction and heterogeneous catalysis. To understand these processes it is of essential importance to understand the adsorption properties of pyrite to organic molecules and water. The structures of the dry (100)-pyrite surface and the interface of the (100)-pyrite surface with water were determined with grazing incidence X-ray diffraction experiments (GIXRD) previously[1,2]. An adsorption model for water at the pyrite surface was determined including three adsorption layers, a transitional zone of partial ordered water molecules and water molecules that occupy defect sites of the topmost FeS_2-layer.

In this study the adsorption model of water at the (100)-pyrite surface near iron and sulphur defect sites is analysed. We present computational experiments, especially molecular dynamic simulations using the forcefield COMPASS. The simulations were carried out with the program package Materials Studio 5.0, particularly the program "Forcite", from Accelrys Inc. The surface cell structure determined previously from GIXRD experimental results was used as a start model for the simulations. A surface area of 3x3 surface cells (16,25Å²) and a depth of 2,5 surface cells (13,5Å) of pyrite was selected. A vacuum slab of 30Å was added as adsorption region to permit enough mobility of the H_2O-molecules during the dynamic runs. Different amounts of iron and sulphur defect sites were introduced to the topmost FeS_2-layers of the surface during different simulation runs. Each dynamic run was calculated for 200ps with a step width of 0.2fs.

The introduction of sulphur defects to the surface model has a stronger influence to the H_2O-adsorption model compared to the effect of iron defect sites. By introducing iron defects the adsorption model shows no differences compared to a surface free of defects. Only a small decrease in the distance between the topmost FeS_2-layer and the first H_2O-adsorption layer about 0.2-0.4Å was determined. The sulphur vacancies show a stronger influence, as H_2O-molecules could be observed in the topmost FeS_2-layers occupying the defect sites of sulphur dimers. Additional adsorbed H_2O-molecules occur between the topmost FeS_2-layer and the first H_2O-adsorption layer that is located 2.0(3)Å above the topmost FeS_2-layer. By adding iron and sulfur defects to the three topmost FeS_2-layers it was possible to obtain results in perfect agreement with the previously determined experimental model from GIXRD-data, with molecular dynamic simulations after Philpott et al.[3] and UHV-experiments carried out by Guevremont et al.[4].

References:

[1] Meis S, Magdans U, Torrelles X, Gies H. *Mineralogical Magazine* (2013), 77(5), 1733

[2] Meis S, Magdans U, Torrelles X. *Acta Cryst.* (2011) A67, C338-C339

[3] Philpott M R, Goliney I Y, Lin T T. *J. Chem. Phys.* (2004), 120(4), 1943-1950

[4] Guevremont J M, Strongin, Schoonen M. A. *Am. Mineral.* (1998), 83, 1246-1255

P079

A Racemic Solution Crystallizing as a Disordered Crystal Structure of Opposite Enantiomers: Pseudoracemates in a Sohncke and a Centrosymmetric Space Group

R. Herrmann[1], B. Braun[1], T. Braun[1]
[1]*Humboldt-Universität zu Berlin, Berlin, Germany*

The crystallization of a racemic solution results mainly in *racemic crystals* (heterochiral) or *conglomerates* (homochiral). The unit cell of a *racemic crystal* contains the (*S*) and the (*R*) enantiomer in an equal amount in a well-ordered manner. Furthermore, *conglomerates* can be obtained as (*S*) and (*R*) handed crystals as a result of spontaneous resolution. Additionally, a rare outcome of the crystallization of a racemic solution is the formation of *pseudoracemic crystals* (heterochiral), in which both the (*S*) and the (*R*) enantiomer are present in the crystal structure in a disordered manner.[1] A *pseudoracemate* is a solid solution of opposite enantiomers which can crystallize in any space group. Their existence was first formulated by *Kipping and Pope*.[2]

Figure 1: Various types of the crystallization of a racemic solution.

All three types of crystalline racemates differ in their melting point diagrams and properties. That is why their accurate characterization is crucial for the pharma industry in the context of chiral drug design. Due to the fact that in general only one of the enantiomers has biological activity, the other is inactive, inhibits the desired effect or is even toxic.

We were fortunate to come across the structure of (*R*)- and (*S*)-*exo*-2-[Pt(C_7H_{11}){Ge(2-$C_6H_4PPh_2)_3$}] which was investigated by X-ray diffraction (see Scheme 1).[3] The compound under discussion crystallizes as a *pseudoracemic* THF solvate in the monoclinic chiral space group $P2_1$ with a small excess of the (*R*)-*exo*-2 enantiomer. The asymmetric unit contains one independent molecule on a general position, whereas no other particularly interesting intra- or intermolecular interactions are present.

Scheme 1: (*R*)- and (*S*)-*exo*-2-[Pt(C_7H_{11}){Ge(2-$C_6H_4PPh_2)_3$}].

Particularly interesting and exceptionally rare is the fact that the norbornyl complex crystallizes as a *pseudoracemate* in a Sohnke space group, therefore giving a heterochiral crystal in a chiral space group.

1. H. D. Flack, *Helvetica Chimica Acta* **2003**, *86*, 905.

2. F. S. Kipping, W. J. Pope, *J. Chem. Soc. Trans.* **1897**, *71*, 989.

3. R. Herrmann, B. Braun, T. Braun, *manuscript in preparation.*

Figure 1

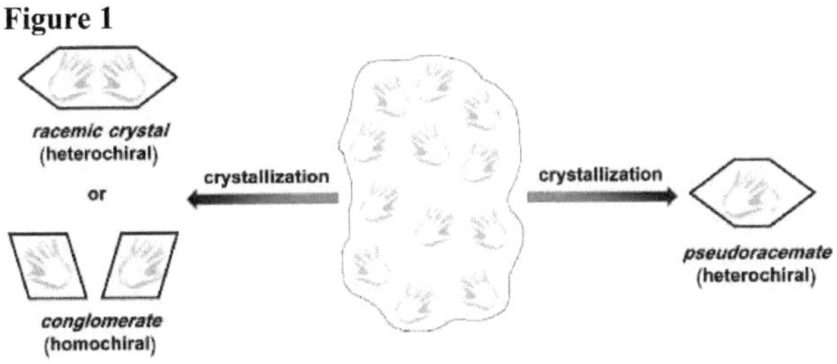

Figure 2

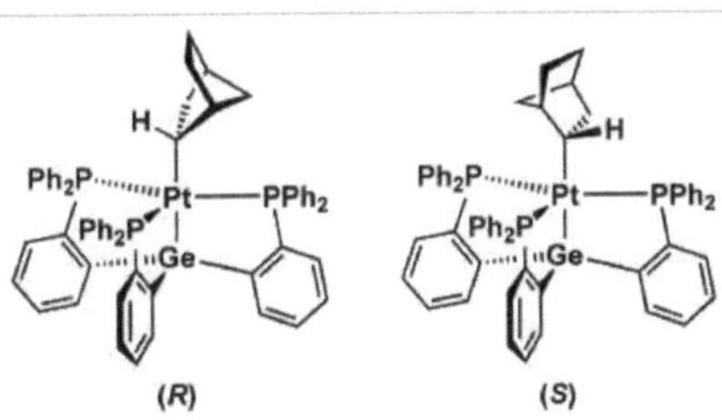

P080

Crystallographic and energetic evaluation tools for the prediction of ionic conduction of Al in oxides

F. Meutzner[1], T. Nestler[1], A. A. Kabanov[2], M. Zschornak[1], W. Münchgesang[1], T. Leisegang[1,2], V. A. Blatov[2], D. C. Meyer[1]
[1]*TU Bergakademie Freiberg, Institut für Experimentelle Physik, Freiberg, Germany*
[2]*Samara State Aerospace University, Samara, Russia, Russian Federation*

An ionic conduction in crystalline oxides is characterisable as jumps of mobile ions between interconnected crystallographic sites in oxygen-coordinated voids. The environment needs to allow passage between jump sites. Considering this theory of ionic conduction, crystallographically motivated evaluation tools can be utilised and combined into a high-throughput identification method for suitable structures for ionic conduction. An energetic consideration in the end allows an in-depth analysis of the ionic conduction process in order to verify the most-promising compounds.

Three approaches can be synergetically used to allow a time-efficient and consistent methodology for the identification of an arbitrarily chosen ion. Voronoi-Dirichlet partitioning (VDP) combined with data mining offers a fast high-throughput screening of crystallographic databases on the basis of a geometrical division of the crystal structure into domains assigned to each constituent of the structure, pointing towards possible void networks.

Bond-valence methodology (BVM) as a chemically motivated evaluation tool assists in the evaluation process as the second step, determining the most probable migration ways by calculating bond-valence sums for each point in the crystal structure. Bond-valence energy landscape calculations then allow for an energetic estimation of jump energies.

The most promising structures are fully electronically modelled by density functional theory (DFT) calculations *ab initio*. Calculating migration step energies subsequently enables a full assessment of jump energies from equilibrated structures as well as electronic conductivities and diffusion coefficients. As it is a very time-consuming method, DFT is applied as the final step.

This work presents the first results of the identification of promising oxidic Al-ion conductors involving all three approaches and their comparison.

This work is funded by the German BMBF (CryPhysConcept: 03EK3029A) and BMUB (BaSta: 0325563B).

P081

Disordered Intermetallic Phases visited by Solid-State NMR and Quantum Mechanical Calculations

B. Mausolf[1], F. Haarmann[1], V. Peters[1], A. Korthaus[1]
[1]*Institute of Inorganic Chemistry at RWTH Aachen University, Arbeitskreis für moderne Strukturanalytik, Aachen, Germany*

Structure-property relations became more and more important for materials and solid state scientists. This is particularly true for disordered materials and materials whose applicability is controlled by incorporated defects. Especially for intermetallic compounds the bonding situation and the atomic structure are in the focus.[1] A combination of XRD, NMR and DFT calculations turned out to be suitable for this purpose during a systematic study of binary intermetallics.[2] Recently, this approach was extended on disordered ternary phases.[3]

The various information about an intermetallic system which are available from quantum mechanical (QM) calculations are in the focus of this contribution. In the presence of a defect, the atomic structure in the vicinity reacts with a change of the atomic arrangement. The atomic shifts due to a structural optimization and along with this the changes of interatomic distances are valuable information which is neither accessible by diffraction methods nor by NMR alone.

Referring to NMR as a local probe, the electric field gradient (EFG) serves as sensitive experimental measure. QM calculations give the EFG resolved for each atom in the structure model and therefore are essential for the interpretation of the experimental results. In addition, the tensor orientation and the sign of the EFG are available by theoretical methods. This offers the possibility to interpret orientation dependent NMR experiments on single crystals or aligned powder samples. Furthermore, the calculated orbital contributions of the EFG give a hint for the change of the bonding situation due to incorporated defects.

Studying the details of the electronic structure leads to an deeper understanding of changes in the bonding situation. For this purpose we analyze the changes of the total, atomic and integrated density of states (DOS). In addition, we focus on theoretical derived formation energies and the efficient use of density functional theory (DFT) codes in this contribution.

We focus on Zintl-like intermetallic materials: These contain metals lying close to the Zintl boarder which form an anionic framework coordinated by alkali- and alkaline-earth metal cations.[4] The applied DFT codes VASP[5] and Wien2k[6] offer the necessary tools and implementations for the above mentioned investigations. For the generation of suitable structure models and super lattice structures we follow the Bärnighausen formalism based on group subgroup relations summarized in the International Tables for Crystallography.[7,8,9]

The system $Cu_{1-x}Al_2$, containing Cu vacancies[10] is currently investigated with our strategy. Details of the atomic and electronic structure in combination with Results from XRD and NMR experiments are presented. Using the high sensitivity of NMR in combination with DFT calculations, we present a systematic way to investigate disordered intermetallic materials.

Literature

[1]Yu. Grin, *Wiss. Z. Techn. Univers. Dresden*, **49**(1), 16, 2000.
[2]F. Haarmann, K. Koch, P. Jeglic, O. Pecher, H. Rosner, Y. Grin, *Chem. Eur. J.*, **17**, 7560 - 7568, 2011.
[3]O. Pecher, B. Mausolf, K. Lamberts, D. Oligschläger, C. Niewieszol, U. Englert, F. Haarmann, *Chem. Eur. J.*, **21**, 1-13, 2015.
[4]F. Laves, *Z. Nat.*, *17*, 244-255, 1941.
[5]G. Kresse, J. Furthmüller, *Phys. Rev. B*, **54**, 11169-11186, 1996.
[6]P. Blaha, K. Schwarz, G. K. H. Madsen, D. Kvasnicka, and J. Luitz, WIEN2k, Vienna *Univ. of Techn.*, 2001.
[7]H. Bärnighausen, MATCH, **9**, 1980.
[8]Int. Tables for Cryst. A, IUCr, 2006.
[9]Int. Tables for Cryst. A1, IUCr, 2006.
[10]F. Haarmann, M. Armbrüster, Yu. Grin, *Chem. Mater.*, **19**, 1147-1153, 2007.

P082

In situ high pressure single crystal X-ray diffraction study of clinoferrosilite

A. Pakhomova[1], L. Ismailova[1], M. Bykov[1], E. Bykova[1], T. Boffa Ballaran[1], L. Dubrovinsky[1]
[1]*Bayerisches Geoinstitut, Bayreuth, Germany*

Pyroxenes, $M1M2Si_2O_6$, are one of the major constituent minerals of the Earth's upper mantle and therefore they have an important influence on its mineralogy, thermodynamics and the geological structure. A well-constrained correlation between chemical compositions, crystal structures and elastic properties of pyroxenes is hence necessary for the modeling their behavior in the Earth's interior. In the present study we report high-pressure single-crystal diffraction (SC XRD) data of the pyroxene Fe end-member: clinoferrosilite (CFs). CFs is known to transform from $P2_1/c$ into $C2/c$ phase between 1.48 and 1.75 GPa (Hugh-Jones *et al.*, 1996), however little is known about its structural behavior at Earth's mantle pressures.

Single crystals of CFs synthesized at 9.5 GPa and 1100 °C were loaded together with a ruby sphere or an Au foil for pressure estimation into diamond anvil cells (DAC). Neon was used as pressure-transmitting medium. *In situ* high-pressure SC XRD experiments were performed at the synchrotron stations P02.2 at Petra III and ID09 of European Synchrotron Radiation Facility.

The $C2/c$ crystal structure of CFs consists of three crystallographically distinct polyhedra: two octahedra $Fe1O_6$, $Fe2O_6$ and a tetrahedron SiO_4. The tetrahedra are connected via a bridging O3 atom forming chains running along the *c* axis. The octahedra form edge-sharing chains parallel to the silicate chains. Tetrahedral and octahedral "layers" alternate along the *a* axis forming a three-dimensional crystal structure by sharing the O1 and O2 atoms of the SiO_4-..-SiO_4 chain. With increasing pressure the crystal structure accommodate the compression by kinking of the tetrahedral chains as reflected in the decrease of the O3-O3-O3 angle from 137.54 to 133.5° up to 30 GPa. Between 30 and 35.8 GPa CFs undergoes a first order phase transformation into a $P2_1/c$ phase associated with volume change of 6%. The $C2/c \rightarrow P2_1/c$ transition is induced by the rearrangement of half of the layers of corner-sharing SiO_4 tetrahedra: in these the Si atoms become octahedrally coordinated and a new type of layer is formed by polymerization of $Si2O_6$ polyhedra via common edges. The remaining chain of corner-sharing tetrahedra $Si1O_4$ is significantly more kinked in the $P2_1/c$ phase as indicated by the small value of the O3-O3-O3 angle of 128.7°.

The isothermal bulk moduli K_0 at zero pressure and its first pressure-derivative K_0' for the $C2/c$ phase were calculated using a third-order Birch-Murnaghan equation of state (BM EoS) and were found to be 113(3) and 6.3(3) GPa, respectively. The axis compressibilities reveal a strong anisotropy: the *a* axis is the stiffest while the *b*-axis is the softest.

Comparison between the structural compression of CFs and those of diopside and hedenbergite will be discussed in order to shed light on the influence of Ca and Mg substitutions on the elastic behavior of pyroxenes.

P083

Millisecond time resolved diffraction study of $SrCO_3$ at high pressures and temperatures

M. Stekiel[1], L. Bayarjargal[1], W. Morgenroth[1], R. Luchitskaia[1], B. Winkler[1]
[1]*Goethe University, Institute of Geoscience, Frankfurt am Main, Germany*

Strontium carbonate, $SrCO_3$, is isostructural to aragonite. At ambient pressure, $SrCO_3$ transforms at 1185 K to a disordered rhombohedral phase [1,2] (calcite-V structure). Upon further heating, $SrCO_3$ transforms into a cubic phase [3] (NaCl structure) at 1690 K. At ambient temperature, $SrCO_3$ undergoes a pressure induced phase transition, but the exact transition pressure is still a matter of discussion, since reported results are in the range of 10 to 35 GPa [4,5,6,7]. Rapoport & Pistorius [8] studied the aragonite-type to calcite-V type transition from ambient pressure up to 4 GPa, and found that the transition temperature increases with pressure up to 1500 K at 4 GPa.

In this study a previously unexplored p-T range of the $SrCO_3$ phase diagram was investigated. Synthetic $SrCO_3$ sample was grown by hydrothermal method.Powder diffraction experiments were performed at the Extreme Conditions Beamline P02.2 at PETRA III,Hamburg, inthe double-sided laser-heated diamond anvil cell (DAC). The sample was heated up to 1600 K at 8 GPa. Then, diffraction images were collected every 2.5 ms using the LAMBDA detector [9], while the two lasers were switched off. This allowed us to observe the time-dependence of the cooling process.

Figure 1 shows representative, time resolved diffractograms. At 8 GPa and 1600 K rhombohedral $SrCO_3$ was observed (t=0 ms), the reflections are labeled with "R" in the figure. Additionally, at these conditions, at least seven reflections (labeled "h") from a previously unknown phase were observed. Indexation gave an orthorhombic unit cell, and we tentatively assign these reflections to a new high (pressure, temperature) phase of $SrCO_3$. Other reflections are from NaCl (labeled "N") which was used as a pressure medium, and Pt and W metals ("M") which were employed as a heat absorber and gasket, respectively. The horizontal L1 and L2 lines indicate the time when the upstream and downstream laser of the double-sided laser heating system were turned off, respectively. The cooling process lasts a few milliseconds only. Surprisingly, upon rapid cooling at 8 GPa $SrCO_3$ does not transform back tothe aragonite structure, stable at ambient conditions, but to another, yet unknown phase with poor crystallinity ("a").

The study is supported byprojects DFG-ANR WI1232/41-1, and BMBF 05K13RF1.

[1] J. J. Lander, *J. Chem. Phys.*, vol. 17, pp. 892-901, 1949.
[2] K. Strømme, *Acta Chem. Scand. A*, vol. 29, p. 105, 1975.
[3] E. H. Baker, *J. Chem. Soc.*, pp. 2525-2526, 1962.
[4] S. Ono, M. Shirasaka, T. Kikegawa and Y. Ohishi, *Phys. Chem. Miner.*, vol. 32, pp. 8-12, 2005.
[5] S. Arapan and R. Ahuja, *Phys. Rev. B*, vol. 82, p. 184115, 2010.
[6] M. Wang, Q. Liu, S. Nie, B. Li, Y. Wu, J. Gao, X. Wei and X. Wu, *Phys. Chem. Minerals*, vol. 42, pp. 517-527, 2015.
[7] C.-C. Lin and C.-g. Liu, *Eur. J. Mineral.*, vol. 9, pp. 785-792, 1997.
[8] E. Rapoport and C. Pistorius, *J. Geophys. Res.*, vol. 72, pp. 6353-6357, 1967.
[9] T. Hatsui and H. Graafsma, *IUCrJ*, vol. 2, pp. 371-383, 2015.

Figure 1

Fig. 1

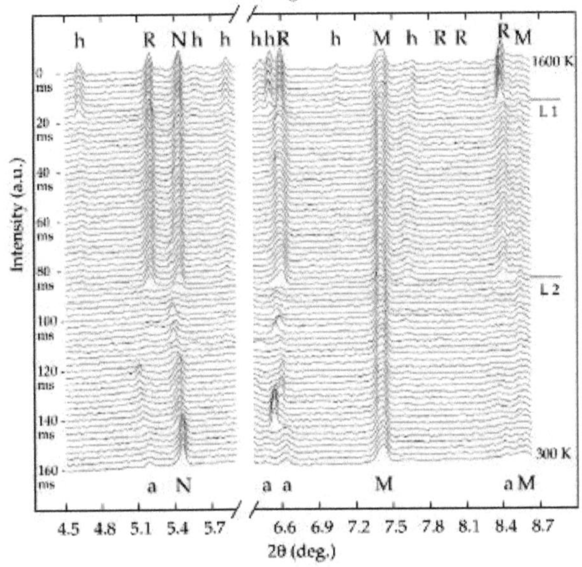

Computational and theoretical crystallography

P084
Delocalization indices and domain-averaged Fermi hole analysis for complex solids

P. Golub[1], A. Baranov[1]
[1]Technische Universität Dresden, Dresden, Germany

Quantum mechanics became a foundation for incessant development of versatile computational methods for analysis of chemical and physical properties of molecules and crystals. Real space methods for chemical bonding analysis, relying upon quantum mechanical observables, have gained much popularity during the last decades.[1] However the practical usage of the methods which belong to aforementioned family often requires a significant amount of time and computer resources. For example such powerful instruments as localization/delocalization indices[2] and domain-averaged Fermi hole orbitals[3] require the evaluation of the so called domain overlap matrices - which is a formidable task for complex and low-symmetry systems.

The elaboration and implementation of algorithm for evaluation of domain overlap matrices for DFT projected augmented-wave method[4] (PAW) is reported. It enables fast computation of delocalization indexes, which can be interpreted as covalent bond orders, and domain-averaged Fermi hole orbitals, providing quantitative and visual representation of chemical bonding in solids [5]. Several complex systems, which were almost inaccessible for such kind of analysis by now, were analyzed employing delocalization indices and domain-averaged Fermi hole orbitals in order to demonstrate the power of real space methods.

Fig. 1. Selected QTAIM basins and delocalization indices in crystalline $Nb_2(Se_2)_2(AlCL_4)_4$.[6]

References

[1] Bader, R. F. W. *Atoms in Molecules: A Quantum Theory*; Oxford University Press: Oxford, 1990.

[2] Bader, R. F. W.; Stephens, M. E. *J. Am. Chem. Soc.*, 1975, 97, 7391.

[3] Ponec, R. J. *Math. Chem.*, 1997, 21, 323. [4] Bloechl, P. E. *Phys. Rev. B*, 1994, 50, 17953.

[5] Baranov, A. I.; Ponec, R.; Kohout, M. in *Chemical Modelling (Eds. M. Springborg and J.-O. Joswig), Royal Society of Chemistry*, 2015, *vol. 12*, 53.

[6] Mueller, U.; Isaeva, A.; Ruck, M. *Z. Anorg. Allg.* Chem. 2014, 640, 1564.

Figure 1

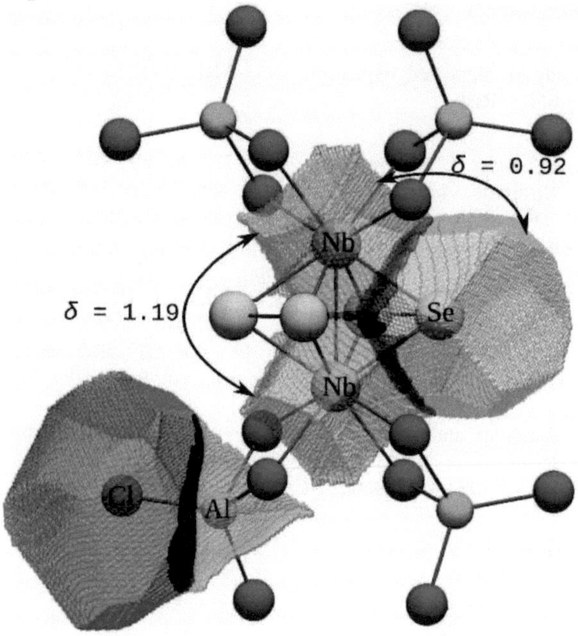

P085
Atomistic simulation study of ternary $(Ca,Sr,Ba)CO_3$ solid solutions

X. Liu[1,2], V. L. Vinograd[3], D. Bosbach[3], B. Winkler[1]
[1]Goethe Universität Frankfurt am Main, Institute of Geosciences, Frankfurt am Main, Germany
[2]Nanjing University, School of Earth Sciences and Engineering, Nanjing, China, Germany
[3]Forschungszentrum Jülich, Institute of Energy and Climate Research 6, Jülich, Germany

The long-term safety of the direct disposal of spent nuclear fuel in deep geological formations depends, in part, on the stability of secondary minerals, which could be formed at the reaction front between the waste and a host rock if water gets into contact with the waste. These minerals may bind the released radionuclides via the solid solution formation. Here we hypothesize that alkaline earth carbonates may become relevant when the content of Ba and Sr in waste is high and when sulfates cannot be formed due to reducing conditions. In this study the $(Ca,Sr,Ba)CO_3$ phase relations are simulated with the force-field model [1].

Experimental studies [2] show that at 823 K and 1 GPa the ternary orthorhombic phase coexists with trigonal (R-3m) disordered solid solutions. At lower temperatures monoclinic barytocalcite occupies the field of the trigonal phase. The aragonite-type solid solution is simulated with the double defect method [3]. A large ternary miscibility gap is predicted. The disordered calcite is modelled with the aid of quasi-random structures prepared from the R-3c calcite. Geometry optimization of these structures leads to a random rotation of CO_3 groups by ±30°. The transition to the disordered calcite is accompanied by an increase in the coordination number and by a decrease in the enthalpy. Thus, the stabilization of the trigonal phase could be promoted by a coupling of the cation disorder with the orientation disorder of the CO_3 groups. Our calculations suggest that at low temperatures barytocalcite, $BaCa(CO_3)_2$, is more stable relative to the trigonal phase and is marginally stable relative to the mixture of aragonite and witherite. Thus in a Ca-rich system the stable assemblage will likely be represented by aragonite and barytocalcite. Additionally, our ab initio calculations point out to an increased stability of the

Ra-analogue of barytocalcite, $RaCa(CO_3)_2$. Consequently, if barytocalcite can be formed at repository conditions, it will be able to efficiently immobilize Ra.

We acknowledge support from the German Federal Ministry of Education and Research (02NUK021) and from the China Scholarship Council.

[1] Raiteri et al. (2015) J. Phys. Chem. C 119: 244447-244458.
[2] Chang, L. (1971) Am. Mineral., 56: 1660-1673.
[3] Vinograd et al. (2009) Phys. Rev. B, 79: 104201-(1-9).

P086

The fast acquisition Laue Camera for neutrons FALCON at neutron source BER II

M. Tovar[1], G. Iles[2], S. Schorr[1]
[1]*Helmholtz-Zentrum Berlin, Berlin, Germany*
[2]*ANSTO, Bragg Institute, Sidney, Australia*

End of 2015 the FALCON Laue diffractometer at the Berlin neutron source BER-II started. It was developed in collaboration with the ILL, Grenoble. The diffractometer is designed for fast neutron data acquisition of single crystals and utilises a white ("pink") thermal neutron beam with wavelength band of about 1-3A. Pattern acquisition is performed by means of a backscattering as well as a transmission detector consisting of four iCCD (intensified charged coupled device) cameras each. The detectors cover an activ area of 400x400 square mm. Acquisition time is about one to three minutes. The sample is mounted on an irelec cradle allowing sample positioning and orienting in the neutron beam. The set-up of the instrument as well as first examples will be shown in detail.

Figure 1

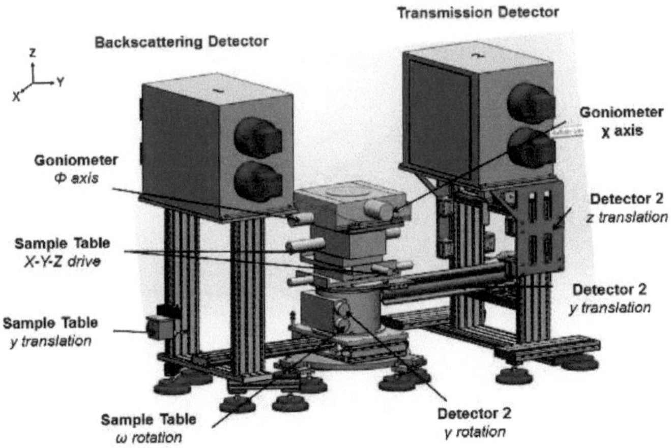

Figure 2

P087

Computational studies of binary and ternary systems and connection to actual synthesis routes

D. Zagorac[1], J. Zagorac[1], K. Doll[2], M. Rosic[1], M. Cebela[1], B. Matovic[1]
[1]*"Vinča" Institute of Nuclear Sciences, Department of Materials Science, Belgrade, Serbien und Montenegro*
[2]*University of Ulm, Institute for Electrochemistry, Ulm, Germany*

In this work, we would like to present computational studies of binary and ternary compounds and their connection to the actual synthesis routes. In the first part, we present two binary systems: lead sulphide (PbS) and barium sulphide (BaS) [1-3]. Each of the investigated compounds was optimized on *ab initio* level using Hartree-Fock, DFT-LDA and hybrid B3LYP functional and the results were in good agreement with the previous theoretical and experimental observations. Furthermore, we have calculated their properties, focusing on structural, electrical and vibrational properties, and we also had good agreement with experimental results [3]. At the end of the first part, we show some additional structure candidates as function of pressure and temperature not previously observed in BaS system and their connection to experimental synthesis.

In the second part, we present two ternary systems: $CaMnO_3$ and $BiFeO_3$. In the $CaMnO_3$ system we have performed a crystal structure prediction study focusing on the structures generated by octahedral tilting according to group-subgroup relations from the ideal perovskite type (), which is the aristotype of the experimentally known $CaMnO_3$ compound in the *Pnma* space group [4]. Furthermore, additional structure candidates have been obtained using data mining. Again, each of the structure candidates were optimized on *ab initio* level using Hartree-Fock, DFT-LDA and hybrid B3LYP functional, and the results obtained using the B3LYP functional showed the best agreement with the existing experimental data. Our results suggest that several of the low-energy structure candidates, mostly with orthorhombic or monoclinic symmetry, may be synthetically accessible at elevated temperatures and/or for slightly non-stoichiometric compositions [4,5]. Finally, we show the results for $BiFeO_3$ compound where we have further combined structure prediction and *ab initio* minimization data mining approach. With the results of this study we were able to understand more about the influence of temperature in $BiFeO_3$ system, and to connect our results to the actual synthesis routes and get additional crystal candidates.

[1] D. Zagorac, K. Doll, J. C. Schön, and M. Jansen, *Phys. Rev. B*, 2011, **84**, 045206.

[2] D. Zagorac, J. C. Schön, K. Doll, and M. Jansen, *Acta Phys. Pol. A,* 2011, **120**, 215.

[3] D. Zagorac, K. Doll, J. C. Schön, and M. Jansen, *Chem. Eur. J.*, 2012, **18**, 10929.

[4] J. Zagorac, D. Zagorac, A. Zarubica, J. C.Schön, K. Djuris, B. Matovic, *Acta Cryst.B,* 2014, **70**, 809.

[5] J. Zagorac, A. Zarubica, A. Radosavljevic Mihajlovic, D. Zagorac, B. Matovic, *Bull. Mater. Sci.*, 2014, **37**, 1.

P088

Electrostatic Potential of Dynamic Charge Densities

C. B. Hübschle[1], S. Van Smaalen[1]
[1]*Universität Bayreuth, Lehrstuhl für Kristallographie, Bayreuth, Germany*

Laboratory of Crystallography, University of Bayreuth, Germany

The computer program PRIOR, which is part of the BayMEM suite [1] is able to calculate a dynamic electron density from a multipole model [2]. Such a density or the result of a maximum entropy method (MEM) calculation can be used to calculate a dynamic electrostatic potential by a modified version of the PRIOR program.

The idea of the algorithm is based on the reciprocal part of an Ewald summation [3]. The application of the U_{ij} parameter to the nuclear density in the above equation improves the convergence of the summation so that a calculation in direct space is unnecessary, which make this approach quite fast.

We will show applications to DL-Serine at different temperatures.

Fig1: Electrostatic potential in the crystal of DL-Serine at 20K mapped on the dynamic electron density isosurface at 0.5 eÅ^{-3}

[1] L. Palatinus, S. van Smaalen and M. Schneider. *Acta Crystallogr.* (2003) **A59** 459-469.

[2] S. Mondal, S. J. Prathapa and S. van Smaalen. *Acta Crystallogr.* (2012) **A68** 568-581.

[3] P. Ewald (1921). “Die Berechnung optischer und elektrostatischer gitterpotentiale.” *Annals Phys.* **64**, 253-287.

Figure 1

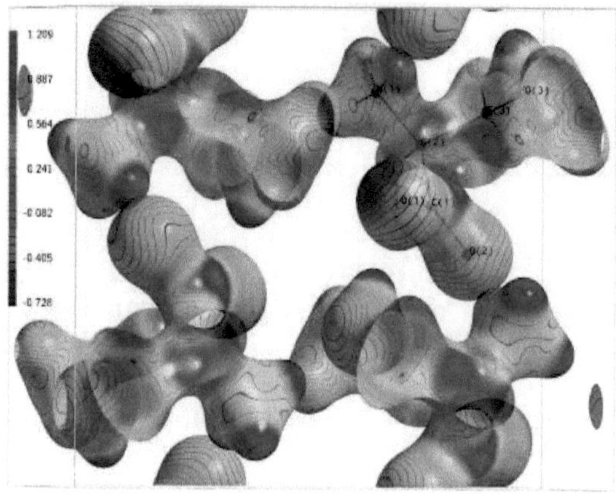

Crystallography, crystal chemistry, and application of layered materials

P089

Crystal Structure of the Ordered Double Perovskite $Sr_2Mn_{1-x}Ni_xTeO_6$

A. zaraq[1]
[1]university hassan II Casablanca, chimestry, casablanca, Morocco

$Sr_2Mn_{1-x}Ni_xTeO_6$ (x=0,25; x=0,5; x=0,75) was synthesized by the solid-state reaction ,the vibrational phonon modes was investigated using optical spectroscopic techniques, for the first time. X-ray diffraction (XRD), Raman and infrared spectroscopies were employed to investigate the structures of these perovskite materials and the results analysed together with group-theoretical predictions. The number and behaviour of the first-order modes observed in both spectroscopic techniques are in agreement with the calculations for a monoclinic I 2/m space group.

Keywords: double perovskites; X-ray diffraction, Raman spectroscopy,infrared spectroscopy and crystal structure.

P090

Structural Characterization of the Two $PbSb_{0.5}Fe_{1.5}(PO_4)_3$ and $Pb_{0.5}SbFe(PO_4)_3$ Phosphates with Nasicon type-structure

A. MARCHOUD[1]
[1]University Hassan II faculty of sciences ben m'sik Casablanca, chimestry, Casablanca, Morocco

the structural characteristics by X-ray powder diffraction (XRD) study using the Rietveld method for $MSb_{0.5}Fe_{1.5}(PO_4)_3$ and $M_{0.5}SbFe(PO_4)_3$ (M = Ca; Sr) phases were realised . All materials exhibit the Nasicon-type structure. In a continuation of our search concerning phosphates with interesting physical applications, the principal objective of the present study was the structural determination, using Rietveld refinement of the XRD patterns, of the two newly synthesised $PbSb_{0.5}Fe_{1.5}(PO_4)_3$ and $Pb_{0.5}SbFe(PO_4)_3$ phosphates. Structure of both phosphates, obtained by conventional solid state reaction techniques, was determined at room temperature from (XRD) using the Rietveld analysis. Results of the structure refinement show that both materials exhibit the Nasicon-type structure. Hexagonal unit cell parameter values are: a = 8.314(1) Å and c = 23.00(1) Å for $PbSb_{0.5}Fe_{1.5}(PO_4)_3$ and a = 8.237(1) Å and c = 22.773(1) Å for $Pb_{0.5}SbFe(PO_4)_3$. Final results of the Rietveld refinement leads to acceptable reliability factors (e.g., R_{wp} = 8.2%; R_B = 4.9% for $PbSb_{0.5}Fe_{1.5}(PO_4)_3$ (S.G.: R-3c) and R_{wp} = 9.0%; R_B = 3.6% for $Pb_{0.5}SbFe(PO_4)_3$) (S.G.:R-3). Both structures are built up from a corner-sharing of $[Fe(Sb)O_6]$ octahedra and $[PO_4]$ tetrahedra. Close structural examination shows clearly a quasi-ordered distribution of Sb^{5+} and Fe^{3+} ions within the $Pb_{0.5}SbFe(PO_4)_3$ Nasicon framework. In $PbSb_{0.5}Fe_{1.5}(PO_4)_3$, the Sb^{5+} and Fe^{3+} ions are statistically distributed within the framework.

Keys Words: Rietveld refinements, Nasicon, Iron and antimony phosphates

P091

Structural features of partially unprotected Aib containing peptides

R. Gessmann[1], H. Brückner[2], K. Petratos[1]
[1]IMBB-FORTH, Iraklion, Greece
[2]University, Giessen, Germany

Introduction and Objectives

The achiral residue α-aminoisobutyric acid (Aib) is an essential residue in peptaibiotics (nonribosomally biosynthesized antibiotic peptides of fungal origin). The conformational space of Aib is severely restricted by the second methyl group attached to the Cα atom resulting in α- or 3_{10}-helical structures for the peptaibols and usually for Aib-containing peptides. Most of the crystal structures containing Aib are N- and C-terminal protected. Deprotection of the N- or C-terminus of peptides may alter the hydrogen-bonding scheme and the structure facilitating thus crystallization. To investigate this effect, we have solved and refined the crystal structures of H-Gly-Aib-Gly-Aib-O*t*Bu and Z-$(Aib)_{10}$-OH.

Materials and Methods

Diffraction data from single crystals of H-Gly-Aib-Gly-Aib-OtBu and Z-$(Aib)_{10}$-OH were collected at 100 K on the microfocus beamline I24 of the Diamond Light Source in Didcot, England at a resolution of 0.72 and 0.65Å, respectively.

Results

Crystal structure of H-Gly-Aib-Gly-Aib-O*t*Bu

The peptide could form an intramolecular hydrogen bond, which is found in all tetrapeptides and N-terminal-protected tripeptides containing Aib with one exception. Here, this hydrogen bond is not observed (distance N4-O1 = 5.88Å). Instead, every molecule is hydrogen bonded to six other symmetry-related molecules with a total of eight hydrogen bonds per molecule (Fig. 1).

Crystal structure of Z-$(Aib)_{10}$-OH

Both molecules in the asymmetric unit form regular left-and right handed 3_{10}-helices. In contrast to the crystal structures of all to date Aib-homopeptides, which are all C-terminal OMe or O*t*Bu protected, here the helical sense of the last residue is not reversed. C-terminal OH-groups form hydrogen bonds to the residues i-3, in part mediated by water molecules (Fig. 2).

Conclusion

Both structures were dramatically changed by deprotection in comparison with the protected peptides. Crystallization of partially protected peptides is facilitated by the possibility to form additional intermolecular hydrogen bonds as described for both peptides. This makes the deprotection of peptides a method of choice in the cases, where a fully protected peptide resists forming crystals suitable for X-ray analysis.

Figure 1

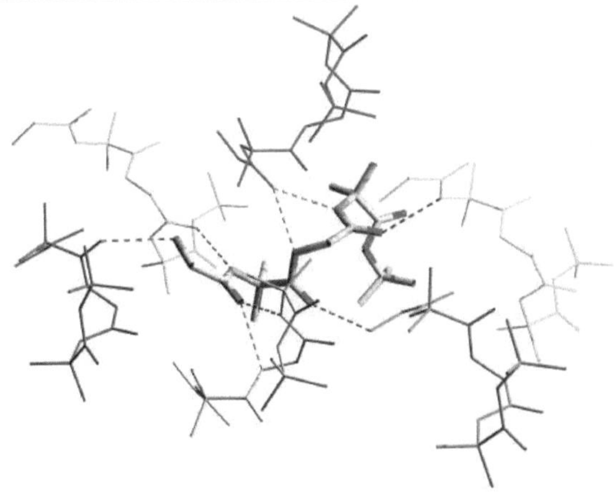

Figure 2

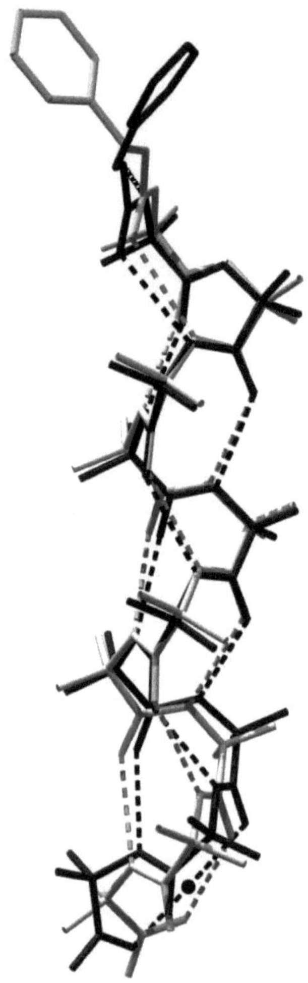

P092

Determination of high pressure and high temperature stability of α- and β-$MgSO_4$ by Raman spectroscopy

M. Herrmann[1], B. Winkler[1], J. D. Bauer[1], L. Bayarjargal[1], K. Refson[2], V. Milman[3]
[1] *Geosience, Crystallography, Frankfurt/Main, Germany*
[2] *Physics, Theoretical Physics, London, Great Britain*
[3] *Physics, Metal physics, Cambridge, Great Britain*

Magnesium sulfate and its hydrates may be important phases on the Martian surface and on Jupiter`s moon Ganymede [1,2]. Three polymorphs of anhydrous magnesium sulfate at ambient pressure and temperatures between 4.2-1300K are known [3,4]. While the crystal structures of α- and β-$MgSO_4$ were solved by Rentzeperis and Soldatos [3] and Weil [4], the crystal structure of γ-$MgSO_4$ is unknown. α- and β-$MgSO_4$ have been found to coexist below 868K [5]. The influence of pressure on the stability of magnesium sulfates is unexplored. Only the pressure dependence of the lattice parameters of α-$MgSO_4$ have been determined by Fortes and Wood [6] at 2.6(2) and 6.4(2)GPa. Wang *et al.* [2] have measured Raman spectra of several magnesium sulfate hydrates and β-$MgSO_4$. However, no Raman data for α-$MgSO_4$ have been published.

Here, we have determined the high pressure and high temperature stability of α- and β-$MgSO_4$ up to 12.7(1)GPa by using in situ Raman spectroscopy in a diamond anvil cell. To investigate the high temperature behavior of α- and β-$MgSO_4$, we have heated a mixture of both phases up to 673K and analysed quenched samples by Raman spectroscopy. We have carried out DFT calculations for a quantitative analysis of our spectroscopic data [7]. The excellent agreement between the experimentally observed and computed Raman spectra (fig. 1) allows an unambiguous assignment of all observed modes.

Our experiments provided new constraints with respect to the stability fields of α- and β-$MgSO_4$. At ambient conditions β-$MgSO_4$ is stable while α-$MgSO_4$ is metastable. We observed a pressure-induced phase transition from β- to α-$MgSO_4$ at 3.4(1)GPa and 298K. The α-phase is stable up to at least 12.7(1)GPa. At ambient pressure and high temperatures the temperature-induced phase transition from α- to β-$MgSO_4$ occurs between 523-573K. β-$MgSO_4$ is stable up to at least 673K.

Fig. 1. Raman spectra of a mixture of α- and β-$MgSO_4$ quenched from high temperatures.

[1] Nakamura, R. and Ohtani, E. Icarus, 211:648-654, (2011).

[2] Wang, A. *et al.*. Geochim. et Cosmochim Acta, 70:6118-6135, (2006).

[3] Rentzeperis, P. J. and Soldatos, C. T.. Acta Cryst., 11:686-688, (1958).

[4] Weil, M.. Acta Cryst., E63, i172, (2007).

[5] Yamaguchi, A. and Kato, E.. Yogyo Kyokai Shi, 80:337-339, (1972).

[6] Fortes, A. D. and Wood, I. G.. ISIS report, 510036, (2006).

[7] Refson, K., Tulip, P. R. and Clark, S. J.. Phys. Rev. B., 73: 155114, (2006).

Financial support from the DFG (grant WI1232) is gratefully acknowledged.

Figure 1

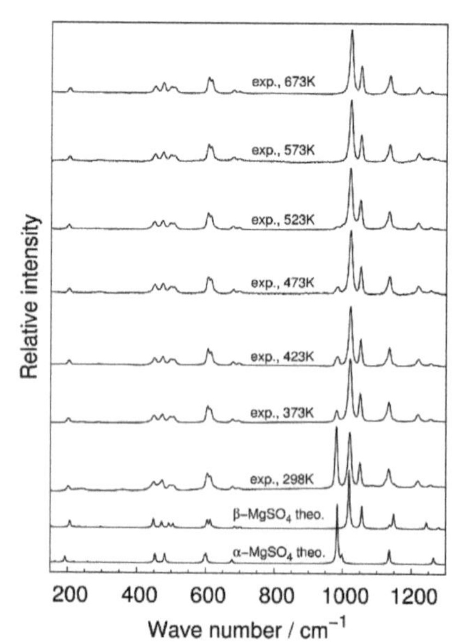

P093

Crystal structure of an anhydrous zinc hydroxide sulfate

L. S. Germann[1], R. E. Dinnebier[1], X. Liu[2], Y. Dong[2], W. Li[2]
[1]Max Planck Institute for Solid State Research, Stuttgart, Germany
[2]Chinese Academy of Science, Qinghai Institute of Salt Lakes, Xi'ning, China

Zinc hydroxide sulfates (ZHS) are ternary compounds with the general formula a $Zn(OH)2 \cdot$ b $ZnSO4 \cdot$ c $H2O$ (abbreviated as abc with a,b $\geq$ 1). The interest in these materials arises due to their possible use as precursors for gas sensors, pigments and fertilizers e.g. Although zinc hydroxide sulfates were intensively studied in the last century, only a few crystal structures are known from single crystal X-ray diffraction studies.

The crystal structure of a previously unknown anhydrous zinc hydroxide sulfate (520 phase) has been determined by high resolution laboratory X-ray powder diffraction (XRPD). It consists of brucite type Zn(OH)2 layers where every 6th octahedral site is vacant. Tetrahedrally coordinated Zn-ions are located above and below these vacancies. Consecutive layers are bridged by sulfate molecules. High temperature XRPD and thermal analysis revealed a multi-phase thermal decomposition with ZnO as final product.

Figure 1 - Left: Crystal structure of 5[Zn(OH)2]·2[ZnSO4] along the a-axis at ambient conditions. Right: measured scattered intensities plotted versus scattering angle 2θ.

Figure 1

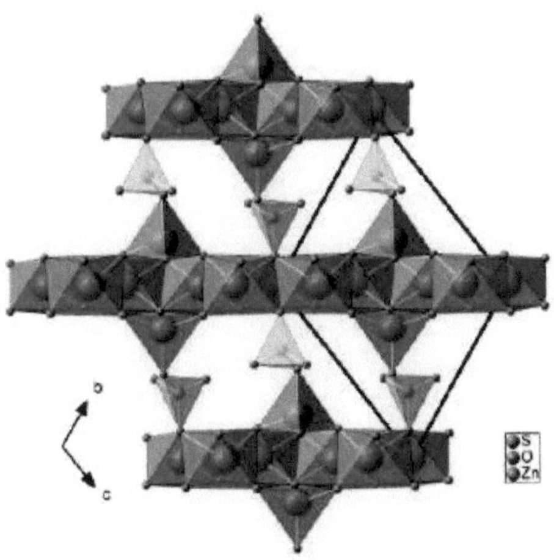

Figure 2

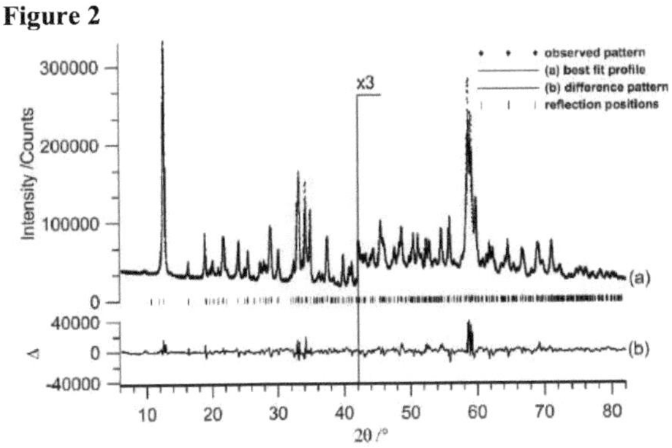

P094

High-pressure synthesis, refractive indices, and long-term structural stability of trigonal B_2O_3 (I) and the polarizability of ^{III}B

M. Burianek[1], J. Birkenstock[1], P. Mair[2], V. Kahlenberg[2], O. Medenbach[3], R. D. Shannon[4], R. X. Fischer[1]
[1]Universität Bremen, FB5-Geowissenschaften, Bremen, Germany
[2]Univ. Innsbruck, Faculty of Geo- and Atmospheric Sciences, Institute of Mineralogy and Petrography, Innsbruck, Austria
[3]Univ. Bochum, Institut für Mineralogie, Bochum, Germany
[4]Univ. Colorado, Geological Sciences/ CIRES, Boulder, CO, Germany

Trigonal B_2O_3 (B_2O_3-I, space group $P3_121$ or $P3_221$) is the stable modification of B_2O_3 at ambient conditions. Refractive indices were determined more than 70 years ago on a 50 µm crystal [1] with hitherto unknown reliability as B_2O_3 is known to be hygroscopic. Accordingly, we redetermined the refractive indices and investigated the crystalline state of the sample.

B_2O_3-I was synthesized in a sealed Pt capsule placed in a piston cylinder apparatus using a pelleted precursor formed by thermal decomposition of H_3BO_3 (Merck, 99.8%). Experiment conditions were 800°C and 1 GPa for 4 h with isobaric cooling of 15°C/h to 500°C within 20 h and followed by quenching at ~30°C/s. After opening the capsules contact to humid air was prevented by immediate embedding in anhydrous oil. Crystals of up to 1 mm were recovered from the sample. Refractive indices were redetermined by the immersion method using a microrefractometer spindle stage yielding n_o = 1.653(3) and n_e = 1.632(3) which corresponds to a total polarizability for B_2O_3 of α = 4.88 Å^3. This value was used to determine the electronic polarizability $\alpha(B)$ = 0.16 Å^3 for 3-coordinated B^{3+}. Because the B polarizability is very small, practically the entire polarizability of B_2O_3 derives from the oxygen ion. This allows a good estimate of $\alpha(O^{2-})$ = 1.52 Å^3.

Single crystal diffraction analyses confirmed previous results [2], no matter if we used crystals with an oil film or if they were oil-free (washed with hexane). Repeated datasets were taken for both types over several months. Surprisingly, although we observed a slight change in the optical behavior in terms of a change in transparency immediately after exposure to air (see fig. 1), we did not find any change of crystal-structure parameters. Even the scale factor of the diffraction data essentially remained unchanged. Accordingly, although B_2O_3 is known to be hygroscopic, our results indicate no change in its crystallinity.

Ref.:

[1] Kracek FC, Morey GW, Merwin HE (1938). Amer J Sci 35A, 143-171
[2] Effenberger H, Lengauer CL, Parthé E (2001). Monatshefte für Chemie 132, 1515-1517

Fig. 1: Changes in transparency of B2O3-I crystal. Micrographs taken
a) 5s, b) 10s, c) 20s, d) several minutes after exposure to air.

Figure 1

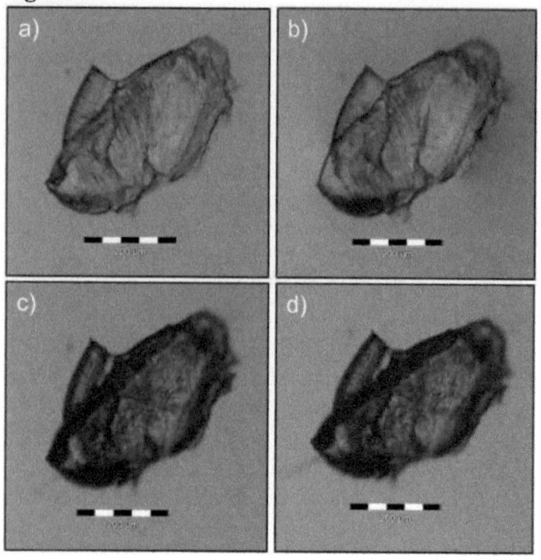

P095

Structure and Properties of ITQ-8, a Hydrous Layer Silicate with Porous Silicate Layers

B. Marler[1], M. Müller[1], H. Gies[1]
[1]*Ruhr University Bochum, Geology, Mineralogy and Geophysics, Bochum, Germany*

Introduction: Hydrous Layer Silicates (**HLS**s) are very interesting compounds since these layered silicates can be used as precursors to produce microporous and mesoporous framework materials by either condensing the silicate layers or by expanding the inter-layer region by covalently bonded linkers. To synthesize HLSs, cations of low charge density which have a strong influence on the layer type being formed, are used as the structure directing agents (SDAs).

Objectives: The knowledge of the crystal structure is of fundamental importance to recognize possible applications of these materials and to develop improved synthesis strategies.

Materials & Methods: The ITQ-8 sample which was kindly provided by Miguel A. Camblor (ICMM-CSIC, Spain) [1] was characterized by SEM, chem. analysis NMR and FTIR spectroscopy. XRD data were recorded from a Siemens D5000 powder diffractometer in modified Debye-Scherrer geometry (λ = 1.5406 Å). The crystal structure was solved by Simulated Annealing (TOPAS 4.2) and subsequently refined using the FullProf 2K program.

Results: ITQ-8 consists of small, colourless, plate-like crystals. The physico-chemical characterization confirmed that ITQ-8 is a hydrous layer silicate. Except for one very weak reflexion the XRD powder pattern could be indexed in monoclinic symmetry with a_0 = 16.032 Å, b_0 = 13.392 Å, c_0 = 17.249 Å and b = 99.69°. Including the additional reflexion, the indexing led to a unit cell of doubled volume with a_0 = 35.509 Å, b_0 = 13.390 Å, c_0 = 16.029 Å, b = 106.74° and possible space group symmetries *Cc* and *C2/c*. The crystal structure was solved using the small unit cell and space group *C2*. A subsequent Rietveld refinement (FullProf 2014) of the structure model converged to residual values of R_{Bragg} = 0.029, R_F = 0.025 and proved that ITQ-8 crystallizes in *C2/c* (larger unit cell). The chemical composition is $[(C_4H_8(C_7H_{13}N)_2]_8$ $[Si_{64}O_{128}(OH)_{16}]$ • 48 H_2O. The structure of ITQ-8 contains silicate layers made up by four-, six-, double six- and eight-rings forming a puckered sheet of interconnected cup-like half-cages which point alternatively up and down. The layer possesses a unique topology and represents a cut through the silicate framework of zeolite levyne. Unlike other HLSs which possess dense silicate layers, ITQ-8 contains 8-rings with a free diameter of 3.5 * 3.4 Å and can be regarded as a "small-pore layer silicate" (Fig. 1). The organic cations, 1,4-diquiniclidiniumbutane, used as SDAs during synthesis are intercalated between the silicate layers. Bands of water molecules which are hydrogen bonded to each other and to the terminal Si-OH/Si-O- groups are located between the org. cations and interconnect the silicate layers (Fig. 2).

Conclusion: The layer topology of IQ-8 is unique among layer silicates. Because of its micro porosity, ITQ-8 is a possible precursor for the synthesis (by topotactic condensation or interlayer expansion) of framework silicates with 3D micro-pore systems which may be useful in applications as e.g. catalysts, catalyst supports or adsorbents.

References: [1] Díaz-Cabañas, M.A. Camblor, Z. Liu, T. Ohsuna, O. Terasaki: Zeolite syntheses using linear diquats of varying length in fluoride media. The synthesis of ITQ-8, ITQ-10, ITQ-14 and high silica Nu-87, J. Mater. Chem., 2002, 12, 249-257.

Fig. 1: The lev type (levyne) layer

Fig. 2: The structure of ITQ-8

Figure 1

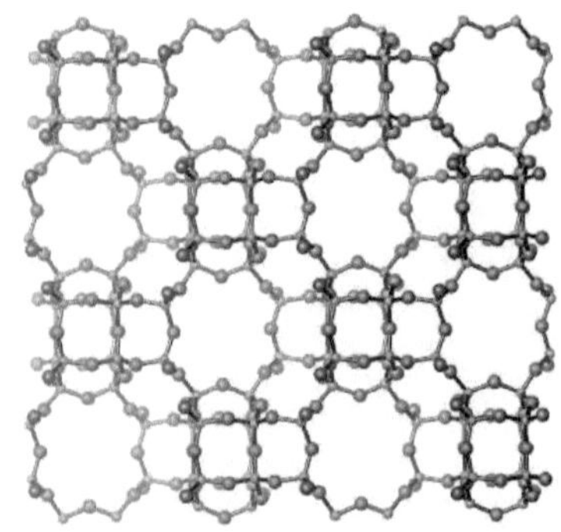

Figure 2

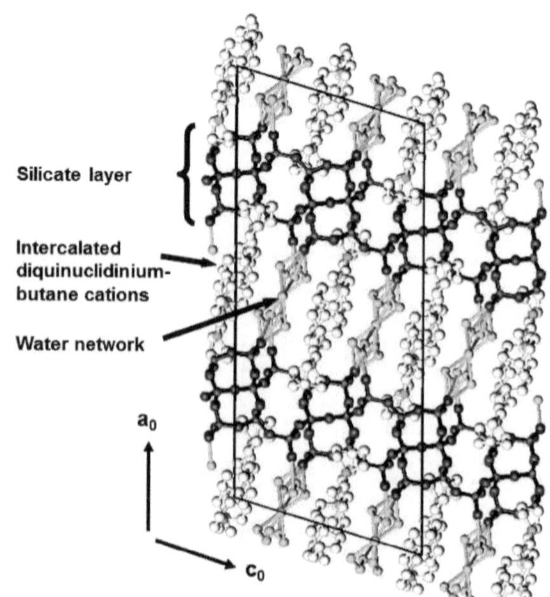

P096

Structure Eludication of the B-Type Lanthanoid Oxide Tellurides Ln_2O_2Te for the Example of Tm_2O_2Te

A. H. Geyer[1], S. Strobel[1], T. Runčevski[2], M. Etter[2], R. E. Dinnebier[2], T. Schleid[1]
[1]*University of Stuttgart, Institute for Inorganic Chemistry, Stuttgart, Germany*
[2]*Max Planck Institute for Solid State Research, Stuttgart, Germany*

The structural properties of the heavy lanthanoid(III) oxide tellurides with the formula Ln_2O_2Te were successfully eludicated for the example of Tm_2O_2Te. Synthesis was carried out within a few days by reaction of the elemental lanthanoids (*Ln* = Ho - Lu) with tellurium dioxide (TeO_2) in evacuated silica ampoules at 750 °C. Investigations on single crystals of Tm_2O_2Te (orthorhombic, *Cmcm; a* = 381.68(2), *b* = 1273.81(6), *c* = 402.06(2) pm; *Z* = 2) turned out futile, so other characterization methods such as X-ray powder diffraction needed to be used. The crystal structure of the thulium oxide telluride Tm_2O_2Te (orthorhombic, *Cmcm; a* = 763.65(2), *b* = 2548.57(9), *c* = 402.13(1) pm; *Z* = 8) was determined by *Rietveld* refinement from synchrotron data and differs from that of the lighter lanthanoid compounds Ln_2O_2Te (*Ln* = La - Nd, Sm - Ho), which crystallize in the A-type structure (tetragonal, *I*4/*mmm*, *Z* = 2) [1] with Nd_2O_2Te as prototype [2]. Both structure types contain layers of edge-sharing $[OLn_4]^{10+}$ tetrahedra with four connecting edges (e) as shown in Figure 1. These cationic sheets of 2D-$\{[OLn_{4/4}^{e}]^{+}\}$ are held together by layers of telluride anions (Te^{2-}) for charge compensation. This results for the A-type structure in eightfold coordination for the Ln^{3+} and Te^{2-} ions. For Tm_2O_2Te with the B-type arrangement [3], both kinds of heavy ions are only sevenfold coordinated. To achieve the smaller coordination spheres, the positions for the Te^{2-} anions have to be half-occupied completely in the smaller and for half of them in the larger orthorhombic *Cmcm* cell (Figure 2). But there are also two monoclinic $P2_1/m$ cells possible: A larger one (*a* = 763.70(3), *b* = 402.14(2), *c* = 1329.84(6) pm, *β* = 106.610(3)°; *Z* = 4) without any need for half-occupation of the Te^{2-} sites, and a smaller one with half the *c* axis (*a* = 763.70(3), *b* = 402.14(2), *c* = 664.92(3) pm, *β* = 106.610(3)°; *Z* = 2), which turned out to be just slightly less valid for the description of this problematic crystal structure.

References

[1] F. A. Weber, Th. Schleid, *Z. Anorg. Allg. Chem.* **1999**, *625*, 1833-1838.

[2] P. M. Raccah, J. M. Longo, H. A. Eick, *Inorg. Chem.* **1967**, *6*, 1471-1473.

[3] M. P. Pardo, J. Flahaut, *Compt. Rend. Acad. Sci. Paris* **1967**, *C 265*, 1254-1256.

Figure 1: Cationic sheets 2D-$\{[OLn_{4/4}^{e}]^{+}\}$ in the A-type (*top*) and the B-type Ln_2O_2Te structure (*bottom*).

Figure 2: The problem of the coordination spheres around the Te^{2-} anions in various possibilities for the B-type structure of Tm_2O_2Te.

Figure 1

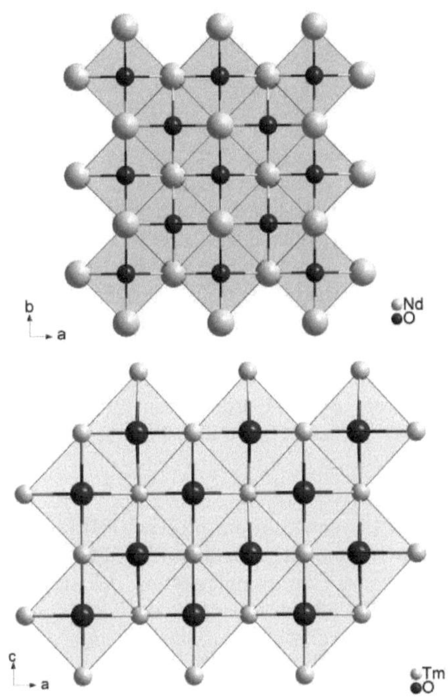

Figure 2

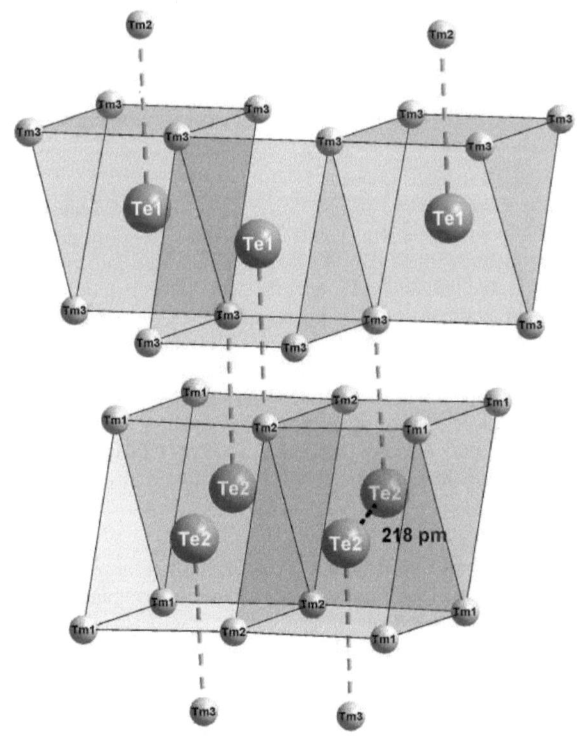

P097

The Structure of Dehydrated PREFER, $(C_9H_{22}N_2)_4[Si_{36}O_{76}]$, the Layered Silicate Precursor of Silica-Ferrierite

M. Müller[1], B. Marler[1]
[1]*Ruhr-Universität Bochum, Bochum, Germany*

Introduction

"PREFER" (precursor of ferrierite) is a hydrous layer silicate which has frequently been used as a precursor to synthesize porous silicates like zeolite ferrierite [1], delaminated ITQ-6 [2] or pillared silicate ITQ-36 [3]. The synthesis and properties of as-made PREFER, $(C_9H_{22}N_2)_4[Si_{36}O_{76}]\cdot 16H_20$, were first described in 1995 [1]. The physico-chemical characterization revealed that PREFER contains fer type silicate layers. However, no structure analysis could be performed, up to the present day, due to ambiguous space group symmetry and a very complex arrangement of disordered guest molecules and water molecules in the inter-layer region.

Objectives

For the understanding of the condensation process (PREFER => ferrierite) a detailed knowledge of the structure of PREFER is very helpful. In order to decrease the complexity of the structure solution process, as-made PREFER was carefully dehydrated while maintaining a good to moderate crystallinity of the material.

Materials & Methods

Dehydrated PREFER was obtained by heating the as-made material in an open glass capillary at 150 °C. *In-situ* X-ray powder diffraction data were recorded from a Siemens D5000 powder diffractometer in modified Debye-Scherrer geometry (λ = 1.5406 Å). The structure of Dehydrated PREFER was determined by simulated annealing (TOPAS 4.2): the fer type layers were fixed at the appropriate position with respect to space group symmetry and unit cell dimensions while the guest molecules were allowed to move freely within the structure as rigid bodies.

Results and Conclusions

The XRD powder pattern could be indexed in orthorhombic symmetry with a_0 = 14.026 Å, b_0 = 26.286 Å, c_0 = 7.426 Å. Surprisingly, the unit cell volume of the dehydrated material hardly decreased in comparison to the hydrous PREFER (a_0 = 7.427 Å, c_0 = 13.983 Å, b_0 = 26.334 Å, β = 90.28°). The analysis of existing reflections led to possible space group symmetries *Pnnm* and *Pnn*2. The structure model was then refined based on *Pnn*2 using Fullprof 2K. Because of anisotropic peak half-widths the Rietveld refinement did not converge to a perfect fit (R_F = 0.044, chi^2 = 5.2) but led, nevertheless, to meaningful interatomic distances and angles.

PREFER possesses two different types of voids to accommodate the organic cations (see Figure 1). Each void is occupied by one org. cation which, however, is statistically disordered about two orientations (related to each other by a 2-fold axis along [001]). In addition to the disordered cation, a slight disorder in the arrangement of layers seems also to be present.

There are neither intra-layer nor inter-layer hydrogen bonds between the terminal oxygen atoms of neighboring silicate layer (O-O-distance: 3.6 Å), instead, hydrogen bonds exist between the silanol groups and the amino groups of the organic cation. These bonds contribute only little to the stability of the structure, heating the sample from 150 °C to 160 °C leads to a considerable decrease of structural order.

[1] L. Schreyeck, Ph. Caullet, J.-Ch. Mougenel, J.-L. Guth, B. Marler: J. Chem. Soc., Chem. Commun. 21, 2187 (1995).
[2] A. Corma, U. Diaz, M.E. Domines, V. Fornés: Angew. Chem. Int. Ed. Engl. 38, 1499 (2000).
[3] A. Chica. A. Corma V. Fornés, U. Diaz, US Patent 6,555,090 (2003).

Figure 1: The structure of Dehydrated PREFER projected on (001).

Figure 1

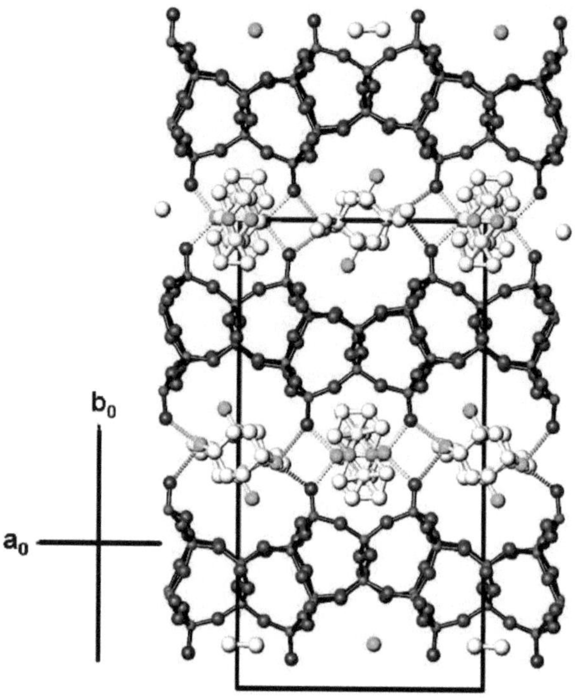

P098

In-situ neutron radiography during tensile testing of bentonite-bonded moulding sand

K. Schiebel[1], G. Jordan[1], W. W. Schmahl[1], A. Kaestner[2], B. Schillinger[3], S. Boehnke[4]
[1]*LMU München, Dept. f. Geo- und Umweltwiss., München, Germany*
[2]*PSI, Villigen, Schweiz, Switzerland*
[3]*FRM II, Garching, Germany*
[4]*S&B Industrial Minerals, Marl, Germany*

Almost any metal part can be casted in bentonite-bonded sand moulds. Bentonite mainly consists of smectite. Depending on the water content, smectite plasticizes and stabilizes the mould. During casting, the sand moisture changes significantly and a temperature gradient develops. Hot sand areas dehydrate. Pore water and smectite interlayer water evaporate. Steam diffuses into colder areas of the mould where it condensates. The high water content in the condensation zone is assumed to decrease the tensile strength of the mould. Additionally, the phase transition of quartz increases the stress in the hot areas. Together, higher stress and lower tensile strength can lead to flaking of sand layers and, therefore, to scabs. The rupture plane in the sand is assumed to be located somewhere between the condensation zone and the dehydrated zone. The exact location, moment, and cause of rupture are insufficiently explored. However, an improved knowledge of the physical background of scabbing could considerably help to improve the usability of bentonite-bonded moulding sand, which is the most environment-friendly mould material. Aim of the project (DFG-Gz JO301/5-1), therefore, is to investigate the effects of the water and heat kinematics on the tensile strength of moulding sand. For this, in-situ neutron radiography is performed during tensile testing of heated moulding sands using a self-constructed tensile test stand (Fig. 1). In the radiographs, the local neutron intensities have to be corrected for beam fluctuations, camera noise, multi scattering, and spectral effects like beam hardening. After applying Beer-Lambert's law, the dynamic water distribution in the sand can be determined with high spatial and temporal resolution (Fig. 2). Initial observations in preliminary experiments with the newly constructed tensile test stand allowed identification of areas with both reduced and increased tensile strength. The quantitative correlation of these areas with the respective momentary water content revealed a parabolic dependency of tensile strength on water content.

Fig. 1 In-situ neutron radiography of a heated moulding sand during tensile testing (T max = 310 °C, the heating plate is positioned above the upper image margin).

Fig. 2 Vertical profile of the water content of the sand before and after rupture.

Figure 1

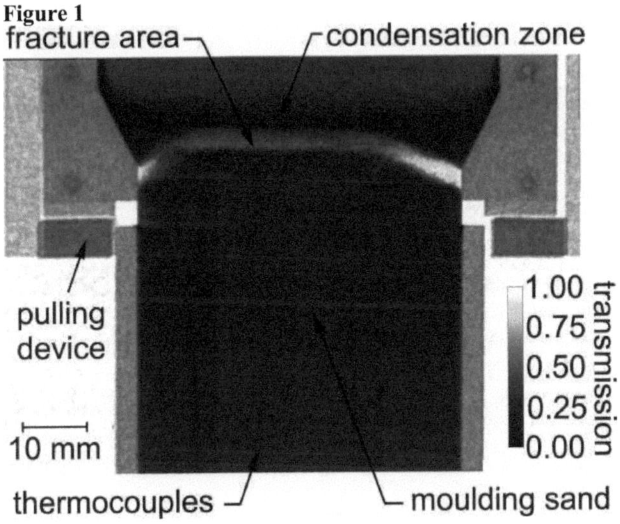

Figure 2

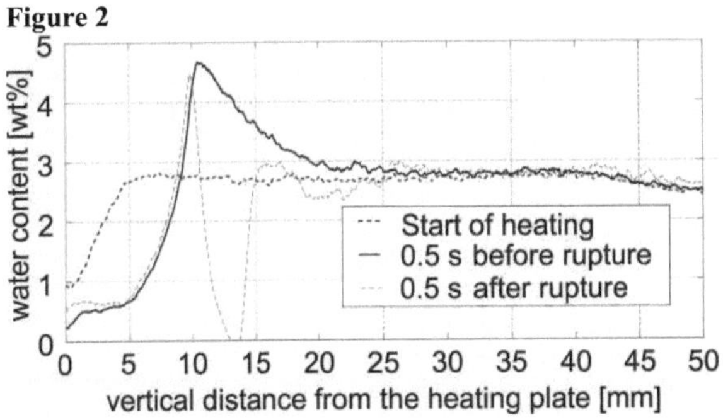

Structural trends in off-stoichiometric $Cu_2ZnGeSe_4$ compound semiconductors

R. Gunder[1], A. Franz[1], S. Schorr[1,2]
[1]*Helmholtz Zentrum Berlin, Crystallography, Berlin, Germany*
[2]*Freie Universität Berlin, Institut für Geologische Wissenschaften, Berlin, Germany*

The research on sustainable solar energy conversion technologies has further branched to another promising alternative, based on kesterite-type absorbers $Cu_2ZnSn(S,Se)_4$ (CZTSSe). In comparison to other compound semiconductor materials kesterites are more favorably because the elements contained in the CZTSSe system are both earth-abundant and, apart from selenium, not hazardous. Also, the optical band gap fits well to the solar spectrum and, additionally, can be tuned by different anion ratios in the solid solution CZTS - CZTSe. However, the difficulties in controlling the anion ratio gave rise to consider alternative means which facilitate the band gap to be designed more easily. Such an alternative is given by the cationic substitution Sn ↔ Ge as done in this study. In dependence of the route of off-stoichiometry (e.g. Cu-poor and Zn-rich), the kesterite phase, then, accommodates particular defect types and defect complexes. Those different kesterite-types, that is, A-, B-, C- and D- type, were introduced by Lafond et al. [1]. In addition, the existence of E- and F- type kesterite has been proposed by our group [2].

Recent investigations on a mixed F- D- type $Cu_2ZnGeSe_4$ (CZGSe) sample [3] carried out by neutron powder diffraction indeed indicate that CZGSe is crystallizing in the kesterite structure (space group), as also predicted by first-principle calculations [4]. According to the different types of off-stoichiometric kesterite proposed by [1] two series of powder samples were synthesized by solid state reaction in a one-zone tube furnace at T_{max} = 700 °C. These sample series comprise Cu-poor A-type ($Cu_{2-2x}Zn_{1+x}GeSe_4$) and Cu-rich C-type ($Cu_{2+2z}Zn_{1-3z}Ge_{1+z}Se_4$) kesterite. The powder samples stoichiometrically are in the range of, respectively, $0 \leq x \leq 0.2$ and $0 \leq z \leq 0.125$. The produced powder material has been pre-characterized by quantitative electron microprobe analysis using WDX spectroscopy as well as X-ray and neutron powder diffraction, respectively. This study aims to provide an overview of the structural response both to the cationic substitution as well as the different off-stoichiometric types, especially in consideration of the extent of deviation from stoichiometry.

[1] Lafond, A., et al., *Crystal Structures of Photovoltaic Chalcogenides, an Intricate Puzzle to Solve: the Cases of CIGSe and CZTS Materials*. Zeitschrift für Anorganische und Allgemeine Chemie, 2012. **638**(15): p. 2571-2577.

[2] Neldner, K., Valle-Rios, L.E., et al., *Existence of off-stoichiometric single phase kesterite*. Journal of Alloys and Compounds, 2015. **657**: p. 408-413.

[3] Gurieva, G., *Crystal structure of Cu2ZnSn(S1-xSex)4 solid solution*. Experimental report, 2014.

[4] Chen, S., et al., *Wurtzite-derived polytypes of kesterite and stannite quaternary chalcogenide semiconductors*. Physical Review B (Condensed Matter and Materials Physics), 2010. **82**(19): p. 195203---195203.

P099

P100

Crystallographic characterization and thermal decomposition behavior of ammonium exchanged zeolites

I. Spiess[1], M. Fischer[1], L. Wang[1], W. H. Baur[2], R. X. Fischer[1]
[1]*University Bremen, FB5-Geowissenschaften, Bremen, Germany*
[2]*International Zeolite Association, Western Springs, IL, United States*

Ammonium-exchanged zeolites, where the NH_4^+ ion represents the charge-compensating cation, play an important role as precursor materials in the production of H-zeolites. The hydrogen zeolites can be used as catalysts with their protons acting as Brønsted acid sites. Here, we present results on ammonium-exchanged zeolites within the framework of a project studying the mechanisms of deammoniation processes in zeolites to produce the H-forms acting as catalysts. Results are reported for zeolite A (synthetic), merlinoite (synthetic and natural samples) and chabazite (natural mineral).

Single crystals and powder samples of synthetic zeolite A (ideally $Na_{12}Al_{12}Si_{12}O_{48} \cdot 27H_2O$) were NH_4 exchanged and investigated by X-ray diffraction methods. In addition, a series of samples with different K/NH_4 ratios was produced by partial exchange. Thermal analyses showed that the zeolite decomposes already at rather low temperatures when about 20% of the K is replaced by NH_4. Ab-initio Molecular Dynamics calculations based on density-functional theory for a hypothetical protonated model of zeolite A indicate that the presence of water in the toc cages (sodalite-type cages) plays a crucial role in the decomposition process.

Ion-exchange experiments were performed with merlinoite (ideally $(K,Na,Ca_{0.5})_{10}Al_{10}Si_{22}O_{64} \cdot 20H_2O$) single crystals in an ammonium nitrate solution. EDX analysis showed that essentially all K was exchanged by NH_4, and powder diffraction experiments indicate a slight change in the lattice parameters upon ammonium exchange. The thermal stability was assessed with high temperature X-ray diffraction experiments. Single-crystal structure refinements are in progress. The exchanged synthetic merlinoite is compared with an NH_4 containing merlinoite mineral.

Single-crystal and powder samples of the natural mineral chabazite (ideally $(Sr,Ca_{0.5},K)_4Al_4Si_8O_{24} \cdot 12H_2O$) were ammonium-exchanged. For short exchange times (1 day), only the Sr^{2+} cations were replaced by ammonium; but a prolonged exposure (3 days) to the solution led to a complete exchange. Further characterization with variable-temperature X-ray diffraction methods is in progress.

P101

Optical properties of tungsten oxide bronzes and misfit layer compounds $(MS)_nTS_2$: Small versus large polaron behaviour

C. Rüscher[1], T. Debnath[2]
[1]*Leibniz University, Mineralogy, Hannover, Germany*
[2]*University Dhaka, Chemistry, Dhaka, Bangladesh*

Polarons are quasi particles formed by electron-phonon coupling and are in general ground states of crystals, therefore [1]. A distinction between large and small polarons could be made in so far as the former preserve the free carrier behaviour whereas the later compromise free carrier aspects and hopping character according to the strong spatial localisation. The coupling must be seen related to the strain tensor, too [2]. The optical reflectivity of hexagonal tungsten bronzes (HTB) of type M_xWO_3, M = K, Rb, Cs and x = 0.2, 0.25, 0.3 were tentatively be described using a Drude free carrier approach [3]. Our more recent results show that these deviations can be explained in favour of a combined Drude-oscillator model suggesting that the carriers behave as small polarons [4]. This somehow new understanding let us search for systems possessing presumably ordinary Drude free carrier effects for comparison, as for example the so called misfit layer compounds (MLC) which are coming into focus again for optimisation of thermoelectric materials [5]. The optical reflectivity and transmissivity spectra of the at that time as new called MLC $(MS)_nTS_2$ reveal ordinary Drude type free electron gas behavior, as shown for compositions with T = Ta, Nb and M = Sn, Pb, Sm, Tb, La (1.08 < n < 1.23) [6, 7]. Their anisotropy is related to the orthorhombic metric of the crystal lattice. The screened free carrier plasma frequencies were found between 3000 and 6000 cm^{-1} for the electrical field E parallel to (001) possessing a small anisotropy related to slightly different effective masses ($m^*[010]/m^*[100] \approx$ 1.04-1.08). The temperature depending damping parameter well approaches values deduced from the dc-resistivity and their linear increase with temperature measured between 150 and 400 K. For E perpendicular to (001) the plasmafrequencies are shifted to below 1500 cm^{-1} due to larger effective masses $m^*_c/m^*_{(001)}$ of about 34 or 6, e.g. for $(PbS)_{1.13}TaS_2$ and $(SmS)_{1.19}TaS_2$, respectively. The volume effect and the volume effect plus doping has been discussed as to be responsible for the decrease in carrier concentration compared to the related phases 2H-TaS_2 and 2H-NbS_2 alone possessing in all cases a dz^2 dominated conduction band. For the misfit layer compounds under investigation a transmission window opens above the plasma edge threshold and becomes closed due to the S^{2-}band edge effect. We argue that MLC may be examples for large polaron behaviour.

[1] A. S. Alexandrov, N. F. Mott: Polarons and Bipolarons, 1996, ISBN 978-981-02-2298-7. World Scientific.

[2] I. G. Austin, N. F. Mott: Polarons in crystalline and non-crystalline materials. Advances in Physics, 18: 71, 41-102 (1969).

[3] A. Hussain, R. Gruehn, C. H. Rüscher: Crystal growth of alkali metal tungsten bronzes MxWO3 (M = K, Rb, Cs), and their optical properties. J. Alloys Compds. 246, 51-61 (1997).

[4] T. Debnath, C. H. Rüscher, A. Hussain: Optical properties of alkali metal tungsten bronzes. Proc. of the Bangladesh Chemical Congress 2008.

[5] D. R. Merril, D. B. Moore, S. R. Bauers, M. Falmbigl, D. C. Johnson: Misfit layer compounds and ferecrystals: model systems for thermoelectric nanocomposites. Materials 8, 2000-2029 (2015).

[6] C. H. Rüscher, C. Haas, S. van Smaalen, G. A. Wiegers: Investigations of the optical reflectivity of misfit layer compounds. J. Phys.: Condens. Matter 6, 2117 (1994).

[7] C. H. Rüscher: Optical reflectivity and transmissivity of misfit layer compounds. phys. stat. sol. b 198, 889-904 (1996).

P102

Biomineralization in *Sepia Officinalis*: Real Structure and Simulation

J. Schneider[1], W. Schmahl[1]
[1]*LMU-München, Earth-and Environmental Sciences, München, Germany*
Biomineralization in *Sepia Officinalis*: Real Structure and Simulation
Julius Schneider & Wolfgang W. Schmahl
LMU München, Department of Earth- and Environmental Sciences, Crystallography Section

Biomineralization has created an enormous variety of solid structures in an amazing range of chemical compositions and in a huge range of length scales [1]. Examples include microfossils like Radiolaria (SiO_2), magnetosomes of submicron-sized magnetite (Fe_3O_4) to help with the orientation of organisms in the earth's magnetic field, ($CaCO_3$)-manifestations ranging from micrometer-sized coccoliths up to coral reefs of 2000 km length (Great Barrier Reef), apatite ($Ca_5(PO_4)_3$, OH) manifestations of small animal bones up to sauriopod bones of tons of weight.

We investigated the aragonite structure of the buoyancy tank (cuttlebone) of *Sepia Officinalis* (cuttlefish) [2]. It consists of a multilayer sandwich structure of highly textured aragonite material, where planes of densely packed platelets alternate with separating sheets of hollow pillars/walls. The structure strongly reminds of a multi-storey car park. Very recent SEM-work on Sepia Officinalis was performed by Checa et al. [3].

Now we are interested in the question on how the biomineralization process can change between these two very different space filling modes. In particular, we focus on the question on how this intricate hollow structure of supporting pillars/walls may be formed.

To visualize the irregular lateral pattern of the pillar/wall arrangement, we produced cuts of this layer of a variety of samples and in varying heights. Amazingly, a pattern formation closely resembling that of mollusk shells [4,5] is observed.

There, observed patterns can be simulated mathematically by means of nonlinear activator-inhibitor formalisms of coupled differential equations. Adapting this formalism, we find that pattern simulations of 1- and 2-dimensional activator-inhibitor schemes nicely reproduce the pattern of the pillar/wall arrangement of *Sepia Officinalis*.

This example thus represents only another special case of the extremely diverse and amazing field of biological pattern formation occurring with many living creatures [6].

[1] Mann S., *Biomineralization: Principles and Concepts in Bioinorganic Materials Chemistry*, Oxford University Press, New York 2001

[2] Schneider J., Griesshaber E., Kaiser-Bischoff I., Merkel C., Macknam K. & W.W.Schmahl, CIMTEC 3rd Int.Conf. ond smart materials, structures and systems, Acireale, Sicily, Italy E HP01, page 106, 2008

[3] Checa A, G., Cartwright J.H.E., Sanchez-Almazo I., Andrade J.P. & Ruiz-Raya F., Sci. Rep. 5 (2015) 11513

[4] Meinhardt H., *The Algorithmic Beauty of Sea Shells,* Springer Verlag Berlin 1995

[5] Meinhardt H. & Klingler M., J.theor.Biol. 126 (1987) 63-69

[6] Meinhardt H., *Models of Biological Pattern Formation*, Academic Press, London 1982

P103

Investigations of the System $[(Cu,Mn)_{1-x}Al_x(OH)_2](CO_3)_{x/2}\cdot nH_2O$ with (Cu,Mn):Al ratios 2:1 and 3:1

T. Linke[1], S. Stöber[1], H. Pöllmann[1]
[1]*Martin-Luther-University Halle-Wittenberg, Mineralogy and Geochemistry, Halle (Saale), Germany*

Many layered double hydroxides are known and used for different applications, e.g. chemical catalysis, agriculture, medicine or for waste decontamination. Otherwise many LDHs are not known or in use. At the moment, different LDHs with chemical compositions similar to the investigated system are applied as electro catalysts or as precursor phases. Recently they are used for active oxygen evolution reaction [1].

In this study different solid phases with the general formula $[(Cu,Mn)_{1-x}Al_x(OH)_2](CO_3)_{x/2}\cdot nH_2O$ were synthesized. Different molar ratios of the di- and trivalent cations were used for the preparation of the layered double hydroxides. Simultaneously the combination of a two and a three-layer endmember is investigated. Whereas the 2H and 3R polytypes are most common in this LDH group [2].

The solid phases were prepared by co-precipitation after Evans & Slade [3] from corresponding metal nitrate solutions at alkaline pH-values. Therefore 1 molar solutions were mixed in the chosen ratio (Me^{II}:Me^{III}) and inserted into an automatic titration device from Mettler-Toledo GmbH. Adding the solution into a reactor with a stoichiometric amount of $NaHCO_3$ with carbonate as interlayer anion produces the LDHs. To keep the pH at a constant value a pH-stat device was used.

After precipitation the suspension is filtrated and the solid residue is instantly analyzed with x-ray diffraction. Afterwards the solid residues are stored in a desiccator with nitrogen atmosphere to dry at 35 % relative humidity. In a next step the dried residues are analyzed with x-ray diffraction. Afterwards the chemical compositions of the single phases are determined by ICP-MS, thermogravimetric analysis, differential thermal analysis and differential scanning calorimetry.

The lattice parameters were refined and investigations of different hydration levels were made with these methods.

The syntheses show first, that it is possible to prepare a single charmarite phase $Mn_4Al_2(OH)_{12}CO_3\cdot nH_2O$ without any accessory phases of manganese oxide or manganese carbonate. Furthermore it was possible to prepare single solid solutions of charmarite and the Cu^{2+} equivalent.

[1] Gong, M. and Dai, H. A mini review on NiFe-based materials as highly active oxygen evolution reaction electrocatalysts. Nano Research 2015; 8: 23-39

[2] Bookin, A., Cherkashin, V. and Drits, V. Polytype Diversity of Hydrotalcite-like minerals II. Determination of the polytypes of experimentally studied varieties. Clays and Clay Minerals 1993; 41 (5): 558-564

[3] Evans, D. and Slade, R. Structural aspects of layered double hydroxides. pp. 187 in: Layered Double Hydroxides. Duan X. and Evans D.G. editors. Structure and Bonding, Springer, Berlin, 2006; 119: 218 pp.

P104

Investigation on the systems $C_3A{\bullet}CaSO_4{\bullet}nH_2O$ - $C_3A{\bullet}Ca(MnO_4)_2{\bullet}nH_2O$ and $C_3A{\bullet}CaCO_3{\bullet}nH_2O$ - $C_3A{\bullet}Ca(MnO_4)_2{\bullet}nH_2O$ — hydration phases of the Mn-CAC

K. M. Dietmann[1], S. Stöber[1], H. Pöllmann[1]
[1]*Martin-Luther-University Halle-Wittenberg, Mineralogy and Geochemistry, Halle (Saale), Germany*

The LDHs investigated in this study are related to the so called AFm-Phases which are formed during the hydration of cementitious materials [1]. Because of their layered structure and the capability to exchange the intercalated anions those minerals can be used as storage minerals [2] and for the fixing of hazardous elements or waste materials [3,4].

Regarding the fixing of manganese in calcium sulfoaluminate cements as an result of the usage of manganese-rich additives the binary systems $C_3A{\bullet}CaSO_4{\bullet}nH_2O$ - $C_3A{\bullet}Ca(MnO_4)_2{\bullet}nH_2O$ and $C_3A{\bullet}CaCO_3{\bullet}nH_2O$ - $C_3A{\bullet}Ca(MnO_4)_2{\bullet}nH_2O$ have been analyzed. Mainly the substitution of structural different anions in the interlayer of layered double hydroxides has been investigated. More precisely the substitution of planar carbonate anions and tetrahedral sulfate anions by equivalent charged tetrahedral permanganate anions. As a side effect the contemporaneous fixing of uncharged potassium permanganate into the interlayer has been observed and investigated.

Therefore different sets of samples with an increasing content of manganese have been prepared for the systems $C_3A{\bullet}(1\text{-}x)CaSO_4 \bullet xCa(MnO_4)_2{\bullet}nH_2O$ and $C_3A{\bullet}(1\text{-}x)CaCO_3{\bullet}xCa(MnO_4)_2{\bullet}nH_2O$ with $0 \leq x \leq 1$.

To guarantee the exclusion of carbonate the preparation took place under nitrogen atmosphere. For three months the samples have been aged at 25 °C under constant agitation. Afterwards the phase composition has been analyzed using X-ray diffraction while the samples have had a relative humidity of 100 % and 35 %. The samples have been dried in a desiccator under nitrogen atmosphere to avoid carbonization. Thus a comparison between the wet and dried samples is possible.

In addition to this tetra calcium aluminate hydrate (TCAH) has been synthesized and aged for two weeks. Afterwards the kinetic behavior of the fixing of permanganate respectively potassium permanganate in the interlayer has been investigated. Therefore time-resolved XRD-measurements have been done.

In a next step the single-phased samples will be investigated using thermo-analytical methods, like thermal gravimetric analysis and differential scanning calorimetry. Chemical investigations on the residue solutions of single-phased samples will be done to quantify their phase content and determine the built-in of cations.

[1] Auer, S. & Pöllmann, H. (1994): Synthesis and characterization of lamellar cadmium aluminium hydroxide salts with SO_4^{2-}, CO_3^{2-}, Cl^-, NO_3^- - Journal of solid state chemistry, 109, 187-196

[2] König, U. (2006): Synthese, Charakterisierung und Eigenschaften von manganhaltigen Layered Double Hydroxides (LDHs) - Dissertation, MLU Halle-Wittenberg

[3] Auer, S. (1992): Bindung umweltrelevanter Ionen in Ettringit und in Schichtstrukturen vom Typus TCAH - Dissertation, Universität Erlangen - Nürnberg

[4] Pöllmann, H. (2007): Immobilisierung von Schadstoffen durch Speichermineralbildung - Berichte aus der Geowissenschaft, 428 S., Shaker Verlag, 2007

P105

In-situ Grazing Incidence Small-Angle-Xray-Scattering Studies Using The Incoatec Microfocus Source IµS

J. Wiesmann[1], A. Stricker[1], J. Graf[1], P. Siffalovic[2], M. Hodas[2], K. Vegsö[2]
[1]*Incoatec GmbH, Geesthacht, Germany*
[2]*Slovak Academy of Science, Institute of Applied Physics, Bratislava, Slovakia*

The Incoatec microfocus source IµS is a low power air cooled X-ray source for diffractometry applications and is available with Cr, Co, Cu, Mo, and Ag anodes. The source is equipped with a two dimensional beam shaping multilayer optics. Therefore, we can form either a highly collimated beam with a low divergence (below 0.5 mrad) or a focusing beam with higher divergence (up to 10 mrad) and very small focal spots (diameter below 100 µm).

Equipped with a collimating optics it can be used for GISAXS, SAXS and X-ray reflectometry studies. When using focusing optics all those experiments can be carried out in transmission geometry, especially in powder diffraction applications. With the Mo-IµS highly absorbing and radiation-damage sensitive materials can be investigated. Consequently, this source is often used for single crystal diffractometry in the chemical crystallography and becomes more and more interesting for investigations of soft matter samples.

In our presentation we will be giving an overview of representative GISAXS setups, which take advantage of the brilliance and outstanding beam quality of this low-maintenance microfocus source.

We will be presenting two applications in more detail, both measured at the Slovak Academy of Sciences:

1) Nanoparticles on a liquid sample were investigated with a special GISAXS setup equipped with a Cu-IµS producing a 5 mrad focused beam and with a Pilatus pixel detector. Ordering phenomena could be observed in-situ during an increase of surface pressure. The particles were transformed from single islands to an almost vertically ordered structure of connecting particles.

2) By using in-situ GISAXS in the home-lab we investigated how multilayer grow during thin film deposition. This kind of experiments is typically done only at synchrotrons. With an IµS it is now also feasible in the home-lab.

P106

Does Stoichiometric Sillenite $Bi_{12}SiO_{20}$ Show Relaxor Behavior or anelastic relaxation?

C. S. Pandey[1], M. Burianek[2], M. Mühlberg[2], J. Schreuer[1]
[1]*Ruhr-Universität Bochum, Institute of Geology, Mineralogy & Geophysics, Bochum, Germany*
[2]*University of Cologne, Institute of Crystallography, Cologne, Germany*

Sillenite Bi12SiO20 (BSO) single crystal have body-centered cubic lattice of symmorphic space group I23 (T3) [1]. Recently it was demonstrated that BSO shows relaxor like behavior in the range of ~ 325 - 480 °C [2]. It is well known that relaxor behavior can be best demonstrated by its temperature dependent elastic properties and thermal expansion studies [3]. Considering this, we have performed thermos-elastic measurements on single crystal BSO up-to ~ 730 °C incorporating an elegent method of the resonant ultrasound spectroscopy. In this presentation we will demonstrate whether BSO really show realxor behavior or an anelastic relaxation effect incorporating the temperature dependent elastic properties and thermal expansion up-to ~ 730 °C of single crystal BSO with high internal consistency. As far as known to the authors, this kind of study has been done for the first time.

(1) Radaev S F, Simonov V I, Kargin Y F, Acta Crystallogr. B **48**, 604–609 (1992).

(2) Hu Y, Sinclair D C, Chemistry of Materials **25**, 48-54 (2013).

(3) Pandey C S, Schreuer J, Burianek M, Muehlberg M, Phys. Rev. B **84**, 174102 (2011).

Storage and battery materials

P107

Pair Distribution Function Analysis of Ammonium Nitrate - EVALUATION FOR ENERGETIC MATERIALS

P. B. Kempa[1], M. Herrmann[1]
[1]*Fraunhofer ICT, Pfinztal, Germany*

Energetic materials are often nanocrystalline, amorphous, liquidor gel structured. Diffraction measurements would not yield sharp Bragg peaks but halos from diffuse scattering.Therefore, it is difficult to obtain structural information from these materials by conventional structure analysis using X-ray diffraction.The use of the total scattering, in which both Bragg and diffuse components of the scattering are analyzed together, is applicable for the study of this kind of problems.The contribution shows the first refinements and the possibility to achieve structural information with the aid of the pair distribution function (PDF) from an energetic material.

The energetic material ammonium nitrate (NH_4NO_3) was chosen as an example for comparative measurements on laboratory measuring systems with different wavelengths. The diffraction data were used for calculating the PDFs and the results are evaluated in order to improve measurement conditions for energetic materials.

Ammonium nitrate (AN) is a well-known energetic material with five phases, used as fertilizer and energetic oxidizer in solid propellants, explosives and gas generator systems. Phase IV, which is stable at room temperature, crystallizes in the orthorhombic space group Pnmm with the lattice constants a = 5.745, b = 5.438 and c = 4.942 Å. The number of formula units Z is 2.

All X-ray measurements were performed in the Bruker laboratory on the diffractometer D8 equipped with Göbel mirror and Lynxeye-detector using a capillary in the range of 3.5 - 150 °2q and the step size of 0.016 °2q for Ag-radiation and 0.02 °2q for Mo-radiation, respectively.

The investigations demonstrate the feasibility of the XRD / pair distribution function analysis for energetic materials. Reasonable PDFs for ammonium nitrate were obtained, with very good agreement of experimental and calculated pattern. The crystallographic parameters bond lengths and lattice parameter, and particularly the particle size (for nanostructured energetic materials) and the static disorder parameter d_2 were identified as highly interesting parameter for future characterizations of the microstructure of energetic material ranging from the highly crystalline to the amorphous state.

Out of various laboratory measurement systems, that with silver radiation was identified as best choice. The findings and data shall be used for the development and refinement of a measurement system and evaluation tools at the Fraunhofer ICT for advances PDF analysis of energetic materials, with e.g. the ammonium nitrate as a reference material for calibration.

P108

Synthesis, Chemical Sodiation & Desodiation of Olivine-Related Compounds $A_yM_xFe_{1-x}PO_4$

J. Bauer[1], R. Haberkorn[1], G. Kickelbick[1]
[1]*Universität des Saarlandes, Anorganische Festkörperchemie, Saarbrücken, Germany*

Energy storage is an increasing demand that can be achieved *via* rechargeable batteries for stationary as well as mobile applications. Li-ion batteries containing cathode materials like $LiFePO_4$ and $LiCoO_2$, are already well established. Nevertheless research in the field of Na-ion batteries gained increasing interest in the last few years.

Our research focuses on olivine-type cathode materials, especially Na-containing materials due to the fact that sodium is less expensive, less toxic and naturally more abundant in the Earth's crust than lithium. While investigations on Li-containing olivine-type compounds are widely described in literature [1-4], research in Na-containing olivine-type compounds is rarely reported [5-7].

Na-containing materials were synthesized by chemical oxidation of $LiM_xFe_{1-x}PO_4$ (M = 3d transition metal) using various oxidation agents like I_2, Cl_2, Br_2 or $K_2S_2O_8$. The as-prepared phosphates then were chemically sodiated using Na_2S in acetonitrile. The materials were characterized with X-ray powder diffraction. The products are compared with materials derived from alternate synthesis routes.

Besides the commonly used bromine for the oxidation of $LiFePO_4$ iodine and chlorine proved to be suitable oxidation agents. Through chemical sodiation several new alkaline phosphate could be obtained. One of those compounds is $NaCo_{0.5}Fe_{0.5}PO_4$ with triphylite structure while the same compound showing maricite structure was yielded *via* solid state reaction. The triphylite structure type is a more promising candidate for technical applications as cathode material [8].

Chemical sodiation and desodiation is a powerful tool avoiding the restrictions from electrochemical processing. Pure phases may be obtained in any quantity. On the one hand this enables more precise characterization of the crystallographic structure and other properties of materials which may not be yielded by high temperature synthesis and on the other hand preprocessing of materials is possible before forming electrodes for technical applications. Expanding the bandwidth of oxidation agents for desodiation or delithiation provides more possibilities in optimization of reaction conditions, sample handling or econimical aspects.

[1] O. Clemens, R. Haberkorn, M. Springborg, H. P. Beck, *Z. Anorg. Allg. Chem.* **2014**, *640, (1),* 173.

[2] R.Qing, M.-C. Yang, Y. S. Meng, W. Sigmund, *Electrochim. Acta* **2013,** *108*, 827.

[3] R. Ruffo, C. M. Mari, F. Morazzoni , F. Rosciano , R. Scotti, *Ionics* **2007,** *13*, 287.

[4] M. Kaus, I. Issac, R.Heinzmann, S. Doyle, S. Mangold, H. Hahn,

V. S. K. Chakravadhanula, C. Kübel, H. Ehrenberg, S. Indris, *J. Phys. Chem.* C **2014**, *118*, 17279.

[5] M. Avdeev, Z. Mohamed, C. D. Ling, J. Lu, M. Tamaru, A. Yamada, P. Barpanda, *Inorg. Chem.* **2013**, *52*, 8685.

[6] P. Moreau, D. Guyomard, J. Gaubicher, F. Boucher, *Chem. Mater.* **2010**, *22*, 4126.

[7] K. T. Lee, T. N. Ramesh, F. Nan, G. Botton, L. F. Nazar, *Chem. Mater.* **2011**, 23, 3593.

[8] P.P. Prosini, C. Cento, A. Masci, M. Carewsk, *Solid State Ionics*, **2014**, *263*, 1.

P109

Electrochemical characterizations of Nasicon type phosphate $Na_3CoZr(PO_4)_3$ as an cathode material for sodium-ion batteries

L. ASMAA[1,2], R. Ismail[1,2], E. B. Abdsslame[1,2], O. Brahim[3]
[1]*university Hassan 2 Ben Ms'ik Casablanca , Casablanca, Morocco*
[2]*Faculté des Sciences Ben M'Sik, Casablanca, Morocco., Laboratoire de Physico-chimie des matériaux appliquésLPCMA, , CASABLANCA, Morocco*
[3]*Universidad del País Vasco, P.O. Box 644, Bilbao, Spain, Departamento de Física de la Materia Condensada, , Bilbao, Spain*

In this paper we report the synthesis, structural, optical and magnetic studies of the Nasicon-type phosphate $Na_3CoZr(PO_4)_3$, which represent a promising candidate as an cathode material for the Sodium-ion batteries thanks to their good energetic performance and structural stability.

The Rietveld refinement of the XRD pattern showed that at room temperature the material has a hexagonal symmetry with the space group (R-3) and the cell parameters $a_h = 8{,}931(3)$Å ; $c_h = 22{,}310(1)$ Å. The structure of $Na_3CoZr(PO_4)_3$ consists of edge sharing [MO_6] octahedra (M = Zr, Co) resulting in chains parallel to [−101]. These chains are linked together via the [PO_4] tetrahedra to form two distinct tunnels in which sodium cations are located.

The optical absorption study showed that Co^{2+} ion is subject to a weak ligand field. It also showed that the Co-O bonds are slightly covalent, which is in good agreement with the experimental distance Co/Zr-O (2.09 Å) which is less than the sum of the ionic radii (2.15 Å).

The magnetic study has shown that the electronic configuration of the Co^{2+} ion is $t^5_{2g}e^2_g$, indicating that oxygen exert a low field on the cobalt site. $Na_3CoZr(PO_4)_3$ follows a paramagnetic behavior of Curie-Weiss type. The low negative value of the Curie temperature pq = - 10 K indicates the existence of weak antiferromagnetic nature of interactions between the Co^{2+} ions, in agreement with the structural study which showed that Co^{2+} ions are distant from each other ($d_{Co^{2+}\text{-}Co^{2+}}$ = 4.58 Å).

The electrochemical properties of $Na_3CoZr(PO_4)_3$ were tested in sodium cells. Ex situ and in situ X-ray diffraction measurements indicate that the intercalation-deintercalation process of Na^+ ions in $Na_3CoZr(PO_4)_3$ is reversible and that the structural framework is maintained during cycling.

P110

Synthesis and characterization of carbon-doped titanium dioxide nanotubes

U. Helbig[1], J. Roudenko[1], K. Herbst[1], J. Helbig[1]
[1]*Technische Hochschule Nürnberg Georg Simon Ohm, Werkstofftechnik, Nürnberg, Germany*

Introduction

Carbon-doped titanium dioxide nanotubes (C-TNTs) are a promising material for electrode materials in PEM fuel cells. They provide a high specific surface, good conductivity and showed improved durability in first electrochemical cell tests.

Objectives

Undoped titanium dioxide nanotubes (TNTs) can be fabricated via a hydrothermal process. The typical inner diameter of the resulting tubes is 3-10 nm and the length can be up to about 500 µm. The structure is suggested to be comparable to Anatase or of titanate type ($H_2Ti_3O_7$). [Liu et al. 2014] Several methods of doping of the tubes are known, including carbon as dopant. However, a challenge is the introduction of the dopant without losing the nanotube morphology and the high specific surface area.

Materials & Methods

For the fabrication of TNTs, commercially available titanium dioxide nanopowder is dispersed in sodium hydroxide solution and treated at temperatures between 100 and 115 °C for up to 20 h. The titanium dioxide nanotubes are subsequently obtained after washing with purified water and hydrochloric acid at room temperature. A following carbothermic treatment causes sintering and loss of specific surface area. For that reason, the fabrication process was modified. The titanium dioxide nanopowder was first doped with carbon and then subjected to the hydrothermal treatment afterwards only. In this modified process, the carbothermal treatment was conducted in a rotary furnace under ethylene/nitrogen gas flow at 700 °C.

Results

The phase content and specific surface area of the pretreated titanium dioxide powder were investigated. It consists of Anatase and Rutile and has a specific surface area of 52 m^2/g. Titanium dioxide nanotubes doped with carbon synthesized from the precursor powder showed a specific surface area as high as 445 m^2/g and an increased electric conductivity compared to undoped TNTs. The material was further investigated by SEM and powder XRD. Typical morphologies and diffraction patterns of TNTs were observed.

Conclusion

The modified synthesis process of C-doped TNTs provides electrically conducting titanium dioxide nanotubes with high specific surface. The investigations done so far suggest that the carbon was successfully introduced into the lattice of the nanotubes.

References

N. Liu, X. Chen, J. Zhang, J. Schwank: A review on TiO_2-based nanotubes synthesized via hydrothermal method: Formation mechanism, structure modification, and photocatalytic applications. Catalysis Today, 225, pp. 34-51, 2014.

Structural chemistry at non-ambient conditions

P111
SUCCESSIVE PHASE TRANSITIONS IN $LaCoO_3$ NANOPARTICLES: AN *IN-SITU* X-RAY POWDER DIFFRACTION STUDY

S. Ortatatli[1], J. Ternieden[1], C. Weidenthaler[1]
[1]*MPI für Kohlenforschung, Heterogeneous Catalysis, Mülheim an der Ruhr, Germany*

Perovskites with the general formula ABO_3 have received great attention as cathode materials in solid oxide fuel cells (SOFCs) and in heterogeneously catalyzed reactions either as catalyst precursors or as catalysts themselves. Among other perovskites, there is an increasing interest in $LaCoO_3$ for being an active catalyst for the decomposition of N_2O. Under non-ambient conditions $LaCoO_3$ crystallizes in different structures [1-3]. Stability of $LaCoO_3$ under reducing and oxidizing atmosphere has been widely investigated since stability is important for fuel cell applications and catalytic oxidation reactions. As a result, deviations from the conventional perovskite structure to oxygen deficient or layered perovskites were observed [3, 4]. Particularly, layered perovskites, so-called Ruddlesden-Popper (RP) type, were tested in SOFCs and found to be promising candidates as electrolytes [5].

In this study, we probed the phase transition mechanism of $LaCoO_3$ (LCO) nanoparticles by using X-ray powder diffraction (XRPD), TG-MS, and chemisorption upon heating as-synthesized LCO from room temperature to 800 °C under N_2 or Ar atmosphere. Nanosized LCO perovskite was synthesized via ball-milling from oxide precursors for 36 h under atmospheric conditions. Recently, we have observed by in situ X-ray diffraction studies that as-synthesized rhombohedral LCO undergoes successive phase transitions starting at temperatures as early as 650 °C to intermediate RP-phases (orthorhombic) and then finally to a tetragonal RP-phase at 800 °C. Upon cooling, another transition from high temperature tetragonal RP- to low temperature orthorhombic RP-phase was detected.

References

[1] Dacquin, J. P., Lancelot, C., Dujardin, C., Da Costa, P., Djega-Mariadassou, G., Beaunier, P., Kaliaguine, S., Vaudreuil, S., Royer, S., Granger, P., *Appl. Catal. B*, 91, **2009**, 596-604.

[2] Wu, Y., Ni, X., Beaurain, A., Dujardin, C., Granger, P., *Appl. Catal. B*, 125, **2012**, 149-157.

[3] Radovic, M., Speakman, S. A., Allard, L. F., Payzant, E. A., Lara-Curzio, E., Kriven, W. M., Lloyd, J., Fegely, L., Orlovskaya, N., *J. Power Sources*, 184, **2008**, 77-83.

[4] Royer, S., Bérubé, F., Kaliaguine, S., *Appl. Catal. A*, 282, **2005**, 273-284.

[5] Tarancón, A., Skinner, S. J., Chater, R. J., Ramírez-Hernández, F., Kilner, J. A., *J. Mater. Chem.*, **2007**, 17, 3175-3181.

P112
Thermal degradation of chalcogenogallate chains $^1_\infty[Ga_2(Q_2)_{2-x}Q_{2+x}^{2-}]$ (x = 0, 1, 2; Q = S, Se) investigated by high-temperature *in situ* PXRD

D. Friedrich[1], M. Schlosser[1], A. Pfitzner[1]
[1]*Universität Regensburg, Regensburg, Germany*

During our investigations of alkali metal group 13 chalkogenometallates, we reported on the compounds $Cs_2Ga_2Q_5$ (Q = S, Se).[1, 2] These compounds feature previously unknown anionic chains $^1_\infty[Ga_2(Q_2)Q_3^{2-}]$ with structural similarities to the parent compounds $CsGaQ_3$[3, 4] and $CsGaQ_2$[5, 6]. The chains can be converted to each other by replacing a dichalcogenide dumbbell Q_2^{2-} with a chalcogenide ion Q^{2-} and *vice versa*.[1] This conversion was previously proposed.[1] The release of gaseous chalcogens upon heating of the compounds $CsGaS_3$ and $Cs_2Ga_2Se_5$ now confirms this assumption. $CsGaSe_3$ exhibits congruent melting at around 600°C.

The thermal decomposition of these compounds was further studied by using *in situ* by high-temperature X-ray diffraction techniques using a STOE Stadi P diffractometer equipped with a high temperature capillary furnace (monochromatic Mo-$K_{\alpha 1}$ radiation). These experiments revealed a two-step decomposition from $CsGaS_3$ to $CsGaQ_2$ with $Cs_2Ga_2Q_5$ as the intermediate compound.

Figure 1: Schematic representation of the stepwise transformation of the anionic chains $^1_\infty[Ga(Q_2)_{2-x}Q_{2+x}]$ (x = 0, 1, 2; Q = S, Se).

[1] D. Friedrich, F. Pielnhofer, M. Schlosser, R. Weihrich, A. Pfitzner, *Chem. Eur. J.* **2015**, *640*, 826-829.

[2] D. Friedrich, M. Schlosser, A. Pfitzner, *Z. Anorg. Allg. Chem.* **2014**, *640*, 826-829.

[3] M. Suseela Devi, K. Vidyasagar, *J. Chem. Soc., Dalton Trans.* **2002**, 4751-4754.

[4] J. Do, M. G. Kanatzidis, *Z. Anorg. Allg. Chem.* **2003**, *629*, 621-624.

[5] D. Schmitz, W. Bronger, Z. Naturforsch. **1975**, *30B*, 491-493.

[6] D. Friedrich, M. Schlosser, A. Pfitzner, *Z. Anorg. Allg. Chem.* **2014**, 640, 2356.

Figure 1

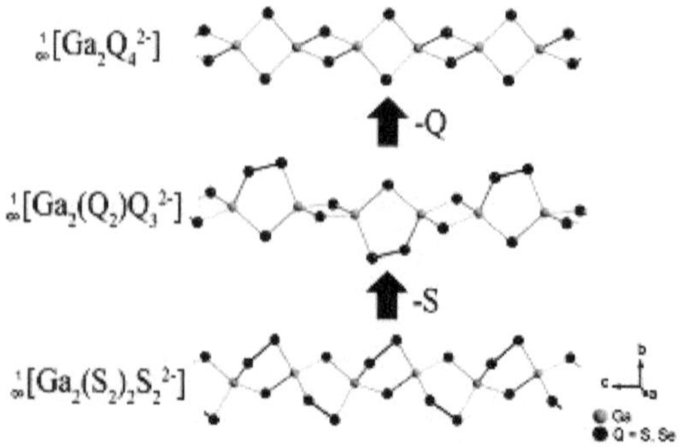

P113

Pressure induced phase transition from calcite to aragonite detected by fluorescence spectroscopy

C.- J. Fruhner[1], L. Bayarjargal[1], B. Winkler[1], R. Luchitskaia[1]
[1]*Goethe Universität, Kristallographie, Frankfurt am Main, Germany*

The pressure-induced phase transition from calcite to aragonite has been the topic of numerous studies and, at ambient temperature, proceeds via intermediate phases (calcite-II, calcite-III, calcite-IIIb) [1].

The goal of this study is to employ pressure dependent fluorescence spectroscopy to further elucidate the calcite to aragonite phase transition and to demonstrate the viability of this approach for the study of phase relations of carbonates at high pressures and temperatures. Eu^{3+} fluorescence can be employed for site symmetry detection [2], and, as in doped samples the calcium sites are partially occupied by europium, pressure-induced changes in the emission spectrum should allow us detect phase transitions.

Europium doped samples of calcite and aragonite were synthesized at T < 500 K. High-pressure experiments were carried out in diamond anvil cells (DAC). DACs were loaded with Eu-doped calcite (calcite:Eu^{3+}) and aragonite (aragonite:Eu^{3+}), ruby chips for pressure determination and argon as pressure transmitting medium. Fluorescence was excited with a blue laser (λ = 473 nm) and measured in the pressure range 0 - 25 GPa at ambient temperature.

The pressure induced shift of the $^5D_0 \rightarrow {}^7F_0$ emission peak of aragonite:Eu^{3+} (Fig. 1) and of calcite:Eu^{3+} (Fig. 2) was determined during pressure increase and decrease. As expected, in aragonite there is no evidence for a phase transition, as there is only a linear, reversible pressure-induced shift of the fluorescence (Fig. 1). In contrast, the fluorescence in calcite:Eu^{3+} (Fig. 2) shows initially a non-linear pressure-dependence on pressure increase. At P > 12 GPa, the pressure dependence changes irreversibly. Upon pressure release, the fluorescence energies are those of aragonite and the pressure-dependence is linear, as in aragonite.

Hence, here we have shown that Eu^{3+} fluorescence allows us to detect the pressure-induced phase transition to aragonite, which occurs at 12 - 15 GPa in our experiments. The formation of calcite II and calcite III/IIIb has not been detected here, which is consistent with the observation that the local environments of Ca in calcite I, II and III are similar. Fluorescence measurements such as those presented here can be carried out at high pressures and high temperatures, and our results therefore imply that this approach is complementary to Raman- and IR spectroscopic studies.

Financial support from the DFG (FOR2125) is gratefully acknowledged.

Figure 1 Pressure-dependence of the fluorescence of aragonite:Eu^{3+}. Filled symbols: P increase, open symbols: P decrease, line: linear fit

Figure 2. Pressure-dependence of the fluorescence of calcite:Eu^{3+}. Filled symbols: P increase, open symbols: P decrease, upper line: Guide to the eye, bottom line: linear fit

References
[1] Merlini et al., *Min. Mag.* **78** 225-233 (2014).

[2] M. M. Fernandes, et al. *J. Colloid Interface Sci.* **321** 323 (2008).

Figure 1

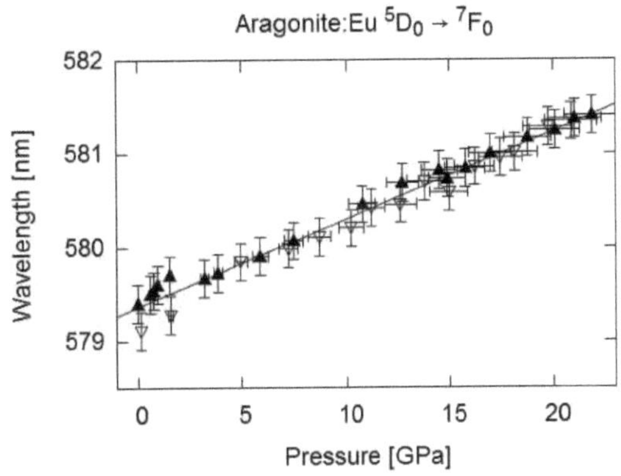

Figure 2

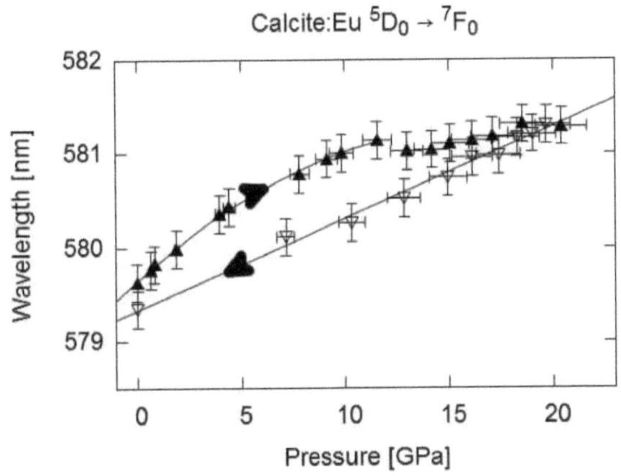

P114

A new micro-furnace for "in situ" high-temperature single crystal X-ray diffraction measurements

M. Alvaro[1], C. Marciano[1], C. Domeneghetti[1], F. Nestola[2], R. Angel[2]
[1]*University of Pavia, Department of Earth and Environmental Sciences, Pavia, Italy*
[2]*University of Padova, Geosciences, Padova, Italy*

The measurement of accurate thermal expansion coefficients of single crystals by diffraction is far more difficult than the measurement of compressibilities. First, the unit cell parameter and volume changes induced upon temperature variation are typically far smaller than those induced by the readily-accessible range of pressure of 0 -10 GPa. Second, temperature stabilization of an open furnace for diffraction is challenging, whereas pressure in a diamond-anvil cell does not change with orientation of the cell.

Here we present a newly-developed H-shaped Pt-Pt/Rh resistance micro-furnace for in-situ high-temperature single-crystal X-ray diffraction measurements. The compact design of the furnace together with the long collimator-sample-detector distance allows us to perform measurements up to $2\theta = 70°$. The micro-furnace is equipped with a water cooling system that allows a constant thermal gradient to be maintained that in turn guarantees thermal stability with oscillations smaller than 5°C in the whole range of operating temperature from room-T to 1200°C. The furnace has been built for use with a conventional 4-circle Eulerian geometry diffractometer equipped with point detector and automated with the SINGLE software (Angel and Finger 2011) that allows the effects of crystal offsets and diffractometer aberrations to be eliminated from the refined peak positions by the 8-position method (King and Finger 1979), and thus maximize precision in unit-cell measurements. The SINGLE software has been modified to reduce chimney effects in the furnace and thus further improve the stability by (i) restricting the χ circle movements to between -90° and +90°, if desired; (ii) optimizing the order of measurements to minimize χ circle movements (iii) imposing a waiting time after large angular movements on χ.

Temperature calibration has been performed iteratively by combining measurements with a standard small diameter thermocouple mounted in the same conditions as the sample together with the lattice parameter determination of materials with known thermal expansion behavior including quartz and pure silicon.

Angel RJ, Finger LW (2011) JAppCryst, 44:247.

King HE, Finger LW (1979) JAppCryst, 12:374.

This work was supported by ERC starting grant 307322, INDIMEDEA, to F. Nestola and SIR grant RBSI140351, MILE DEEp project, to M. Alvaro.

P115
Anomalous compression of cristobalite-like phosphorus oxonitride

M. Bykov[1], E. Bykova[1], D. Baumann[2], W. Schnick[2], M. Hanfland[3], L. Dubrovinsky[1], N. Dubrovinskaia[4]
[1] *University of Bayreuth, Bayerisches Geoinstitut, Bayreuth, Germany*
[2] *Ludwig Maximilian University, Department of Chemistry, Munich, Germany*
[3] *ESRF, Grenoble, France*
[4] *University of Bayreuth, Laboratory of Crystallography, Bayreuth, Germany*

The pseudo-binary system P_3N_5/P_2O_5 has been investigated intensively because the properties of related ceramic materials are promising for industrial applications. A mid-member of this system is phosphorus oxonitride (PON), whose chemical stability is essential for its use as an insulator or fireproofing material. This compound has attracted significant attention as a ternary base compound of electrolytes for rechargeable thin-film Li/Li-ion batteries. Phosphorus oxonitride is an isoelectronic analogue of silica (SiO_2) with the charge-balanced substitution $P^{5+} + N^{3-} = Si^{4+} + O^{2-}$. The crystal structures of the polymorphic forms of SiO_2 and PON are built of tetrahedral SiO4 and PO2N2 units, respectively. At present, five modifications of PON have been identified. Four of them are isostructural to known silica polymorphs, α-quartz [1], β-cristobalite-[2, 3], moganite- [4] and coesite-type [5,3]. The fifth one, δ-PON [6], has a structure type different from any of the silica modifications.

In this contribution we will discuss the details of the compression mechanism of cristobalite-like PON (*cri*-PON). The high-pressure behavior of *cri*-PON has been studied by means of single-crystal X-ray diffraction in a diamond anvil cell. The analysis of the structural changes in a pressure range from 0.0001 to 55 GPa revealed that the compression mechanism of cri-PON is nontrivial. First of all, cri-PON possesses negative linear compressibility along the *c*-axis. Furthermore, the volume-pressure data cannot be adequately described by any of the available third-order equations of state (EoS). The compression of *cri*-PON can only be described by an EoS of at least fourth-order. Refined parameters for the fourth-order Birch-Murnaghan EoS are K_0 = 59.8(7) GPa, K' = 1.78(17), K'' = 0.054(5) GPa^{-1}. The anomalous positive value of K'' means that the rate, at which *cri*-PON becomes stiffer, increases with increasing pressure, which is unusual for standard materials.

At ambient conditions *cri*-PON has a distorted β-cristobalite structure. Upon increasing pressure, the topology of cri-PON continuously changes towards a so-called collapsed cristobalite structure (Figure 1). Below 25 GPa, the compression involves only tetrahedral tilting with negligible changes of the intertetrahedral P-(O,N) distances. Only above 25 GPa, changes in the P-(O,N) distances start to play a noticeable role in the overall compression. At 53 GPa, the total change in the volume of P-(O,N)4 tetrahedra is 5%, while the total change of the unit cell volume is 35%. The compression of *cri*-PON is an example of the collapse of a framework structure to a close-packed structure form in a continuous manner without a phase transition. Such polyhedra tilting mechanisms are of great importance in determining the physical and thermodynamic properties of solids and result in continuous rather than abrupt changes of these properties. Such processes may also play a role in the densification of glasses and amorphous solids.

Figure 1. Illustration of the compression mechanism of *cri*-PON.

[1] Leger, J. *et al.* J. Phys. Chem. Solids, 60, 145-152 (1999).

[2] Leger, J. *et al.* Phys.Chem. Miner. 28, 388-398 (2001).

[3] Bykov, M. *et al.* Acta Cryst. E71, 1325-1327 (2015)

[4] Chateau, C. *et al.* Am. Mineral. 84, 207-210 (1999).

[5] Baumann, D. *et al.* Angew. Chem. 54, 4388-4391 (2015).

[6] Baumann, D. *et al.* Angew. Chem. 51, 4707-4709 (2012).

Figure 1

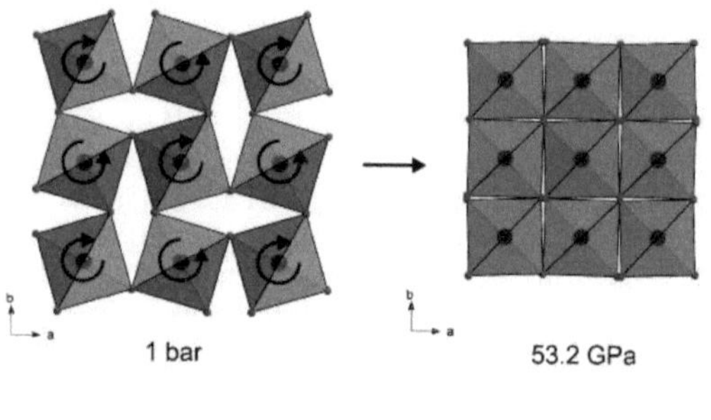

P116
Low-temperature phase-transitions of |$Na_6(H_2O)_8$| $[ZnPO_4]_6$

L. Robben[1], T. M. Gesing[1]
[1] *Uni Bremen - FB2 - Institut für anorganische Chemie und Kristallographie, Chemische Kristallographie fester Stoffe, Bremen, Germany*

The only known sodalite with a zincophosphate framework is the hydrosodalite |$Na_6(H_2O)_8$| $[ZnPO_4]_6$. It crystallizes in space group P-43n , is easy to synthesize by several different synthesis routes and has an extraordinary low thermal stability. This sodalite decomposes in a two step-process by gradually losing its water molecules to $NaZnPO_4$ [1]. The comparatively small lattice parameter of 883.613(4) pm at room-temperature makes symmetry changes due to framework distortions likely but up to now the temperature-dependent behaviour of this sodalite below room-temperature was not examined.

Samples of |$Na_6(H_2O)_8$|$[ZnPO_4]_6$ were investigated by differential scanning calorimetry (DSC) from 113 K to 300 K and Raman spectroscopy from 83 K to 300 K. The DSC shows an exothermic signal in the cooling curve between 181.1(1) K (onset) and 164.4(1) K (offset) with maxima at 172.1(2) K and 175.5 (1) K corresponding to an energy loss of 0.732(1) kJ/mol. Upon heating an endothermic signal (-1.141(2) kJ/mol) between 176.4(1) K (onset) and 201.7(1) K (offset) with minima at 188.8(2) K and 198.8(1) K could be observed. Both signals indicate a two-step process, the one in the heating curve more pronounced than the one in the cooling curve (Fig.1). The signal shape as well as the hysteresis indicates a displacive phase transition. The Raman spectra of the asymmetric T-O-T framework vibration and its autocorrelation analysis show a reduction of the sodalites symmetry while cooling between 173 K and 163 K and between 103 K and 93 K (which is not accessible in the DSC) by multiple degeneration. Fitting the strongest band shows a softening of the mode with decreasing temperature and a splitting into two (153 K) respectively three modes (103 K) (Fig. 2), which is indicative for displacive phase transition [2].

The contribution presents a detailed analysis of the spectroscopic data of |$Na_6(H_2O)_8$|$[ZnPO_4]_6$ at low temperatures and suggestions for possible structure models, which are able to describe this displacive phase transitions and thus the spectroscopic results.

Figure 1: Low-temperature DSC curves of |$Na_6(H_2O)_8$|$[ZnPO_4]_6$ (left), fitting of the cooling (middle) and of the heating curve (right) with two peak functions.

Figure 2: Temperature-dependent peak position of the strongest asymmetric T-O-T framework vibration and its neighbouring peaks after splitting. The red line is a linear fit ($\omega(T)=965.9(2)\ cm^{-1}+0.014(1)\ K^{-1}*T$).

References

[1] L. Robben, T.M. Gesing, Temperature-dependent framework-template interaction of |Na6(H2O)8|[ZnPO4]6 sodalite, J. Solid State Chem. 207 (2013) 13-20. doi:10.1016/j.jssc.2013.08.022.

[2] Dove, M.T., Structure and Dynamics - An atomic view of materials, Oxford, 2003, pp. 247-272

Figure 1

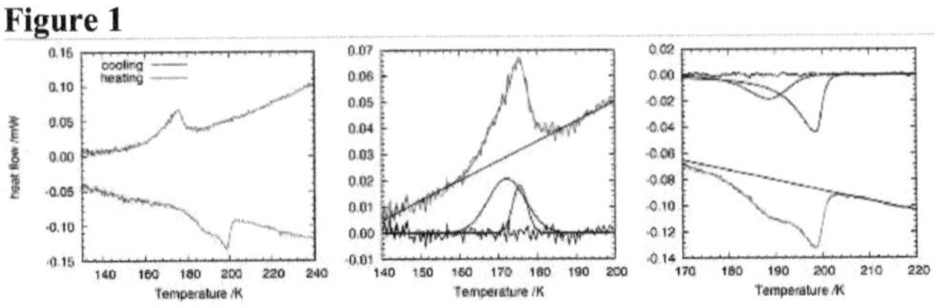

be shown in order to give an overview of the capabilities of this beamline particularly with respect to in situ diffraction experiments, which benefit a lot from the high photon energy, if large/thick sample environments have to be penetrated.

P118

CoMg olivine: cation partitioning, thermal expansion and structural variation studied by in situ neutron and synchroton powder diffraction

P. Schmid-Beurmann[1], H. Kroll[1], A. Kirfel[2], P. Sutanto[2], A. Sell[1], J. Büscher[1]
[1]*Institut für Mineralogie, Universität Münster, Münster, Germany*
[2]*Steinmann Institut, Universität Bonn, Bonn, Germany*

Olivine group minerals are orthorhombic orthosilicates $(A,B)_2[SiO_4]$, (SG *Pbnm*), where A and B denote octahedrally coordinated, divalent cations like Mg, Mn, Fe, Co, Ni. These cations reside in two non-equivalent sites, M1 (I) and M2 (m), so that the smaller M1 and the larger M2 octahedra differ with respect to size, site symmetry and distortion. Depending on cation species and temperature, the structure accommodates a wide range of different distributions of A and B over M1 and M2, ranging from ordered to disordered to anti-ordered arrangements. When the larger A cation resides in the larger M2 site, the configuration is termed "ordered" so that "anti-ordered" denotes B in that site.

The objectives were: (1) Investigation of the Co,Mg order/disorder behaviour. (2) Modelling of the Co,Mg equilibrium distribution using electron density derived parameters. (3) Description of the thermal expansion by isothermal equation-of-states (EoS) combined with the thermal Mie-Grüneisen EoS.

The CoMg olivine sample was prepared at 1450°C by solid-state reaction, then quenched. Three powder diffraction experiments were carried out on two instruments (A. Kirfel and P. Sutanto) , two on the angle dispersive TOF-diffractometer ROTAX at the pulsed neutron spallation source ISIS, Rutherford Appleton Laboratories, Chilton, UK, and another one on the powder diffractometer installed at the synchrotron radiation beam-line B2, HASYLAB, Hamburg. Measurements were performed at temperatures between 23°C and 1050°C.

During quench, the olivine acquired a partially anti-ordered Co,Mg site occupancy, $X_{M1,Co} > X_{M2,Co}$, which upon ensuing heating persisted up to 550 °C due to slow exchange kinetics. Above 600°C, internal equilibrium was readily obtained. The temperature variation of the Co,Mg site occupancies was modelled by $RT\ln(K_D) = -12080(\pm620) - 1.60(\pm0.58)T$ [J/mol]. The pronounced negative exchange enthalpy strongly favours the anti-ordered state, as does the positive exchange entropy so that CoMg olivine will remain in that very state up to the melting point.

From procrystal electron density distributions and ensuing topological analyses A. Kirfel (in Kroll et al., 2006) derived bonded radii and electronegativities, which allow modelling exchange enthalpies of intermediate olivines. Likewise, octahedral face distortions were found to measure exchange entropies. For CoMg olivine, the equations of Kroll et al. predict $\Delta H^0_{exch} = -11.0(\pm1.4)$ kJ/mol and $\Delta S^0_{exch} = 1.62(\pm0.01)$ $Jmol^{-1}K^{-1}$, respectively, in agreement with the experimental values given above.

The anisotropy of the thermal expansion was analysed using isothermal EoS's coupled with the thermal Mie-Grüneisen EoS. This coupling allowed for relating the anisotropic expansivity to an anisotropic thermal pressure. Both properties increase from Mg via CoMg to Co and Fe olivine. Expansivity and thermal pressure are linked to the temperature-dependent rotations of the M1 and M2 octahedra.

The Kumar and Birch-Murnaghan EoS's were found to deliver virtually indistinguishable results upon modelling thermal expansion.

References

Kroll H, Kirfel A, Heinemann R (2006) Order and anti-order in olivine II: Thermodynamic analysis and crystal-chemical modelling. Europ J Mineral 18, 691-704

P119

Laboratory Based x-Ray Powder Diffraction Down to Liquid Helium Temperatures

R. K. Kremer[1], S. Höhn[1], R. E. Dinnebier[1], F. Adams[1]
[1]*Max Planck Institute for Solid State Research, Stuttgart, Germany*

We describe a newly constructed versatile cryostat system that allows collecting laboratory based x-ray powder diffraction patterns in Debye-Scherrer geometry on air sensitive samples enclosed in capillaries down to liquid Helium temperatures. First test measurements on systems with magnetoelastic distortions at low temperatures are discussed.

P120

Neon-bearing ammonium metal formates: effect of inclusion at high pressure

I. Collings[1], E. Bykova[2], M. Bykov[2], S. Petitgirard[2], M. Hanfland[3], D. Paliwoda[3], N. Dubrovinskaia[1], L. Dubrovinsky[2]
[1]*University of Bayreuth, Laboratory of Crystallography, Bayreuth, Germany*
[2]*University of Bayreuth, Bayerisches Geoinstitut, Bayreuth, Germany*
[3]*European Synchrotron Radiation Facility, Grenoble, France*

The formation of neon compounds is extremely rare due to the inert nature of the neon atom. A different way to synthesise neon-bearing compounds is by neon encapsulation and insertions into network solids at specific temperature and pressure conditions.[1-5] The formation of neon-bearing network solids can give several insights on the properties of the empty and filled network states. For example, pore dimensions can be rigorously evaluated,[4] the thermodynamic stability may be modified,[2] and the mechanical properties of the filled and empty network states can be drastically different.[6, 7]

We study the high-pressure behaviour of dense metal-organic frameworks, namely ammonium iron and nickel formates [Figure 1(a)], using different noble gas atoms (neon and argon) as the pressure-transmitting medium to investigate (*i*) the pressure-induced structural evolution of these compounds and (*ii*) whether the small porous metal formate pockets (~6 Å) are of the right size to allow selective entry of neon. High-pressure single-crystal X-ray diffraction measurements were performed at the beamline ID09A (ESRF, Grenoble) up to ~4 GPa, and high-pressure Raman spectroscopy was measured at the Bayerisches Geoinstitut up to 10 GPa. Analysis of the X-ray diffraction data revealed that neon had entered into ammonium metal formates without any symmetry changes [Figure 1(b)], and that the neon occupancy could be refined close to 1 after ~1.5 GPa giving the chemical formula $Ne_x[NH_4][M(HCOO)_3]$, where x = occupancy. In contrast, by using argon as the pressure-transmitting medium, there was no inclusion in the ammonium metal formates upon pressurisation. The inclusion of neon has a drastic impact on the pressure-dependent behaviour of the ammonium metal formates: in particular it maintains the structural stability of the frameworks up to ~6 GPa, while the compression of the empty ammonium metal formates gives rise to structural distortions as low as 0.5-1.4 GPa.

Figure 1 Crystal structures of (a) $[NH_4][Fe(HCOO)_3]$ at ambient conditions and (b) $Ne_{0.4}[NH_4][Fe(HCOO)_3]$ at 0.4 GPa. The $[Fe(HCOO)_3]_5$ cages are illustrated next to each structure. Octahedral iron coordination is represented by polyhedra, oxygen atoms are in red, carbon in black, nitrogen in blue, hydrogen in pink and neon in grey.

References

[1] M. Saunders, H. A. Jimenez-Vazquez, R. J. Cross, S. Mroczkowski, M. L. Gross, D. E. Giblin, R. J. Poreda, *J. Am. Chem. Soc.* **1994**, 116, 2193-2194. [2] X. Yu, J. Zhu, S. Du, H. Xu, S. C. Vogel, J. Han, T. C. Germann, J. Zhang, C. Jin, J. S. Francisco, Y. Zhao, *PNAS* **2014**, 111, 10456-10461. [3] T. Plisson, G. Weck, P. Loubeyre, *Phys. Rev. Lett.* **2014**, 113, 025702. [4] Y. Finkelstein, A. Saig, A. Danon, J. E. Koresh, *J. Phys. Chem. B* **2003**, 107, 9170-9174. [5] R. Barrer, D. Vaughan, *T. Faraday Soc.* **1967**, 63, 2275-2290. [6] J. Haines, O. Cambon, C. Levelut, M. Santoro, F. Gorelli, G. Garbarino, *J. Am. Chem. Soc.* **2010**, 132, 8860-8861. [7] T. Sato, N. Funamori, T. Yagi, *Nat. Comm.* **2011**, 2, 345.

Figure 1

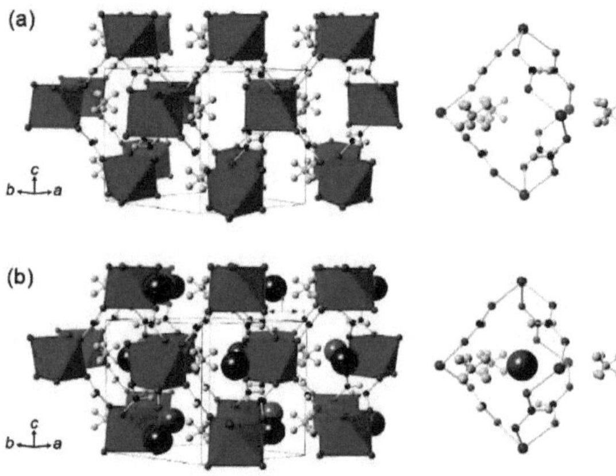

Molecular structural chemistry: Syntheses, structure and applications

P121
Charge Density Study of Distorting Tetrahedral and Octahedral Cu (II) Complexes

A. Wang[1], U. Englert[1]
[1]*RWTH Aachen University, Inorganic Chemistry, AACHEN, Germany*

How to understand the intermolecular interactions (especially the hydrogen bond and halogen bond) in the context of crystal packing and the utilization of such understanding in the design of new solids with desired physical and chemical properties is one of the most flourishing fields in crystal engineering [1]. Based on the high resolution X-ray data, experimental charge density provides us one perspective to study the interaction existing between two molecules. Furthermore, for those transition metal complexes, which electrons of *d*-orbital is not fully occupied, experimental charge density also leaves people some kind of clues, which can help to figure out how the distribution of *d*-orbital's configuration in different coordination situation is [2]. In this work, we will present different charge settings for central metal and refinement strategies based on the different data banks in aspheric atomic density expansion [3] for distorting tetrahedral and octahedral coordinating surrounding Cu^{2+}.

[1]Desiraju, G. R. Crystal Engineering: The Design of Organic Solids, Elsevier, 1989, Amsterdam.
[2]Mallinson, P. R., Koritsanzsky, T., Elkaim, E., Li, N., and Coppens, P. *Acta Cryst.* A44, 336(1988).
[3]Hansen, N. K. & Coppens, P. *Acta Cryst.*, **A34**, 909-921(1978).

Figure 1

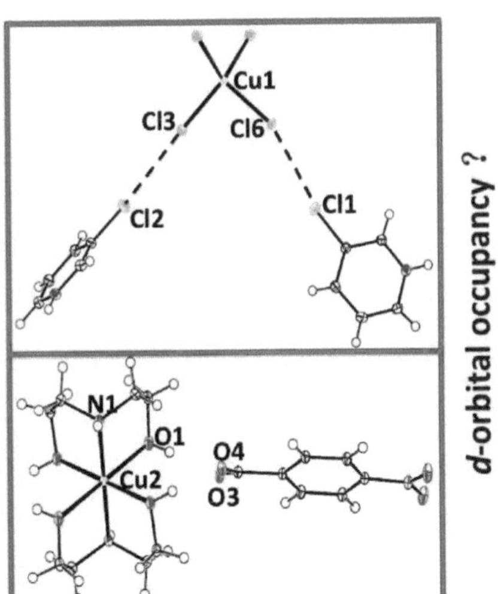

P122
in situ Crystallisation technique - an approach to circumvent crystal structures with Z'>1

G. Lipinski[1], B. Mallick[1], K. Merz[1]
[1]*Ruhr-University Bochum, Department of Inorganic Chemistry I, Bochum, Germany*

The Z'>1 structures are ones in which a significant molecular moiety occurs more than once in the asymmetric unit.[1] Desiraju defines structure with Z'>1 as crystals "on the way" in the sense of being frozen in some high energy kinetic forms.[2] He suggests further that the probability of appearance of a high Z' structure will depend largely on the crystallisation temperature. In those cases where kinetic and thermodynamic forms are one and the same, Z'>1 is not possible.[3] To analyze these phenomenon, Brock & Duncan have extensively surveyed a great range of alcohols and phenols.[4] According to their explanation, the tendency of the hydroxy compounds (ROH) to form cooperative hydrogen bonded chains of the O-H···O-H type is countered by the steric demands of the R-groups, and high Z' is one of the outcomes.

One suitable approach to circumvent crystal structures with Z'>1 could be the usage of *in situ* crystallization technique directly on the diffractometer.[5] To test this method we investigated the polymorphic crystallization behaviour of $(CF_3)_2CH$-OH. The fluorinated alcohol crystallizes in the monoclinic space group $P2_1$ with Z'=8.[6] Furthermore we studied the ability of $(CF_3)_2CH$-OH to form co-crystals with liquid co-former.[7]

[1] P. A. Wood, Cambridge Crystallographic Data Centre, 12 Union Road, Cambridge, CB2 1EZ, UK, **2002**.

[2] G. R. Desiraju, *CrystEngComm*, **2007**, 9, 91.

[3] K. Merz, V. Vasylyeva, *CrystEngComm*, **2010**, **12**, 3989.

[4] C. P. Brock, L. L. Duncan, *Chem. Mater.*, **1994**, 6, 1307.

[5] K. Merz, M.V. Evers, F. Uhl, R.I. Zubatjuk, O.V. Shishkin, *Crystal Growth & Design* **2014**, 14, 3124.

[6] A.Berkessel, J.A.Adrio, D.Huttenhain, J.M.Neudorfl, *J.Am.Chem.Soc.*, **2006**, 128, 8421.

[7] G. Lipinski, B. Mallik, K. Merz, *Crystal Growth & Design* in prep

Figure 1

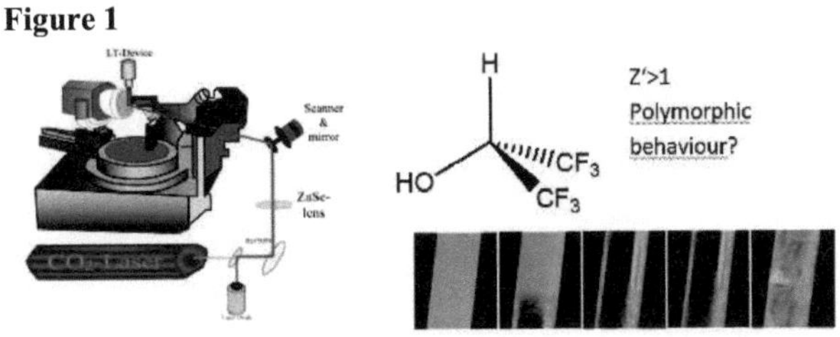

P123

Synthesis of New Functional, Catalytic Materials via Thermal Decomposition of Ag(I)/Yb(III) Bimetallic Coordination Networks

K.- N. Truong[1], M. Konkol[2], A. Schwedt[3], U. Englert[1]
[1]*RWTH Aachen University, Inorganic Chemistry, Aachen, Germany*
[2]*New Chemical Syntheses Institute (INS), Puławy, Poland*
[3]*RWTH Aachen University, Gemeinschaftslabor für Elektronenmikroskopie, Aachen, Germany*

Bimetallic coordination networks facilitate the preorganization of the two cations in three-dimensional space. The nanoscale, predefined distribution of the different metal cations is used to design new catalytic materials. Thermal decomposition of those bimetallic networks, *e.g.* removal of the organic linker, results in a bimetallic metal/metal oxide compound. Those decomposition compounds have been analysed by X-ray powder diffraction, SEM-SEI, SEM-BSE, EDX and XRF.

In previous works, a non-commercial acetylacetone derivative has been used successfully as an organic linker [*Dalton Trans.* **2012**, *41*, 4664-4673]. The decomposition compound containing silver and ytterbium oxide exhibited catalytic activity towards the decomposition reaction of nitrous oxide at 500 °C (commercially-available catalyst: 600 °C). The aims of the new project were to optimize the catalyst system by using cheaper, commercially-available ligands, such as EDTA (see scheme), and try to decrease the reaction temperature.

Figure 1

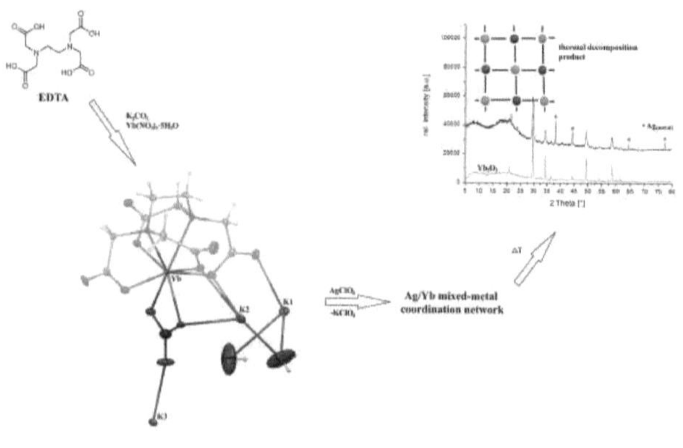

P124

Crystals giving birth to other crystals

C. Döring[1], L. Pinkert[1], C. Taouss[1], P. G. Jones[1]
[1]*Technische Universität Braunschweig, Institut für Anorganische und Analytische Chemie, Braunschweig, Germany*

We are interested in the structure and packing of urea and thiourea adducts with organic solvents such as morpholine, 1,4-dioxane, lutidines or pyrazines that contain nitrogen. Several structures with varying compositions have been obtained, in which the residues are connected by hydrogen bonds.[1-3] By recrystallization of these adducts from simpler organic solvents the range of composition can be extended; lowering the effective concentration of the nitrogenous component should also lower its mole fraction in the new adduct.

In the case of thiourea and 2,5-dimethylpyrazine, the composition of the first adduct was 4:3. From methanol, a solvate of composition 1:1:1 was obtained. These single crystals lose methanol at room temperature (under inert oil) and thereby convert to single crystals of the already known 4:3 adduct (Fig. 1).

Both structures were determined. In the 1:1:1 adduct, the thiourea molecules form chains of a well-known type involving $R^2_2(8)$ rings (Fig. 2); these are connected on the one side directly via pyrazine molecules and on the other by pyrazines and methanols, forming a layer structure. The 4:3 adduct has a more complex three-dimensional structure.

Further investigations will show if this phenomenon of solvate formation with subsequent loss of solvent can be observed with other urea or thiourea adducts.

Fig. 1 Conversion of the methanol solvate to the methanol-free adduct.

Fig. 2 Packing of the methanol solvate, layer parallel to (0-12).

[1] C. Taouss, P. G. Jones, *CrystEngComm.* **2014**, *16*, 5695-5704.

[2] C. Taouss, L. Thomas, P. G. Jones, *CrystEngComm.* **2013**, *15*, 6829-6836.

[3] P. G. Jones, C. Taouss, N. Teschmit, L. Thomas, *Acta Cryst.*, **2013**, *B69*, 405-413

Figure 1

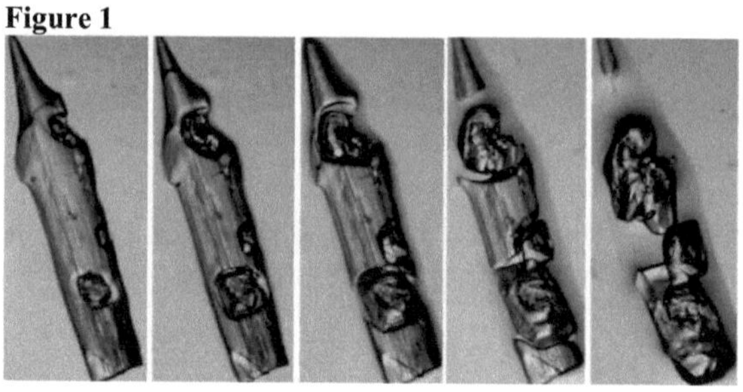

Figure 2

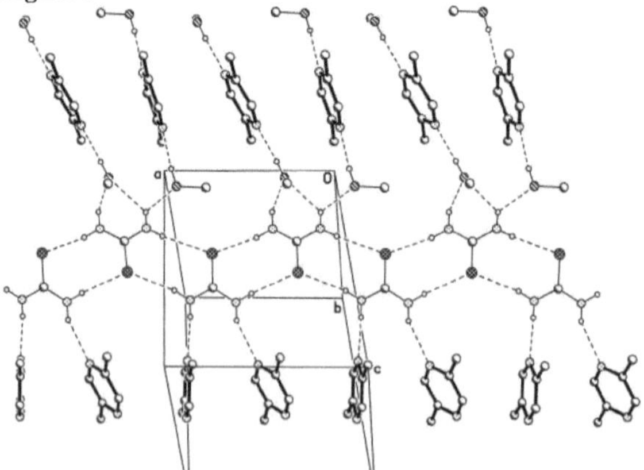

P125

Giant Pentaphosphaferrocene-Based Supramolecules: Supramolecular Synthons

E. Peresypkina[1], A. Virovets[1], M. Scheer[1]
[1]*University of Regensburg, Lst. Prof. M. Scheer, Regensburg, Germany*

The inorganic analog of ferrocene, pentaphosphaferrocene [$Cp^RFe(\eta^5-P_5)$] (R = alkyl or aryl), is a versatile building block in supramolecular chemistry. Coordination of Cu^+ atoms to a rigid and chemically stable pentagonal P_5 ring results in the formation of giant supramolecules with hollow inorganic cores of fullerene and fullerene-like topology [1]. The [$Cp^*Fe(\eta^5-P_5)$] and [$Cp^{Et}FeP_5$] (Cp^{Et} = η^5-C_5Me_4Et) give rise to a series of the supramolecules with 80- and 90 non-carbon atoms in the inorganic core, [$(CuX)_{20-n}(Cp^*Fe(\eta^5-P_5))_{12}$] and [$(Cp^*Fe(\eta^5-P_5))_{12}(CuX)_{25}(MeCN)_{10}$] (n = 0-4, X = Cl, Br), respectively. In addition, 110-vertex nano-capsule [Cp^*FeP_5]@[$\{Cp^*FeP_5\}_9\{CuCl\}_{10}$] and 99-vertex supersphere C_{60}@[$Cu_{26}Cl_{26}(H_2O)_2\{Cp^*FeP_5\}_{13}(CH_3CN)_9$] were obtained. Surprisingly, the giant supramolecules of 2.2-3.2 nm in diameter form perfect single crystals of 0.1-0.5 mm in size suitable for laboratory X-ray structural analysis. Our study on the crystal packing demonstrates the applicability of the concept of supramolecular synthons [2] even to so large yet rigid molecules.

The outer surface in the supramolecules is confined by the organic cyclopentadienyl ligands. In the crystals of the supramolecules, directed intermolecular interactions of two types are observed in addition to van der Waals interactions: π-π interactions between Cp* or Cp^{Et} ligands and σ-π interactions between terminal halides X and Cp^R ligands. They occur from structure to structure, and their geometrical characteristics are reproducible. Thereby, two supramolecular synthons for the giant supramolecules, halogen-π and π-π (Fig.), are found.

The halogen-π synthon is the most frequent and is observed in all crystal structures of 90-vertex and in almost all structures of 80-vertex supramolecules. The typical Cl-π and Br-π distances amount to 3.20 - 3.61 and 3.29 - 3.60 Å, respectively. The other opportunity of agglomeration of supramolecules in the crystal is π-stacking interactions, i.e. the formation of a π-π synthon. Two types of these were observed with classical face-to-face and slipped geometry. Typical interplanar Cp*-Cp* distances 3.34-3.52 Å are common for a π-stacking. This synthon plays a rather supporting role, because no crystal structure is based solely on it. The analysis of the crystal packing undoubtedly proves that despite the huge size the supramolecules do aggregate in the crystal *via* specific patterns of halogen-π and π-π synthons.

This work was supported by ERC grant AdG339072-SELFPHOS.

[1] E. Peresypkina, C. Heindl, A. Schindler et al (2014)*Z. Kristallogr.*, **229**, 735, and references therein.
[2] G. Desiraju (1995), *Angew. Chem. Int. Ed.*, **34**, 2311.
[3] C. Janiak (2000), *Dalton Trans.*, 3885.

Figure 1

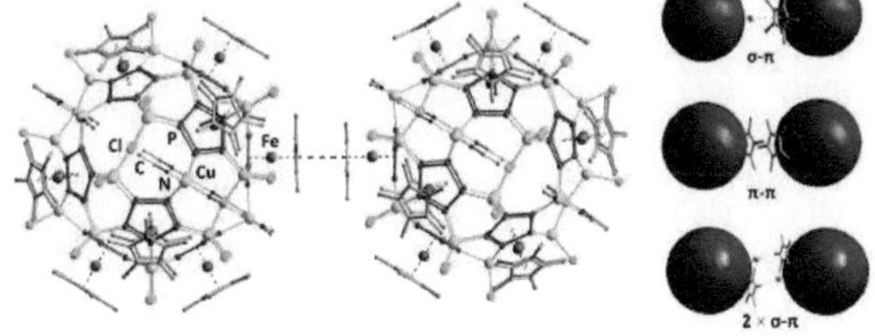

P126

B_2neop_2 revisited: A Simple Compound with an Intricate Structure Determination.

C. Kleeberg[1], T. B. Marder[1,2]
[1]*Technische Universität Braunschweig, Institut für Anorganische und Analytische Chemie, Braunschweig, Germany*
[2]*Universität Würzburg, Institut für Anorganische Chemie, Würzburg, Germany*

Introduction

The diborane(4) derivative B_2neop_2 (neop: neopentylglycolato, $(OCH_2)_2C(Me)_2$) was first described in 1998 by *Norman, Marder, Green* and co-workers.[1a] Since then this compound has become a widely used reagent in borylation reactions.[2] At the time a single crystal X-ray structure determination on B_2neop_2 (at 160 K) was conducted but severe crystallographic problems were encountered. Whilst the molecular structure obtained confirmed the connectivity it represented a chemically not plausible conformation and connected with that it exhibited whole-molecule disorder and apparently pyramidalized boron atoms (see Figure). The authors recognised that this were artefacts due do the fact that the B_2neop_2 molecule is situated on a centre of inversion as well as a three-fold axis (in *R*-3*m*) and stated "*pyramidalization at boron are undoubtedly artefacts and not genuine*", but no resolution of these problems could be provided.[1b]

Objectives

We reinvestigated the solid-state structure of B_2neop_2 and can finally provide a well-behaved single crystal structure analysis of B_2neop_2 as well as a rationalisation of the problems encountered.[3]

Material and Methods

Single crystals of B_2neop_2 were studied by X-ray diffraction at variable temperatures between 290 K and 100 K.

Results

From 290 K to 140 K data in agreement with the earlier results were obtained leading to a structure with the above mentioned problems. However, upon cooling to 120 K the *c* axis is doubled (now space-group *R*-3*c*) and as the B_2neop_2 molecules are no longer residing on a centre of inversion a more plausible molecular structure exhibiting a planar environment at the boron atoms is found, albeit whole-molecule disorder due to the boron atoms situated on a three-fold axis is still present. Cooling to 100 K effects again a change in the cell metrics (doubling of all axes) and the diffraction pattern still possesses an apparently three-fold symmetry. However, closer analysis of the data revealed that they are better described by a monoclinic cell ($P2_1/n$) together with three-component twinning.[

Conclusion3]

In conclusion a detailed, temperature depended, X-ray diffraction study lead to a (re-)determination of the molecular structure of B_2neop_2 and, moreover, to an understanding and resolution of the problems encountered in a previous structure determination (see Figure).

Figure: Apparent molecular structures, including disorder (thin lines), of B_2neop_2 at 160 K, 120 K and 100 K. Hydrogen atoms omitted for clarity; B: dark grey, O: medium grey, C: light grey.[1b]

References:

[1] (a) F. J. Lawlor, N. C. Norman, N. L. Pickett, E. G. Robins, P. Nguyen, G. Lesley, T. B. Marder, J. A. Ashmore, J. C. Green, *Inorg. Chem.* **1998**, *37*, 5282-5288. (b) W. Clegg, M. R. J.

Elsegood, F. J. Lawlor, N. C. Norman, N. L. Pickett, E. G. Robins A. J. Scott, P. Nguyen, N. J. Taylor, T. B. Marder, *Inorg. Chem.* **1998**, *37*, 5289-5293.

[2] See for example: (a) T. Ishiyama, N. Miyaura, *Chem. Record* **2004**, *3*, 271-280. (b) I. A. I. Mkhalid, J. H. Barnard, T. B. Marder, J. M. Murphy, J. F. Hartwig, *Chem. Rev.* **2010**, *110*, 890-931.

[3] C. Kleeberg, T. B. Marder, *manuscript in preparation.*

Figure 1

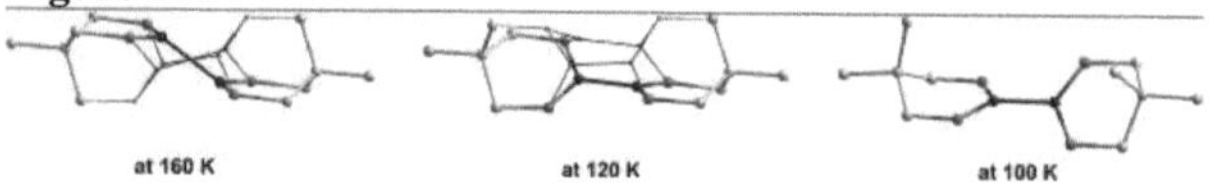

P127

Cu-Ni bimetallic compound with chain-like crystal structure

I. Kočanová[1], J. Kuchár[1], M. Orendáč[2], J. Černák[1]
[1]*P. J. Safarik University in Kosice, Faculty of Science, Institute of Chemistry, Department of Inorganic Chemistry, Kosice, Slovakia*
[2]*P. J. Šafárik University in Košice, Faculty of Science, Institute of Physics, Košice, Slovakia*

Alternating spin chains represent an interesting class of magnetic materials. There exist a theoretical prediction for spin chains, in which integer and half spins alternate, that these at sufficient low temperature $T << J$ will display ferromagnetic properties despite the presence of antiferromagnetic interactions [1]. Such alternating spin chains can be chemically modelled by linking Cu (S=1/2) and Ni (S=1) atoms into a structurally one-dimensional system. Previously we have studied and characterized several Cu-Ni compounds with chain-like crystal structures, e.g. Cu(*cyclam*)Ni$(CN)_4$ (*cyclam* = 1,4,8,11-tetraazacyclodecane) [2], but these comprised diamagnetic nickel atoms. As a continuation of our study on copper-nickel complexes [3] we have synthetized complex $[Cu(cyclam)Ni(NCS)_4(H_2O)_2]_n$ under mild chemical conditions. Its chain-like crystal structure is built up of Cu(II) and Ni(II) atoms linked by bridging NCS^- groups. Cu(II) atom exhibits hexacoordination with the *cyclam* ligand bound in the equatorial plane and the axial positions are occupied by S-bonded bridging thiocyanato ligands. The Ni(II) atom is also hexacoordinated by four N-bonded NCS^- groups and two aqua ligands in *trans* fashion. Weak hydrogen bonding interactions of the O-H···S type links the formed chains into supramolecular structure. The magnetic studies performed in the temperature range 1.8—300 K revealed that the studied complex obeys the Curie-Weiss law and the fitted magnetic parameters are $\theta = 0.54$ K, C = 1.64 K and $g = 2.19$.

References

1. Furusaki A., Sigrist M., Lee P. A., Tanaka K., Nagaosa N.: Phys. Rev. Lett. 73 (1994) 2622.

2. Černák J., Kuchár J., Stolárová M., Kajňaková M., Vavra M., Potočňák I., Falvello L. R., Tomás M.: Trans. Met. Chem. 35 (2010) 737.

3. Černák J., Kočanová I., Orendáč M.: Comm. Inorg. Chem. 33 (2012) 2.

P128

The application of silver X-ray's in single crystal diffraction

A. Griffin[1]
[1]*Rigaku Oxford Diffraction, Oxford, Great Britain*

Rigaku Oxford Diffraction has a range of high-flux, low maintenance micro-focus sources to suit even the most challenging of samples. Weakly diffracting proteins and small molecule crystals can be studied using **Nova** (Cu) sources. The Nova source is ideal for absolute structure determination on pure organic compounds. Samples that suffer from absorption can be looked at with the **Mova** (Mo) source, which also opens up the possibility to perform charge density experiments, as well as high pressure studies.The new Rigaku Oxford Diffraction micro-focus silver source, **Silva,** (Fig. 1) extends the ability to study a wider range of samples than using traditional copper or molybdenum wavelengths.

Figure 1. The Silva Source

The shorter wavelength of the silver source (0.56 Å compared to 0.71 Å for molybdenum) means that it is now possible to measure very highly-absorbing samples, obtain better completion on high pressure setups, and choose to either push the resolution limits of charge density experiments, or perform single-theta charge density measurements. Here we describe a variety of experiments with the Silva source and its enhancements over other wavelengths with the results showing that, in many cases, silver radiation outperforms in terms of data quality and data collection time over other available wavelengths.

Figure 1

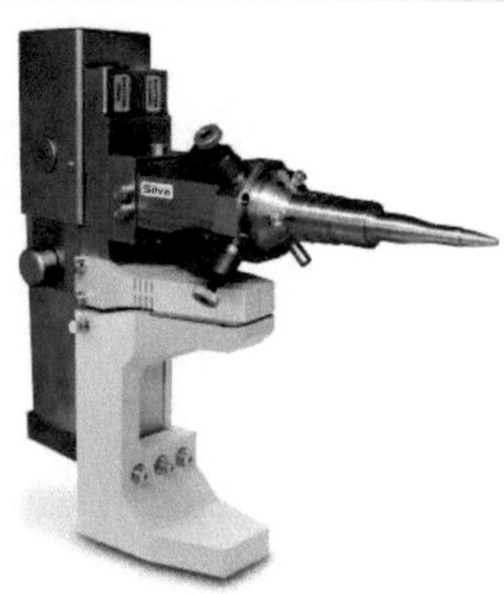

P129

6-Propyl-2-thiouracil vs. 6-methoxymethyl-2-thiouracil: enhancing the hydrogen bonded synthon motif by replacement of a methylene group with an oxygen atom

W. M. Hützler[1], E. Egert[1]
[1]*Goethe-University Frankfurt am Main, Institute of Organic Chemistry and Chemical Biology, Frankfurt am Main, Germany*

A major objective of crystal engineering is the search for reliable non-covalent intermolecular interaction motifs ("synthons") that can be applied for the design of new solids with desired properties.[1] The antithyroid drug 6-propyl-2-thiouracil[2] (PTU) and the structurally related compound 6-methoxymethyl-2-thiouracil (MOMTU) both contain an *ADA* hydrogen bonding site (A = acceptor, D = donor). Therefore, both compounds should be capable of forming a mixed *ADA*/*DAD* N—H···O/N—H···N/N—H···S synthon consisting of three different hydrogen bonds with suitable coformers like 2,4-diaminopyrimidine (DAPY) or 2,4,6-triaminopyrimidine (TAPY). As MOMTU comprises an additional oxygen atom the hydrogen bonding pattern should be enhanced compared to PTU. Cocrystallization experiments yielded one solvated cocrystal salt of PTU with DAPY, (I). In contrast, for MOMTU each one salt with DAPY, (VI), and one cocrystal-salt with TAPY, (VII), were obtained (Fig. 1). In order to analyze the preferred conformation of the side chain of MOMTU crystallization experiments yielded four solvates, (II) – (V), showing anti periplanar conformations for both torsion angles ω and φ (Fig. 1 and Table 1). All four solvates show the formation of dimers in the crystal packing either by two N—H···O or two N—H···S hydrogen bonds. Comparing their hydrogen bond patterns, PTU showed an *AA*/*DD* motif in (I), while, in contrast, MOMTU formed the desired *ADA*/*DAD* synthon in the cocrystal salts (VI) and (VII). Moreover, the conformation of the side chain of MOMTU in (VI) and (VII) deviated strongly from its preferred conformation and the oxygen atom O8 formed hydrogen bonds in both structures.

[1] Desiraju, G. R. (2013). *J. Am. Chem. Soc.* **135**, 9952-9967.

[2] Visser, T. J., Van Overmeeren, E., Fekkes, D., Docter, R. & Hennemann, G. (1979). *FEBS Lett.* **103**, 314-318.

Fig. 1: Structures of the compounds used, composition of the crystals and definition of the torsion angles ω and φ.

Table 1: Selected torsion angles of (I) - (VII).

Figure 1

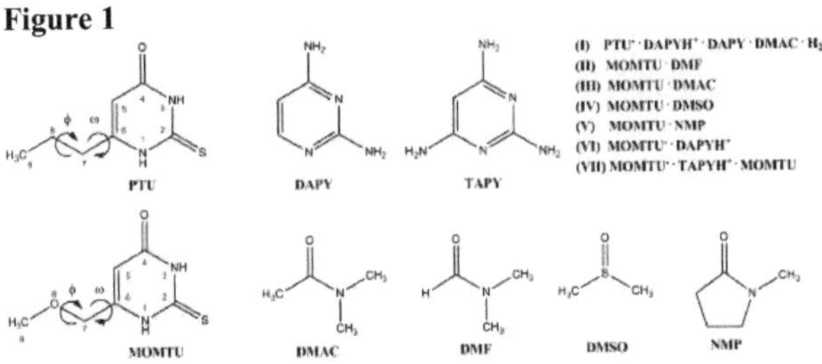

Figure 2

Crystal	ω [°]	ϕ [°]
(I)	172.48 (14)	179.29 (15)
(II)	−176.14 (14)	−176.91 (14)
(III)	167.01 (14)	171.95 (15)
(IV)	−179.58 (15)	−178.12 (17)
(V)	177.35 (15)	179.35 (15)
(VI)	−153.8 (3)	**79.4 (4)**
(VII)	**15.5 (3)**	180.0 (2)

P130
"in situ" Crystal growth and low temperature structure determination of Meerwein's boron trifluoride-carboxylic acid-adducts

S. Parizek[1], W. Frank[1]
[1]*Heinrich-Heine-Universität Düsseldorf, Lehrstuhl für Material- und Strukturforschung, Düsseldorf, Germany*

Adducts of boron trifluoride and carboxylic acids have been known since their first description by Meerwein et al. 1927 and Bowlus et al. 1931.[1,2] Although these highly acidic compounds offer numerous applications as catalysts in organic synthesis[3,4], for example the oligomerization of olefins[5], *Fries*-rearrangements[4], or acylations[6], their crystal structures had yet to be determined.

By using inert low temperature crystallization and preparation techniques [7], we succeeded in obtaining crystals suitable for single-crystal X-ray diffraction. Thus, the structures of boron trifluoride—acetic acid (1/1) (**1**), boron trifluoride—propionic acid (1/1) (**2**), boron trifluoride—valeric acid (1/1) (**3**) and boron trifluoride—caprylic acid (1/1) (**4**) could be determined.

The compounds were prepared by direct action of gaseous boron trifluoride on the corresponding acids. The liquid product (compounds **2**, **3** and **4**, respectively) was sealed in an argon filled thin walled glas capillary and cooled to a temperature slightly below or above its melting point, usually between -10° C to -70° C. Modifying a procedure introduced by Renaud and Fourme[8], a cooled or warmed copper wire was applied to the capillary for the induction and control of the growth of single crystals.

In all compounds boron trifluoride coordinates to the carbonylic oxygen of the carboxylic acid. These adducts form dimers via hydrogen bonding, a single fluorine atom acting as a hydrogen bond acceptor. Figure 1 shows the dimers of **3**. Geometric parameters of the hydrogen bond in this compound are d(O···F) 2.616(5) Å, d(O-H) 0.89(5) Å, d(H···F) 1.80(5) Å, angle(O-H···F) 151(5)°. The crystal packing is influenced to a great extend by the alkyl chains of the carboxylic acids. The crystal data of all compounds are given in table 1.

References:

[1] H. Meerwein, *Liebigs Ann. Chem.* **1927,** *455*, 227-253.
[2] H. Bowlus, J. A. Nieuwland, *J. Am. Chem. Soc.* **1931.** *53*, 3835-3840.
[3] M. Dawidowski et al., *Tetrahedron* **2012**, *68*, 8222-8230.
[4] J. L. Boyer et al. *J. Org. Chem.* **2000**, *65*, 4712-4714.
[5] J.A. Brennon, US Patent 3769373 **1973**
[6] P. Carty, M. Dove, *J. Org. Chem.* **1970**, *21*, 195-192.
[7] M. Veith, W. Frank, *Chem. Rev.* **1988**, *88*, 81-92
[8] M. Renaud, R. Fourme, *Bull. Soc. Fr. Mineral. Cryst.* **1966**, *89*, 243

Table 1: Crystal Data

Compound	**1**	**2**	**3**	**4**
Formula	$C_2H_4BF_3O_2$	$C_3H_6BF_3O_2$	$C_5H_{10}BF_3O_2$	$C_8H_{16}BF_3O_2$
M [g/mol]	127.86	141.88	169.94	212.02

Crystal system	monoclinic	triclinic	triclinic	triclinic
Space group	*C2/c*	$P\bar{1}$	$P\bar{1}$	$P\bar{1}$
a [Å]	9.757(2)	6.2236(12)	6.3164(12)	5.797(1)
b [Å]	7.8800(16)	6.4486(13)	6.3184(13)	6.746(2)
c [Å]	13.178(3)	8.1905(16)	11.640(2)	15.514(4)
α	90°	80.97(3)°	99.97(3)°	97.93(3)°
β	101.98(3)°	77.42(3)°	101.19(3)°	100.68(3)°
γ	90°	84.29(3)°	105.23(3)°	91.63(3)°
V [Å^3]	989.1(3)	316.12(12)	427.21(18)	589.6(3)
Z	8	2	2	2
R(I >2σ I)	R_1 = 0.0344, wR_2 = 0.0746	R1 = 0.1044, wR2 = 0.2313	R1 = 0.0552, wR2 = 0.1584	R1 = 0.1079, wR2 = 0.2287
R(all Data)	R_1 = 0.0564, wR_2 = 0.0780	R1 = 0.1846, wR2 = 0.2528	R1 = 0.1580, wR2 = 0.1811	R1 = 0.3136, wR2 = 0.2616
GooF	0.896	0.973	0.890	0.964

Figure 1

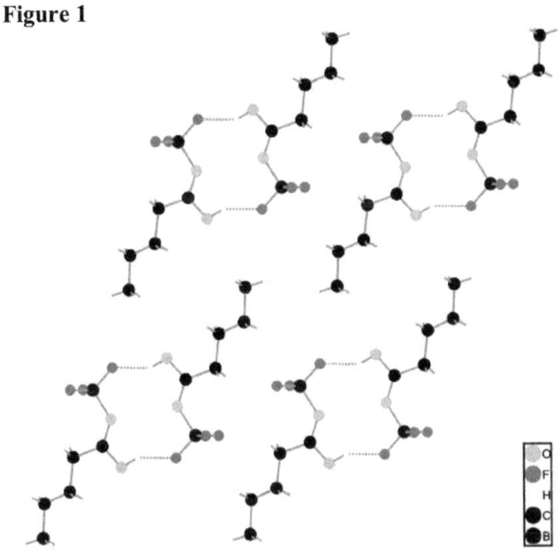

[5] D.A. Haynes, W. Jones, W.D.S. Motherwell, *CrystEngComm* **2006**, 8, 830.

[6] M. J. A. Bowker, P.H. Stahl, C.G.Wermuth, *Handbook of Pharmaceutical Salts*, **2002**,VHCA,Wiley-VCH: New York.

[7] R. Bhogala, S. Basavoju , A.Nangia, ***CrystEngComm*, 2005, 7**, 551.

[8] K. Molčanov, B. Kojić-Prodić, *CrystEngComm* **2010**, 12, 925.

[9] D. Enkelmann, G. Lipinski, K. Merz, Cryst. Growth Des., **in prep.**

Figure 1

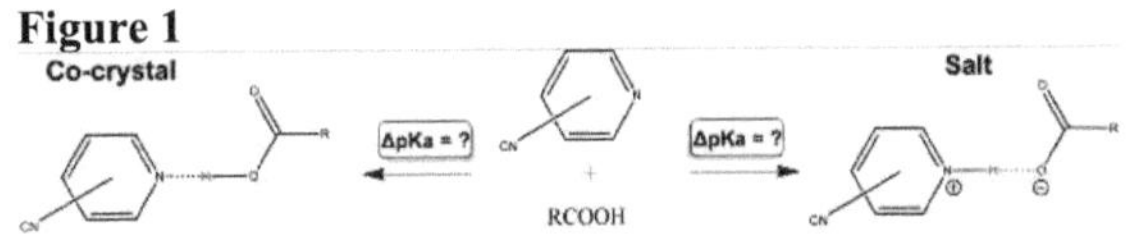

P131

Co-crystal Systems of Cyanopyridines and Carboxylic acids

D. Enkelmann[1], G. Lipinski[1], K. Merz[1]
[1]*Ruhr-Universität Bochum, Bochum, Germany*

In the recent past, multicomponent crystals like co-crystals, salts and solvates became more and more interesting due to their upcoming applications in pharmaceutical use and materials science.[1] The understanding of the formation of those compounds is an essential part of crystal engineering to achieve more knowledge about multicomponent crystal aggregation and subsequently to use this information to tune chemical and physical properties of active pharmaceutical ingredients (API) like solubility, bioavailability, melting point and stability.[2, 3]

In some cases it can be preferable to synthesize a co-crystalline compound instead of a salt, for example based on the poor predictability of salt structures in respect of their chemical and stoichiometric composition.[4] To select suitable compounds for targeted co-crystal growth the pK_a-rule is a helpful tool. The ΔpK_a of a two component system (defined as $\Delta pK_a = pK_{a[base]} - pK_{a[acid]}$) can give reliable information concerning co-crystal or salt formation.[5-8]

In order to study the applicability of this method, selected compounds were chosen for a co-crystal screening. Different cyanopyridines, acting as bases with relatively low pK_a-Values were intended to be formed into co-crystals via solution crystallization with selected carboxylic acids as co-crystal-former.

In our studies, the pK_a-rule turned out to be a very accurate instrument for specific co-crystal approach. Depending on this rule we were able to design various co-crystals consisting of pyridine derivatives and carboxylic acids.

[1] D. Yan *et al., Angew. Chem. Int. Ed.*, **2011**, 50, 12483.

[2] Vishweshwar *et al., J. Pharm. Sci.*, **2006**, 95, 499.

[3] Ning *et al., Int. J. Pharm.*, **2011**, 419, 1.

[4] C.B. Aakeröy, M.E. Fasulo, J. Desper, Mol. Pharm., **2007**, 4, 317.

P132

Synthesis, Crystal Structure, Spectroscopic Studies and Biological activity of 1-(5-bromo-2,3-dimethoxybenzylidene)-2-(pyridine-2-yl) hydrazine Schiff Base Molecule

T. Tunç[1], S. A.Ali Noma[2], M. Erzengin[2,3]
[1]*Aksaray University, Science/(Physics), Aksaray, Turkey*
[2]*Aksaray University, Aksaray, Turkey*
[3]*Aksaray University, Chemistry, Aksaray, Turkey*

Synthesis, Crystal Structure, Spectroscopic Studies and Biological activity of 1-(5-bromo-2,3-dimethoxybenzylidene)-2-(pyridine-2-yl) hydrazine Schiff Base Molecule

In the first part of this study, 1-(5-bromo-2,3-dimethoxybenzylidene)-2-pyridine-2-yl) hydrazine Schiff Base Molecule was synthesized. In order to prepare this hydrazone derivative; solution of 2-hydrazinopyridine was gradually added to the solution of 5-bromo-2,3-dimethoxybenzaldehyde, and a pale-yellow color crystal product was obtained. Later, the structure of the compound has been elucidated by elemental analysis, H-NMR, FTIR, UV-VIS scanning and XRD techniques. Diffraction measurements were performed at room temperature on a Bruker SMART BREEZE CCD diffractometer [1]. The structures were solved by direct methods using SHELXS-97, and refined by a full matrix least-squares procedure using the program SHELXL-97 [2]. All nonhydrogen atoms were refined using anisotropic displacement parameters and hydrogen atoms were included in their idealized positions and refined isotropically, except for H8B, H8C and H2. The structure of B-2,3-MBPyH crystallizes into a triclinic lattice with space group P-1. In the molecule, the benzene and pyridine rings and the hydrazine bridge are practically coplanar. Molecular geometry calculations were performed by CCDC-Mercury [3]. Atom-labeling scheme is shown in figure 1. Molecules of the title compound adopt an E-configuration about the azomethine C=N double bond, with an N2-N1-C9-C6 torsion angle of 178.81(19)°. The N2-C10 and N1-N2 bond distances are between 1.372(3)°A and 1.362(3) °A, respectively. The N1-C9 double bond is 1.272(3) °A.
In the second part of the study, synthesized compound was subjected to screening for their free radical scavenging activity by DPPH method. DPPH free radical scavenging activity is one of the methods for assay of antioxidant activity. The radical scavenging activity of hydrazone derivative was determined by a spectrophotometric method based on the reduction of ethanol solution of DPPH (2,2-diphenyl-1-picryl hydrazyl) using the method of Blois [4]. The hydrazone compound was dissolved in

DMSO, and various concentrations (1, 2, 3, 4 and 5 mg/mL) of each compounds were prepared. A similar concentration of BHT (butylated hydroxytoluene) was used as a control. The compound providing 50 % inhibition (IC50) was calculated from the graph plotting inhibition percentage against tested hydrazone derivative concentration (Fig.2). Antioxidant activity of BHT was investigated to compare with the hydrazone derivative compound as control.

In the third part of the study, human serum paraoxonase 1 (hPON1) was purified using two-step procedures, namely ammonium sulphate precipitation and Sepharose-4B-L-tyrosine-1-naphthylamine hydrophobic interaction chromatography. SDS–polyacrylamide gel electrophoresis of the purified enzyme have been performed. Using the paraoxon as a substrate, we determined the KM and Vmax values of the purified enzyme. Finally, in vitro inhibition effect of synthesized novel compound on purified hPON1 was investigated by using paraoxon as a substrate.

REFERENCES

[1] Bruker (2012). APEX2, SAINT and SADABS. Bruker AXS Inc. Madison, Wisconsin, USA.
[2] Sheldrick, G. M. (2008). Acta Cryst. A64, 112–122.
[3] C. F. Macrae, P. R. Edgington, P. McCabe, E. Pidcock, G. P. Shields, R. Taylor, M. Towler, J. van De Streek, J. Appl. Cryst. 2006, 39, 453–457.
[4] Blois M.S., Antioxidant determinations by the use of a stable free radical, Nature, 1958; 181: 1199-1200.

Figure 1 View of the crystal packing of 1-(5-bromo-2,3-dimethoxybenzylidene)-2- (pyridine-2-yl) hydrazine.

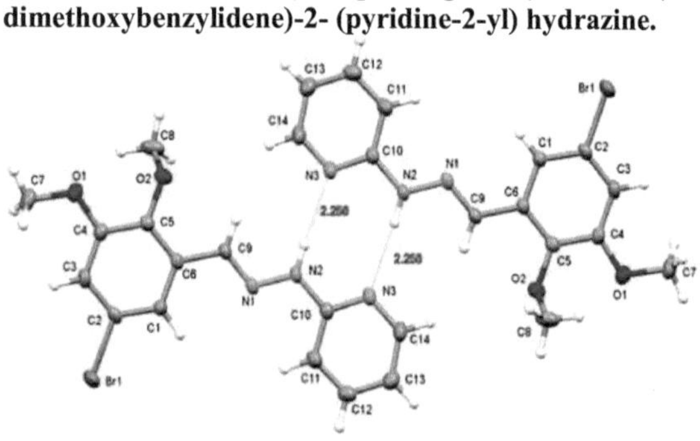

Figure 2 Scavenging activity of 1-(5-bromo-2,3-dimethoxybenzylidene)-2- (pyridine-2-yl) hydrazine on DPPH radicals.

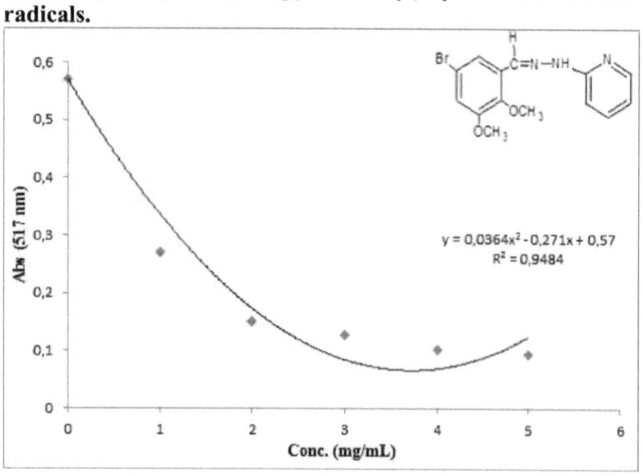

P133

Gemfibrozil chain conformation and packing requirements

C. Schwalbe[1,2], M. Ramirez[1], B. Conway[3], P. Timmins[4]
[1] *Aston University, School of Life & Health Sciences, Birmingham, Great Britain*
[2] *Cambridge Crystallographic Data Centre, Cambridge, Great Britain*
[3] *University of Huddersfield, Pharmacy, Huddersfield, Great Britain*
[4] *Bristol-Myers Squibb, Moreton, Great Britain*

Gemfibrozil [Gem, systematic name 5-(2,5-dimethylphenoxy)-2,2-dimethyl-pentanoic acid] is a drug used to reduce lipid levels. For efficient packing, the highly polar carboxyl group at one end of Gem and the bulky nonpolar dimethylphenyl group at the other end must both find suitable interaction partners. They can be well separated if the intervening alkoxy chain has an all-*trans* conformation.

Structures of Gem free acid (XAPVOA, [1]) and four salts (GICHEG, GICHIK, GICHOQ, GICHUW [2]) appear in the Cambridge Structural Database. Additionally, we have determined the structures of the adamantylammonium (GAdam), benzylammonium (GBenz), cyclobutylammonium (GCbut), cyclohexylammonium (GChex), and triethanolammonium (GTea) salts. The carboxyl groups in paired molecules of the free acid form $R_2^2(8)$ dimers. In (aryl)alkylammonium salts the ionic groups form ladders of hydrogen-bonded rings, either $R_4^3(10)$ or alternating $R_4^2(8)$ and $R_4^4(12)$. Cations with one or two OH groups build these into the chains of rings; but in the Tris salt the cations form hydrogen bonded layers [2], still with two hydrogen bonds to one carboxylate O atom and one to the other.

In the Ph-O-$(CH_2)_3$-CMe_2-COO linkages of the free acid, the first four are *trans* and the last two are *gauche*. This arrangement persists in most of the salts, creating ether O...carboxyl O distances in the range 5.17-5.89 Å; but with Tris the Gem anions lengthen one such O...O distance to 6.32 Å. In its complex with human transthyretin [3], with Gem in the thyroxine T4 binding site, the O...O distances are stretched by almost a further 1 Å. In these cases the carboxyl group is *trans* to the rest of the chain. The prevalence of the 4-*trans*,2-*gauche* conformation suggests that it has low energy. In the Tris salt there is no extra hydrogen bonding to COO^-, nor to COOH in the transthyretin complex, so any energy needed to alter the Gem conformation must be supplied by packing interactions.

We thank the U.K. National Crystallography Service for data collection.

[1] Bruni, B., Coran, S., Di Vaira, M. & Giannellini, V. (2005). *Acta Cryst.* E**61**, o1989-o1991.

[2] Cheung, E. Y., David, S. E., Harris, K. D. M., Conway, B. R. & Timmins, P. (2007). *J. Solid State Chem.* **180**, 1068-1075.

[3] Iakovleva, I., Brännström, K., Nilsson, L., Gharibyan, A. L., Begum, A., Anan, I., Walfridsson, M., Sauer-Eriksson, A. E. & Olofsson, A. (2015). *J. Med. Chem,* **58**, 6507-6515.

P134

New X-ray Sources and X-ray Detectors for Fast and Efficient in-House Experiments

M. Adam[1], T. Stürzer[1], H. Ott[1]
[1]*Bruker AXS, Marketing, Karlsruhe, Germany*

Collecting better data — faster. The availability of modern systems operating the most powerful microfocus X-ray sources, more sensitive, larger, CPAD technology based X-ray detectors and last but not least, more comprehensive, professionally developed and maintained software suites are the proven driving forces in single crystal X-ray diffraction (SC-XRD) and has led to a tremendously increased number of published structures.

However, up to now a typical diffraction experiment was performed following the workflow which was established for the first automated goniometers with scintillation detectors: a pre-experiment to establish the orientation matrix is followed by the complete data collection and subsequent structure evaluation. Often the crystal screening step has been relatively time-consuming, taking several minutes and revealing little more than the unit-cell dimensions. However, the latest generation of X-ray detectors and microfocus X-ray sources, such as the PHOTON II and the IμS 3.0, used in the second generation of the D8 QUEST and D8 VENTURE (Figure 1) allows crystallographers to retire the traditional experiment design for the majority of samples. Shutterless mode used by modern X-ray detectors has eliminated read-out overhead times and reduced overall data collection times by up to 75 %. This overhead free data collection opens the opportunity to quickly collect complete data sets, often in less than 60 s.

Data sets collected using this new approach have significant advantages, as they provide a huge amount of information, and of course considerably more reflections for indexing than traditional matrix scans. The information gained can already be used for a detailed analysis of the diffraction pattern, e.g. to establish twinning. The new APEX3 software suite offers the tools required for this analysis. These tools also help in revealing crystal defects or other sample deficiencies, which may suggest the merit of picking another crystal. From the quick data collection typically the structure determination can be performed, completely revealing the chemical constitution of the crystal. This facilitates confident decision-making on whether to discontinue or continue the experiment, and provides more information to better plan the more meticulous data collection.

Typical day-to-day examples will be presented, taking advantage of the D8 QUEST and D8 VENTURE second generation, featuring PHOTON II detector and the IμS 3.0, the first ever microfocus source designed for X-ray diffraction.

Keywords: microfocus X-ray source (IμS); CPAD X-ray detector; fast - high quality data collection

Figure 1

P135

Monomers, Dimers, Tetramers, Clusters - the Manifold Appearances of Functionalised Cycloheptatrienide-Pd Complexes

C. Jandl[1], A. Pöthig[1], K. Öfele[2], F. E. Kühn[2], W. A. Herrmann[2]
[1]*Technische Universität München, Catalysis Research Center, Garching, Germany*
[2]*Technische Universität München, Lehrstuhl für Anorganische Chemie I, Garching, Germany*

In recent time we have intensely studied the class of imidazolium-substituted cycloheptatrienide-Pd complexes which are not only highly dynamic and versatile in their coordination modes but also in the motifs the Pd-cores adopt together with coordinating halides. The basic structure is monomeric due to the zwitterionic organic ligand requiring two halides for an even charge balance, but exchanging one halide for a non-coordinating anion forces the complexes to form oligomers.[1] Although these mostly appear as dimers, there is an equilibrium with higher oligomers in solution and in one case we were able to co-crystallise the dimer and the tetramer of a complex.[2] The core of the dimeric compounds consists of two Pd(II) ions bridged by two halides which can either be in plane creating a Pd_2Br_2-square or bent out of plane allowing for a Pd-Pd-interaction.[1,2] The core of the tetramer is an eight-membered ring with alternating Pd- and halide ions and it furthermore displays two different coordination isomers of the substituted cycloheptatrienide rings two of which coordinate in the β-position while the others coordinate in the γ-position.[2] The crystal structure therefore captures images of two dynamic processes in solution: the fluxionality of the organic ligand and the dynamic equilibrium of different oligomers. Finally, the ligand-system can also support clusters as has been shown by the serendipitous discovery of a Pd_3-sandwich complex formed upon decomposition of a sample of monomeric compound.[3]

[1] C. Jandl, K. Öfele, F. E. Kühn, W. A. Herrmann, A. Pöthig, *Organometallics*, **2014**, *33*, 6398-6407.

[2] *in preparation.*

[3] K. Mantas-Öktem, K. Öfele, A. Pöthig, B. Bechlars, W. A. Herrmann, F. E. Kühn *Organometallics* **2012,** *31*, 8249-8256.

Figure 1: Core-fragments of the molecular structures of various imidazolium-substituted cycloheptatrienide-Pd complexes.

Figure 1

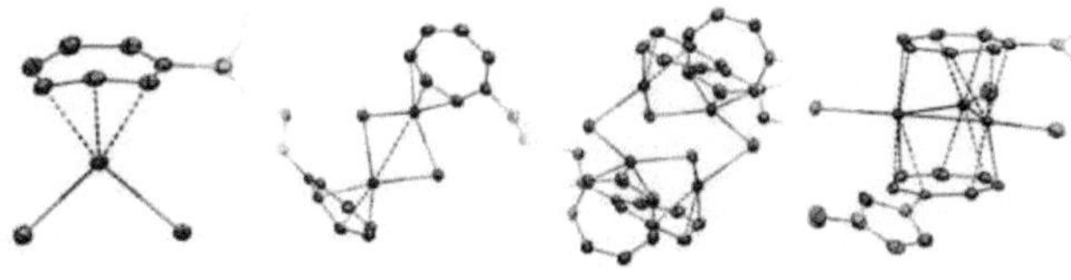

P136

Synthesis of 1-(5-bromo-2,4-dimethoxybenzylidene)-2-(pyridine-2-yl) hydrazine Schiff Base Molecule, Evaluating of its Free Radical Scavenging Activity and Determination of *its, in vitro, Inhibitory Effect on Purified Human Serum Paraoxonase-1*

M. Erzengin[1], S. A. A. Noma[1], T. Tunç[2]
[1]*Aksaray University, Chemistry, Aksaray, Turkey*
[2]*Aksaray University, Physics, Aksaray, Turkey*

In this study, a new hydrazone derivative namely 1-(5-bromo-2,4-dimethoxybenzylidene)-2-(pyridine-2-yl) hydrazine was synthesized and its structure has been elucidated by XRD, H-NMR, FTIR, UV-VIS scanning and elemental analysis techniques. The following figure illustrates the chemical synthesis of this compound.

Figure 1. Synthesis of 1-(5-bromo-2,4-dimethoxybenzylidene)-2-(pyridine-2-yl) hydrazine

In order to prepare this compound, a solution of 2-hydrazinopyridine prepared in hot methanol was gradually added to a solution of 5-bromo-2,4-dimethoxybenzaldehyde prepared in hot acetone at 298-303 K, with constant stirring. Since the condensation reaction is carried out in acidic conditions, synthesis was made at about pH 4-5. The procedure was completed in approximately 30 min.The yellow color crystal was left inside the hood for 3 days and then filtered. The yellow precipitate formed was dissolved in hot methanol, refluxed for 3 hours, and kept in a laboratory fume hood for 2 days for recrystallization and then filtered.

The free radical scavenging activity of the newly synthesized compound was measured by using 1,1-diphenyl-picryl-hydrazil (DPPH). Because, DPPH radical scavenging is a widely used method to evaluate antioxidant activities in a relatively short time compared with other methods.The effective concentration of the compound required to scavenge DPPH radicals by 50 % (IC_{50} value) was obtained by plotting percantage inhibition versus different concentrations of the studied compund. The synthesized compound was found to be significantly active against DPPH radicals with the IC_{50} value of 0.185 mg/mL . On DPPH assay, higher radical scavenging activity is associated with a lower IC50 value.

Paraoxonase-1 (PON1: EC 3.1.8.1) is a calcium-dependent enzyme associated with high-density lipoproteins (HDLs) and has a protective effect against oxidation of low-density lipoproteins (LDLs) in mammals. In this study, human serum paraoxonasel (hPON1) was purified using two-step procedures, namely ammonium sulphate precipitation and Sepharose-4B-L-tyrosine-1-naphthylamine hydrophobic interaction chromatography. SDS-polyacrylamide gel electrophoresis of the purified enzyme showed a single band with an apparent MW of 43 kDa. The purified enzyme had a specific activity of 21.22 U/mg. The overall purification fold and yield were found to be % 17.484 and 561.375 respectively. Furthermore, using the paraoxon as a substrate, we determined the K_m and V_{max} values of the purified enzyme, as 0.018496 mM and 114.955 U/mL, respectively. In this study, *in vitro* inhibition effect of the newly synthesized *compound* on purified hPON1 were also investigated by using paraoxon as a substrate.The result showed that hydrazone derivative inhibited the hPON1 enzyme activity in a concentration-dependent fashion. Studied hydrazone derivative, showed a good inhibitory effect on hPON1 activity, with the low IC_{50} value of 0.0215 mg/mL. The present study has demonstrated that hPON1 activity is very highly sensitive to studied hydrazone derivative.

References

[1] Tunc T., Tezcan H., Ahin E., and Dilek N. (2012). Synthesis, Crystal Structure, and Spectroscopic Studies of N-(4-Bromobenzylidene)-N-(2-Pyridyl) Hydrazine Schiff Base Molecule. Molecular Crystals and Liquid Crystals, 552:1, 194-208.
[2] Blois M.S. (1958). Antioxidant determinations by the use of a stable free radical. Nature, 181: 1199-1200.
[3] Erzengin M., Basaran I., Cakir U., Aybey A., Sinan S. (2012). In vitro inhibition effect of some dihydroxy coumarin compounds on purified human serum paraoxonase 1 (PON1). Appl Biochem Biotechnol, 168:1540-1548.

Figure 1

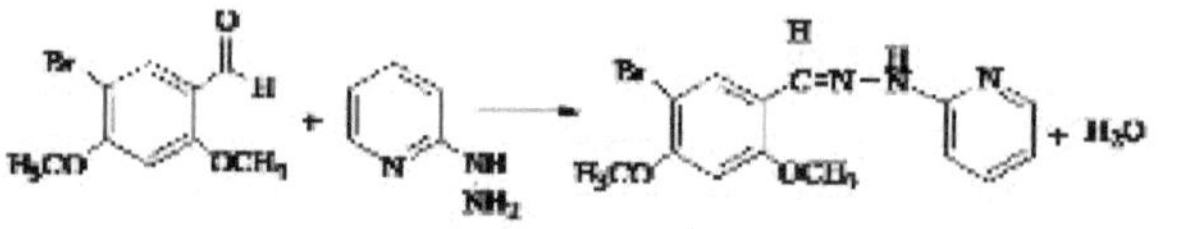

P137

Transition metal complexes with defined cavities for supramolecular applications

A. Pöthig[1]
[1]*TU München, Catalysis Research Center, Garching, Germany*

Supramolecular strategies including molecular catalysts are on the rise in the field of homogeneous catalysis. Generally, tailored ligands are applied which e.g. place several metal centers in a certain distance and surrounding to interact with substrate molecules, as it is the case for the so-called "Pacman" complexes.[1] Another approach is to regioselectively introduce functional groups into the ligand backbone, which can interact with the substrate and prealign/preactivate it towards the metal center as it is the case for the "Hangman" complexes.[2] We recently reported the synthesis of a new family of cyclophanes based on alkylene-bridged imidazolium- and pyrazole-moieties.[3] These macrocyclic compounds can act as a ligand precursors for dinuclear bowl-shaped transition metal complexes. We now extended the synthetic scope towards ligand and metal variation and observed different coordination modes. The resulting complexes possess a defined spatial arrangement featuring differently shaped and sized cavities and can act as host molecules for selective self-assembly with different guests. All compounds were analyzed by means of standard characterization methods as well as single crystal XRD experiments, which are presented.

[1] Love et al., *Chem. Commun.*, **2009**, 3154-3165.
[2] Nocera et al., *ChemSusChem* **2013**, *6*, 1541 - 1544.
[3] Philipp J. Altmann, Christian Jandl, Alexander Pöthig* *Dalton Trans.*, **2015**, *44*, 11278-11281.

P138

High-Brightness Microfocus Sources for Chemical Crystallography

J. Graf[1], A. Kleine[1], J. Wiesmann[1], C. Michaelsen[1]
[1]*Incoatec GmbH, Geesthacht, Germany*

Modern microfocus X-ray sources define the state-of-the-art for single crystal diffraction and small-angle scattering. These sources are usually combined with multilayer X-ray mirrors which are excellent X-ray optical devices for beam shaping and preserving the source brightness.

Microfocus rotating anode systems deliver flux densities in the range of 10^{11} photons/s/mm^2 at power loads of a few tens of kW/mm^2. However, these sources are expensive and need regular and sometimes time-consuming maintenance. Low power microfocus sealed tube sources, such as the Incoatec Microfocus source IµS, represent an interesting low-maintenance alternative to rotating anode generators. Power loads of several kW/mm^2 in anode spot sizes of < 50 µm deliver a small and bright beam. Flux densities of up to 10^{10} photons/s/mm^2 can be achieved in a focused beam suitable for most protein crystals and poorly diffracting small molecule samples. The IµS is available for Cu, Ag and Mo and delivers a beam with a FWHM in the range of 100 µm. Since its launch in 2006 more than 600 IµS were sold for a large variety of applications in diffractometry for biology, chemistry, physics and material science.

The latest generation of the IµS, the IµS 3.0, is the first microfocus X-ray source that is designed for X-ray diffraction resulting in a gain in intensity of about 30% compared to the IµS High-Brilliance. We will be discussing the main parameters of the IµS and demonstrate how they affect the performance of the newest generation of IµS. We will be presenting selected results to demonstrate the impact of these modern microfocus X-ray sources on the data quality for applications in chemical crystallography.

P139

$Na_2(H_2O)_8[B_{12}Br_{12}]$: Disodium Dodecabromo-*closo*-Dodecaborate Octahydrate - An *anti*-$CdCl_2$-Type Crystal Structure with an Unusual $[Na_2(H_2O)_8]^{2+}$ Dimer Cation

I. Tiritiris[1], T. Schleid[1]
[1]*University of Stuttgart, Institute for Inorganic Chemistry, Stuttgart, Germany*

Introduction

The synthesis of salt-like compounds with $[B_{12}X_{12}]^{2-}$ (X = Cl - I) anions is known in literature for more than forty years [1]. In the last 15 years a few perhalogenated *closo*-dodecaborates mainly with the heavy alkali-metal cation Cs^+, such as $Cs_2[B_{12}Cl_{12}] \times SO_2$ [2], $Cs_2[B_{12}X_{12}]$ (X = Cl - I) [3, 4] and $Cs_2[B_{12}X_{12}] \times 2\ CH_3CN$ (X = Cl - I) [4, 5] were fully characterized by single-crystal X-ray diffraction. With the lighter alkali metals, except for $[Li_2(SO_2)_8][B_{12}Cl_{12}]$ [6], much less structural data are available in literature [4]. Compounds containing the weakly coordinating $[B_{12}X_{12}]^{2-}$ (X = F - I) anions are useful salts, because they can be utilized for the stabilization of new and unusual electrophilic cations in the solid state.

Structure Description

The new compound $Na_2(H_2O)_8[B_{12}Br_{12}]$ crystallizes in the monoclinic space group $P2_1/c$ with a = 942.03(5) pm, b = 1646.31(9) pm, c = 1024.42(6) pm and β = 99.810(3)° (Z = 4). All atoms reside at the general *Wyckoff* position 4*e*. The crystal structure can be described best as an *anti*-$CdCl_2$-type structure arrangement (Figure 1). The sodium cations are building up a cubic close packing, in which the *quasi*-icosahedral $[B_{12}Br_{12}]^{2-}$ anions (Figure 2, *right*) reside in the octahedral voids of every alternating layer. Every Na^+ cation is thus surrounded trigonal non-planar by three $[B_{12}Br_{12}]^{2-}$ anions. Characteristic in the crystal structure of $Na_2(H_2O)_8[B_{12}Br_{12}]$ is the centrosymmetric dimer cation $[Na_2(H_2O)_8]^{2+}$ ($[(H_2O)_3Na(H_2O)_2Na(H_2O)_3]^{2+}$, d(Na-O) = 233 - 251 pm, d(Na×××Na) = 395 pm; Figure 2, *left*), which is stabilized in the solid state by the weakly-coordinating dianion $[B_{12}Br_{12}]^{2-}$. Thus the coordination sphere around sodium can be described as a distorted trigonal bipyramid of water molecules (Figure 2, *left*). The apical Na-O distances are longer (d(Na-O) = 250 - 251 pm) than the equatorial ones (d(Na-O) = 233 - 241 pm). A similar dimer cation was observed in the crystal structure of the lithium salt $[Li_2(SO_2)_8][B_{12}Cl_{12}]$ [6]. In addition, two $[B_{12}Br_{12}]^{2-}$ anions are coordinating via edges of bromine atoms at the sodium cations of the dimers (d(Na-Br) = 359 pm), resulting in a distorted octahedral coordination environment for Na^+ (5 ′ O + 1 ′ Br). The cluster bond lengths are in typical ranges (d(B-B) = 176 - 182 pm, d(B-Br) = 194 - 196 pm) and thus similar to the data for $Cs_2[B_{12}Br_{12}]$ [3, 4]. The crystal structure is stabilized by O-H^{d+}×××$^{-d}$O-H hydrogen bonds between the water molecules and O-H^{d+}×××$^{-d}$Br-B hydrogen bonds between the water molecules and the $[B_{12}Br_{12}]^{2-}$ anions.

References

[1] W. H. Knoth, H. C. Miller, J. C. Sauer, J. H. Balthis, Y. T. Chia, E. L. Muetterties, *Inorg. Chem.* **1964**, *3*, 159-167.
[2] V. Geis, K. Guttsche, C. Knapp, H. Scherer, R. Uzun, *Dalton Trans.* **2009**, *2009*, 2687-2694.
[3] I. Tiritiris, Th. Schleid, *Z. Anorg. Allg. Chem.* **2004**, *630*, 1555-1563.
[4] I. Tiritiris, *Doctoral Thesis*, Univ. Stuttgart **2004**.
[5] I. Tiritiris, Th. Schleid, *Z. Anorg. Allg. Chem.* **2001**, *627*, 2568-2570.
[6] C. Knapp, C. Schulz, *Chem. Commun.* **2009**, *2009*, 4991-4993.

Figure 1. View at crystal structure of $Na_2(H_2O)_8[B_{12}Br_{12}]$ along [001].

Figure 2. $[Na_2(H_2O)_8]^{2+}$ dimer cation (*left*) and *quasi*-icosahedral $[B_{12}Br_{12}]^{2-}$ anion (*right*) in the crystal structure of $Na_2(H_2O)_8[B_{12}Br_{12}]$.

Figure 1

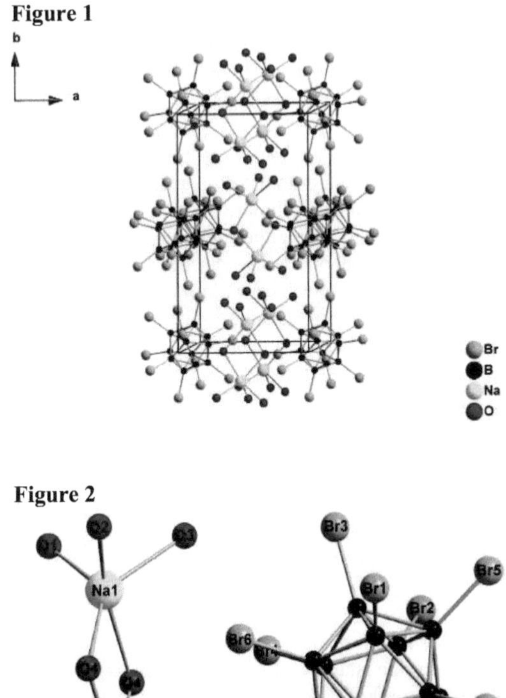

Figure 2

Time resolved crystallography

P140

Phase transitions of SiO_2 under dynamic compression and up to 900 °C

E.- R. Carl[1,2], A. Danilewsky[2], H.- P. Liermann[3], U. Mansfeld[4], F. Langenhorst[4], L. Ehm[5], T. Kenkmann[1]
[1] *University of Freiburg, Geology, Freiburg, Germany*
[2] *University of Freiburg, Crystallography, Freiburg, Germany*
[3] *DESY, Photon Science, Hamburg, Germany*
[4] *Institut für Geowissenschaften, Mineralogie, Jena, Germany*
[5] *Stony Brook University, Mineral Physics, Stony Brook, United States*

SiO_2 as α-quartz is one of the most abundant minerals in the Earth's crust and an important engineering material. Under dynamic and non-hydrostatic conditions, its phase diagram becomes intricate and exhibits many different crystalline phases. Due to this great complexity, previous investigations of SiO_2 at high pressure and temperature have left many unanswered questions concerning the behavior of the material.

Our time-resolved X-ray diffraction experiments under dynamic compression provide insights into the kinetics of high-pressure phase transitions of α-quartz making use of fast data acquisition now available at synchrotron sources.

The experiments were conducted at the Extreme Conditions Beamline at PETRA III at DESY, Hamburg, Germany, in a membrane-driven diamond anvil cell. A mixture of α-quartz powder and Au flakes for internal pressure calibration was used. The maximal pressure was 66 GPa and different compression rates up to 3.0 GPa/s were applied at room temperature. During compression and decompression, diffractograms were collected every 1-10 seconds. Additional experiments were conducted at elevated temperature up to 900 °C and low compression rates using a graphite resistance heater. After surveying all experimental data, single potential diffractograms were selected for further analysis with the Rietveld method.

All experiments reveal that α-quartz transforms to stishovite during compression at about 25 GPa. Stishovite remains stable upon reaching maximal pressure and during the time while the maximal pressure is kept for more than one hour. During decompression, stishovite reflections remain observable, while no reflections of α-quartz appear in the patterns. Coesite could not be identified in the analyzed data at any stage of the experiments.

In the course of compression, new single reflections occur that can be attributed to the monoclinic post-quartz phase. At elevated temperatures, the reflections of the monoclinic post-quartz phase occur independently during heating.

The formation of stishovite was confirmed by TEM analysis of some recovered samples.

The experiments show that α-quartz transforms to stishovite after amophization, whereas the stability field of coesite is skipped. Additional heating up to 900 °C does not influence the formation of high-pressure phases.

P141

Monitoring synthesis reactions using laboratory *in-situ* X-ray powder diffraction techniques

M. Müller[1], H. Gies[1]
[1] *Ruhr-Universität Bochum, Bochum, Germany*

Introduction

Commonly, the reaction processes which occur during chemical synthesis within the reaction vessel remain unknown. However, for many reactions a detailed knowledge of these processes can be beneficial. Via reaction monitoring precursor species and intermediate reaction products might be detected. The obtained findings can also be useful in order to isolate intermediates and optimize the reaction conditions, for example the minimal reaction time, which can be particularly beneficial in case of long lasting reactions e.g. for hydrothermal synthesis of porous materials.

Objectives

The determination of the reaction products without interruption of the reaction process can give new insights into the evolution of the reaction. Among the methods which can be used to follow reaction processes without affecting them, we choose x-ray scattering. Even synthesis in liquid media can be followed using x-ray powder diffraction techniques, as the formation of crystalline phases will be visible within the recorded data.

Methods

Based on modifications of a previously developed *in-situ* reaction cell for x-ray scattering measurements [1], we adapted such a cell in order to fit on a Huber Guinier camera G670 (Figure 1). Two resistivity heaters, which are placed in proximity to a glass capillary containing the reaction mixture, allow heating the sample to the reaction temperature (for the present setup up to about 180 °C). The use of an image plate detector allows recording a complete diffraction pattern at the same conditions which is important for the analysis of the reaction process, especially if the investigated reaction is fast.

Results

Using this reaction cell we are able to monitor hydrothermal synthesis reactions occurring inside a regular glass capillary. The results of initial measurements which provide a first impression of the application of such a reaction cell will be presented.

Conclusion

The proposed set-up allows following reactions which occur during hydrothermal synthesis of porous materials. Depending on the investigated reaction new insights into the reaction process can be obtained. Further applications, e.g. gas-flow studies monitoring interactions of the adsorbate with a porous sample are planned.

[1] P. J. Chupas, *et al.*: A versatile sample-environment cell for non-ambient X-ray scattering experiments. J. Appl. Cryst. (2008) 41(4) 822-824.

Figure 1: *In-situ* reaction cell placed on the Huber Guinier camera.

Figure 1

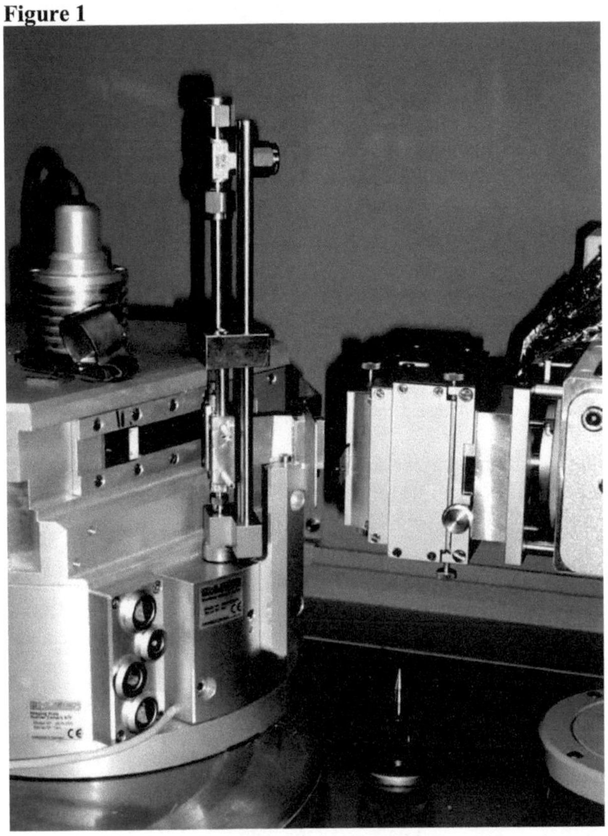

Bragg peaks, which give a first insight to the evolution of the domain shape with time inversion.

[1] M. Baum, J. Leist, Th. Finger, K. Schmalzl, A. Hiess, L. P. Regnault, P. Becker, L. Bohatý, G. Eckold, and M. Braden, Physical Review B 89,144406 (2014).

P143

Pink beam at the Extreme Conditions Beamline P02.2 at PETRA III

W. Morgenroth[1], H.- P. Liermann[2], B. Winkler[1]
[1] *Uni Frankfurt, Frankfurt, Germany*
[2] *Deutsches Elektronen-Synchrotron DESY, Hamburg, Germany*

The Extreme Conditions Beamline (ECB) P02.2 at PETRA III [1] is a beamline for micro X-ray diffraction studies of matter at simultaneous high-pressure and high/low-temperature in the diamond anvil cell (DAC). In 2015 two important upgrades have been commissioned to strengthen time-resolved X-ray diffraction experiments:

a) pink beam with full harmonic and

b) fast GaAs based LAMBDA detector.

Pink beam at the ECB offers a broad energy band and high flux and will be used to study rapid reactions (*in situ* research), small-molecule partial structure determination in the laser heated DAC without sample rotation and μ-fluorescence spectroscopy experiments which benefit from the high flux.

Commissioning - and future use - of pink beam was subdivided into two separate steps. First, the beam of the central undulator cone (3rd harmonic at 25.6 keV) was steered to the sample position in the experimental hutch. Focusing this beam with the Kirkpatrick-Baez mirror system down to 3×3 μm (h × v, FWHM) was straightforward and successful. Secondly, the off-axis beam from the undulator cone will be used to achieve a broad energy band in the pink beam.

One fast LAMBDA detector, a 2M GaAs bonded area detector from *X-Spectrum GmbH*, is available at the beamline to conduct time-resolved powder and single-crystal diffraction experiments up to 2 kHz (Fig. 1). A second detector will be used to collect diffraction data in forward and perpendicular direction in the future.

Results using one LAMBDA detector and monochromatic synchrotron radiation can be found in contribution “Millisecond time resolved diffraction study of $SrCO_3$ at high pressures and temperatures” by M. Stekiel et al. at this conference.

Within this presentation we will describe the current and upcoming capabilities of the Extreme Conditions Beamline to conduct time-resolved X-ray diffraction experiments.

The authors gratefully acknowledge financial support from the BMBF (project 05K13RF1).

[1] H.-P. Liermann et al., “The Extreme Conditions Beamline P02.2 and The Extreme Conditions Science Infrastructure at PETRA III”, JSR 22, 908-924, (2015)

Figure 1:

P142

Stroboscopic diffraction studies on the domain inversion in multiferroic $TbMnO_3$ and $DyMnO_3$ using polarized neutrons

J. Stein[1], T. Cronert[1], A. Komarek[1], T. Fröhlich[1], J. Leist[2], K. Schmalzl[2], G. Eckold[2], M. Braden[1]
[1] *Univerisität zu Köln, Köln, Germany*
[2] *Universität Göttingen, Göttingen, Germany*

Controlling magnetic order with an external field is the most likely application of multiferroic materials characterized by the coexistence and coupling of ferroelectric and magnetic order. Polarized neutron diffraction is the ideal tool to study this domain inversion in the so-called type-II multiferroic materials, where ferroelectric order arises from a chiral magnetic structure, because chiral magnetic components can be directly and efficiently determined. We have studied the rise times of multiferroic switching in $RE$$MnO_3$ by applying a periodically modulated electric field and by registering the chiral response with a stroboscopic technique as described in earlier work on $MnWO_4$ [1]. $TbMnO_3$ exhibits a simple activation law for the rise times, which can be followed over more than five decades in time from ms to several 100 s, similar behavior is also seen in $DyMnO_3$ but the temperarture intervall is strongly reduced due to the lower transition temperatures. This simple activation law contrasts with the more complex behavior in proper ferroelectrics and points to a well-defined pinning of multiferroic domains. In total eight decades in time can be studied with this stroboscopic neutron scattering technique. Close to the multiferroic transitions the switching becomes much faster reaching the experimental limit of our set up. In general we may deduce from these measurements that fast control of multiferroic domains will well be possible in multiferroic materials with higher transition temperature, in particular above ambient temperature. Using an automated setup we could register full time-resolved scans across the magnetic

LAMBDA 2M GaAs detector (*X-Spectrum GmbH*) at the Extreme Conditions Beamline P02.2 at PETRA III.

Figure 1

diffracted intensity can be used and assigned to a specific state of the sample.

The experiment shows that microstructural changes can be followed on this time scale and reveal the tiny differences in position and intensity between reflections concerning their evolution in time and with electric field.

M. Hinterstein, M. Hoelzel, J. Rouquette, J. Haines, J. Glaum, H. Kungle, M. Hoffman
Interplay of strain mechanisms in morphotropic piezoceramics
Acta Materialia 94 (2015) 319-327

H. Choe, S. Gorfman, M. Hinterstein, M. Ziolkowski, M. Knapp, S. Heidbrink, M. Vogt, J. Bednarcik, A. Berghäuser, H. Ehrenberg, U. Pietsch
Combining high time and angular resolutions: time-resolved X-ray powder diffraction using a multi-channel analyser detector
J. Appl. Cryst. (2015). 48, 970-974

M. Hinterstein, J. Rouquette, J. Haines, Ph. Papet, J. Glaum, M. Knapp, J. Eckert, M. Hoffman
Structural contribution to the ferroelectric fatigue in lead zirconate titanate ceramics
Phys. Rev. B 90 (2014) 094113

M. Hinterstein, J. Rouquette, J. Haines, Ph. Papet, M. Knapp, J. Glaum, T. Granzow, H. Fuess
Structural description of the Macroscopic Piezo- and Ferroelectric properties of Lead Zirconate Titanate (PZT)
Phys. Rev Lett., 107 (2011) 077602

P144
High resolution PD at 60 keV on microsecond timescale using stroboscopic methods

A. Schökel[1], A. Berghäuser[2], H. Ehrenberg[1], M. Etter[1,3], S. Gorfman[4], H. Choe[4], M. Vogt[4], M. Hinterstein[1], M. Knapp[1]
[1]*Karlsruher Institut für Technologie, Karlsruhe, Germany*
[2]*Universität Hamburg, Hamburg, Germany*
[3]*Deutsches Elektronensynchrotron, Hamburg, Germany*
[4]*Universität Siegen, Siegen, Germany*

The P02.1 beamline at the PETRA III synchrotron (DESY, Hamburg) is dedicated to X-ray powder diffraction. Here we present the novel high-resolution multi analyzer crystal detector (MAD) together with a custom-built FPGA based stroboscopic data acquisition system. Using this instrument, powder diffraction profiles with an angular resolution of $2.5 \cdot 10^{-4}$ degree and time resolution down to 10 ns can be collected in stroboscopic measurements at the X-ray energy of 60keV. We illustrate the capability by an in-situ stroboscopic study of a ferroelectric ceramics under 10 kHz unipolar alternating electric field.

Under variable electric field ferroelectric materials at the morphotropic phase boundary typically show a complex interplay between the intrinsic lattice strain, extrinsic domain switching and texturing leading to a macroscopic change in dimension. The individual contributions can be separated by measuring full diffraction patterns at different sample orientations [Hinterstein 2015]. To investigate the temporal behavior of these contributions high resolution PD is necessary on the relevant time scale of the process simultaneously measured on different reflections [Hinterstein 2011, 2014].
Lead-free piezoelectric ceramics are of high industrial relevance. One promising material is the solid solution of calcium doped barium titanate (BCT) and zirconium doped barium titanate (BZT). The sample used in this study had a composition of 40% BCT and 60% BZT. At room temperature this system has a rhombohedral symmetry. With increasing temperature the symmetry changes to orthorhombic, tetragonal and cubic above ~350 K. In our measurement data was taken with the maximum time resolution of 10 ns and afterwards binned to 1 μs to enhance statistics. The reading of each single detector (analyzer crystal plus scintillator) is thereby stored in an individual channel with the mentioned time resolution. After integration of sufficient intensity at a certain angle the whole MAD is moved to the next position. This way the whole

Aperiodic and periodic complex materials

P145

Structure Review of the $R_2T\mathrm{Si}_3$ family

M. Nentwich[1], M. Zschornak[1], M. Sonntag[1], D. C. Meyer[1]
[1]*TU Bergakademie Freiberg, Experimentelle Physik, Freiberg, Germany*

The family of $R_2T\mathrm{Si}_3$ compounds is in focus of recent research, especially due to their large variety of magnetic properties, but also because of the numerous variants of lattice symmetry and superstructures. Here, the R elements are alkaline earth metals, rare earth metals or actinoides and the T elements are transition metals or silicon. All $R_2T\mathrm{Si}_3$ compounds are crystallizing in structures derived from either AlB_2 or $ThSi_2$ type symmetry. The structures based on hexagonal AlB_2-type symmetry (space group P6/mmm (191)) exhibit a wide range of superstructures with several stacking variants all following a common pattern (see Figure 1) of alternating concentric rings of either Si or T atoms. Thereby, the symmetry is reduced to orthorhombic and even monoclinic. In contrast, the tetragonal $ThSi_2$ (space group $I4_1$/amd (141)) based composites possess only one further version, being orthorhombic $GdSi_2$ (space group Imma (74)). An overview of the aristotypes and lower symmetric variants will be given in a Bärnighausen diagram.

To find the key to the formation of one or the other crystal symmetry, an extensive literature research has been performed based on more than 400 structure reports of the $R_2T\mathrm{Si}_3$ family. Experimental values like the lattice parameters, their ratio and the multiplicity of the superstructure and the thermal treatment are collected and analyzed. In addition to these purely experimental parameters, density, atomic packing factor and shortest Si-T bond length have been calculated using listed values for atomic diametres and atomic mass. Furthermore, the theoretical parameters ratio of atomic radii, valence electron concentration and coordination number of the anion sublattice were determined.We found that the thermal treatment has a massive effect to the formation of superstructures. Furthermore, there are two different kinds of hexagonal $R_2T\mathrm{Si}_3$ compounds being ionic or metallic, depending on the R element. Additionally, the main influence to the variation of the Si-T bonds is the electronic interplay between R element and Si lattice rather than the R radii. The interplay among the preceeding parameters was analyzed by using correlations and mathematical boxplots to resolve the introductory question.

Figure 1

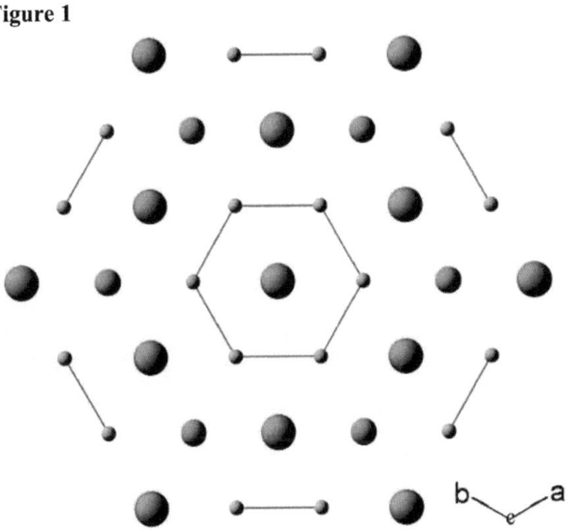

P146

New giant (pseudo)hexagonal structures in intermetallic compounds Sr(Cd/Hg/Zn)$_{\sim 5}$

C. Röhr[1], M. Wendorff[1]
[1]*Universität Freiburg, Freiburg, Germany*

Alkaline earth mercurides and cadmides exhibit a fascinating crystal chemistry, especially in the composition range $AM_{\sim 6}$, where iQC and their approximants are observed [1]. The partial exchange of Cd/Hg against smaller Zn atoms allows to modify the complex cubic structures and e.g. fill the empty cubes or eliminate the disorder of M_4 tetrahedra in the YCd_6-, Eu_4Cd_{25}- or $Ba_{20}Hg_{103}$-type structures [2].

In the same chemical systems, at slightly reduced M contents, related (pseudo)hexagonal complex structures have been investigated:

♦ Starting from the border phases '$Sr_{13}M_{58}$' (M=Cd, Hg) [3], which have been assigned from powder data to form the $Gd_{13}Cd_{58}$-type structure [4] (*hP*142, *P*6_3*/mmc*, $a\approx1500$, $c\approx1500$ pm), a series of ternary strontium Cd-Zn, Hg-Zn and Cd-Hg compounds were synthesized and their single crystal data were examined carefully: As already indicated for '$Dy_{13}Zn_{57}$' [5], the correct orthorhombic symmetry of all members of this type is hidden by a multidomain hexagonal twinning. Fortunately, the correct structure/composition could be determined from some untwinned Zn-containing crystals (*e.g.* $Sr_{13}Cd_{56.8}Zn_{2.5}$: *oP*288, *Pnma*, a=1588.6, b=1593.8, c=2774.7 pm, $R1$=3.9%,) and the twinned data sets could be refined afterwards, using the twin law derived by the group-subgroup relation *P*6_3*/mmc*→t3→*Cmcm*→k2→*Pnma*. Crystal-chemically, $^1/_3$ of the M_8 cubes, which are arranged in kagome nets (KN, fig. 1) are empty (→$A_{13}M_{57}$). The empty cubes are squeezed, leaving space for two further M atoms (→$A_{13}M_{59}$). Additionally, one Cd site is statistically substituted by Zn_2 dumbbells (→$A_{13}M_{59+x}$). The relation of the structure to the cubic approximants of iQCs as well as the dominating structure elements [filled cubes, tetrahedra stars (TS) etc.] are discussed.

♦ At a distinct Zn content of ≈22 %, a related complex hexagonal phase, also forming a new structure type (e.g. $SrCd_{4.67}Zn_{1.04}$, *hP*468, *P*-62*m*, a=2775.45, c=1575.55 pm, $R1$=7%), was obtained. Herein, the space inbetween the very similar KN layers of now completely filled cubes are filled with further building blocks, including M_8 cubes and TS.

♦ At a Zn content of ≈14 % (and also in the system Ba/Hg/Cd), a third complex hexagonal structure appears, which nevertheless contains comparable cube KN layers (e.g. $SrCd_{4.42}Zn_{0.60}$, *hP*638, *P*6_3*/mmc*, a=1604.84, c=6841.92 pm, $R1$=9.1 %).

[1] M. Wendorff, C. Röhr, *J. Alloys Compd.* **546**, 320 (2013).
[2] M. Wendorff, C. Röhr, *Acta Crystallogr.* **A71**, s418 (2015).
[3] G. Bruzzone, *Gazzetta Chim. Ital.*, **102**, 234-242 (1972).
[4] G. Bruzzone, M. L. Fornasini, F. Merlo, *J. Less-Common Met.* **30**, 361-375 (1973).
[5] C. P. Gomez, S. Lidin, *Solid State Sci.*, **4**, 901-906 (2002).

Fig. 1: Pseudohexagonal layers in the structure of $Sr_{13}Cd_{56.8}Zn_{2.5}$

Figure 1

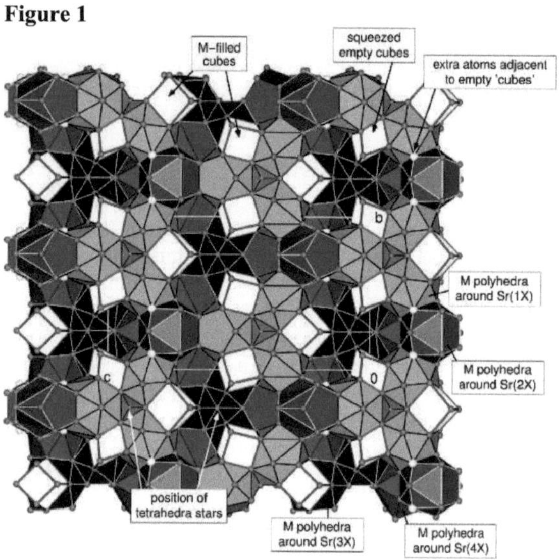

P147

Charge-density waves in CuV_2S_4

S. Ramakrishnan[1], N. H. An Bui[1], F. Feulner[1], A. Schönleber[1], S. van Smaalen[1], D. Chernyshov[2]

[1] *University of Bayreuth, Laboratory of Crystallography, Bayreuth, Germany*

[2] *SNBL ESRF, Grenoble, France*

The symmetry of the spinel compound CuV_2S_4 at room temperature is described by the space group *Fd 3m* [1]. At 90 K the di raction pattern shows a CDW with a modulation wave vector q=(3/4 + δ,3/4 + δ,0) with δ= 0.06 [2]. The X-ray powder di raction shows a lowering of the lattice from cubic symmetry to tetragonal at temperature below 90 K [3]. At 70 K, the X-ray powder di ffraction has revealed a splitting of the main reflections and the weak super lattice reflections with modulation wave vector q =0.7391(5)*b** and the symmetry is orthorhombic with the superspace group *Imm*2(0*β*0) [4]. Further phase transitions at lower temperatures were proposed but not characterized. We have studied CuV_2S_4 by single-crystal X-ray di raction pattern at temperatures down to 5 K, and we will present the low-temperature phase diagram of this 3-dimensional CDW compound.

References

[1] L. S. R. Kumara , M. Hidaka, N. Tokiwal, J. Awaka , T. Hagino, S. Nagata, and Y. J. Park, Phys. Stat. Sol. (b) 245, No. 11, 2539{2549 (2008).

[2] R. M. Fleming, F. J. DiSalvo, R. J. Cava, and J. V. Waszczak: Phys. Rev. B 24 ,2850 (1981).

[3] H. Okada, K. Koyama and K. Watanabe J.Phys.Soc.Jpn 73,3227 (2004).

[4] S. Kawaguchi, H. Ishibashi, N. Tsuji, J. Kim, K. Kato, M. Takata, and Y. Kubota, J. Phys. Soc. Jpn. 82, 064 603 (2013)

Materials: Properties and applications

P148
New alkali metal iron chalkogenido salts with cluster anions Fe_4Ch_8 and Fe_3Ch_7

P. Stüble[1], A. Berroth[1], C. Röhr[1]
[1]*Albert-Ludwigs-Universität Freiburg, Institut für Anorganische und Analytische Chemie, Freiburg i. Br., Germany*

In the course of a systematic study of alkali metal iron chalcogenido salts containing clusters with a Fe_4Q_4 cubane core resembling the $[Fe_4(\mu_3\text{-}S_4)]^+$ cluster, *e.g.* in the active site of ferredoxins, a series of new mixed-valent potassium and rubidium selenido and tellurido ferrates (II/III) were discovered [1]. Therefore, pure elements in stoichiometric composition were heated carefully in Al_2O_3 crucibles under an argon atmosphere up to maximum temperatures of 800 to 900°C. Crystal structures were determined by means of X-ray single crystal diffraction.

The mixed-valent $Fe^{II/III}$ tellurido ferrates $A_7[Fe_4Te_8]$ form three different structure types. All structures contain tetramers of four edge sharing $[FeTe_4]$ tetrahedra, which are connected by common edges to form only slightly distorted tetrahedral $[Fe_4Te_8]^{7-}$ anions with a Fe_4Te_4 cubane core. In all cases, the tellurido ferrate anions are surrounded by 26 alkali cations, which are located at the eight corners, the six faces and the twelve edges of a cube. The three crystal structures can be thus described by three different packings of this type of cubes:

The monoclinic rubidium compound $Rb_7[Fe_4Te_8]$ (space group *C*2/c, a = 2000.16(7), b = 897.79(3), c = 1768.12(6) pm, β = 117.4995(10)°, Z = 4, $R1$ = 0.0296), is isotypic to both the cesium tellurido and sulfido ferrate $Cs_7[Fe_4(S/Te)_8]$ [1,2].

Depending on the temperature, $K_7[Fe_4Te_8]$ forms two different but closely related new structure types: The tetragonal r.t. modification (space group $P4_2/nmc$, a = 1222.25(14), c = 872.1(2) pm, Z = 2, $R1$ = 0.0583) was preliminary mentioned in a poster abstract [3]. It crystallizes in a supergroup of the newly discovered orthorhombic l.t. (100 K) form (space group *Pbcn*, a = 1715.5, b = 866.76(3), c = 1715.50(7) pm, Z = 4, $R1$ = 0.0160).

Attempts to synthesize the corresponding selenium compound $K_7[Fe_4Se_8]$ resulted in the formation of the likewise mixed-valent compound $K_6[Fe_4Se_8]$. Despite the modified composition, the new orthorhombic structure (space group *Pbcn*, a = 1632.62(6), b = 821.10(3), c = 1592.75(6) pm, Z = 4, $R1$ = 0.0540) is next to isotypic to the structure of the low-temperature form of $K_7[Fe_4Te_8]$, the only difference is one missing alkali site.

According to the mentioned abstract [3], the new compound $K_5Fe_2Te_5$ (cubic, space group *Pa*-3, a = 1709.02(5), Z = 4, $R1$ = 0.0594) crystallizes in a new structure type. Pursuant to $K_5Fe_2Te_5$ = $K_{15}[Fe_3Te_7]_2(Te)$, its structure contains mixed-valent trimers $[Fe_3Te_7]^{(6/7)-}$ and an isolated tellurid ion which is surrounded by a cube of K^+ cations.

[1] P. Stüble, A. Berroth C. Röhr. *Z. Naturforsch.* (submitted)
[2] M. Schwarz, C. Röhr. *Inorg. Chem.* **54**, 1038-1048 (2015).
[3] W. Bronger, C. Bomba et. al. *J. Less Common Met.* **167**, 161-167 (1990).

[4] K. O. Klepp, H. Boller in: *XI. Int. Conf. On Solid Compounds of Transition Elements, Münster.* 129 (1991).

Fig. 1: *Four edge sharing tetrahedra $FeCh_4$ form the anion cluster Fe_4Ch_8 (left) like e.g. in r.t. $K_7Fe_4Te_8$. The structurally related anion cluster Fe_3Ch_6 (right) formed by only three tetrahedra can be found in $K_5Fe_2Te_5$.*

Fig. 2: *Similar to most of the compounds discussed, cluster anions in r.t. $K_7Fe_4Te_8$ show a slightly distorted cubic coordination by alkali metal cations.*

Figure 1

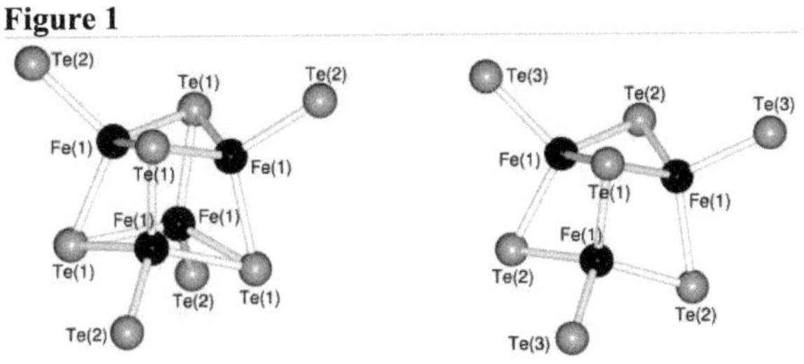

Figure 2

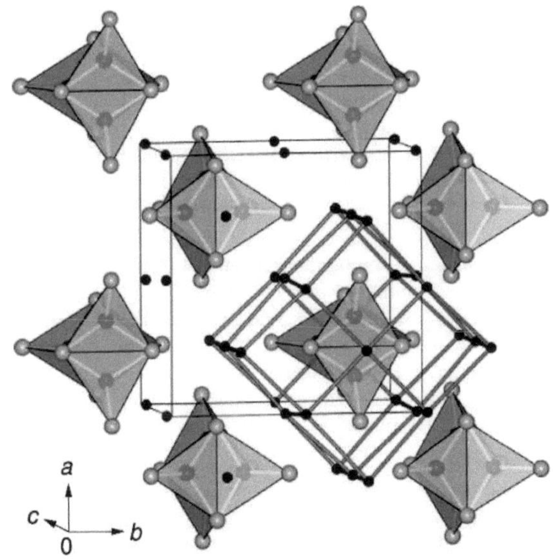

P149
Dye degradation catalyzed by piezoelectric materials

M. Sonntag[1], E. Mehner[1], H. Stöcker[1], B. Störr[1], T. Leisegang[1], D. C. Meyer[1]
[1]*TU Bergakademie Freiberg, Freiberg, Germany*

The prospective global energy transition demands new opportunities for energy conversion and storage. Therefore, research is focussing on harvesting electric energy using physical phenomena like pyro-, thermo- or piezoelectricity. But also the direct conversion of heat or mechanical vibration into better storable chemical energy is promising.

When mechanical stress (e. g. ultrasonic vibration) is applied to a piezoelectric material, a change of its polarization results, which in turn changes the amount of surface charges. These charges may then react with the environment to form radicals. It was already shown that this so-called piezocatalysis can be used for hydrogen generation [1, 2] and dye degradation [3]. Nevertheless, details about the occurring catalytic processes at the surfaces of the piezoelectric materials are rare.

Hence, we investigate the piezocatalytic degradation of aqueous solutions of indigo carmine using single crystalline lithium niobate powder (space group *R*3*c*) in a low frequency ultrasonic bath (35 kHz, 160 W) with long reaction times. Strontium titanate (space group *Pm*-3*m*) is used as non-piezoelectric reference. Fixation of test tubes and a high working temperature of 75 °C ensures a minimization of external pertubations. The degradation of indigo carmine is monitored by UV-Vis spectroscopy.

Indigo carmine is decomposed both by piezocatalysis and sonocatalysis, with the degradation processes following zero order kinetics. We show that the piezocatalytic rate exceeds the sonocatalytic one. In contrast to literature, no upper limit for catalyst loading is found. Larger particle sizes cause higher degradation rates. These results identify piezocatalysis as a promising improvement for ultrasonic triggered energy conversion as well as cleaning of waste water.

[1] K.-S. Hong, H. Xu, H. Konishi, X. Li, J. Phys. Chem. Lett. 2010, 1, 997.
[2] M. B. Starr, X. Wang, Sci. Rep. 2013, 3, Art. 2160.
[3] K.-S. Hong, H. Xu, H. Konishi, X. Li, J. Phys. Chem. C 2012, 116, 13045.

P150

Schottky contribution to the heat capacity of monazite type (La,Pr)PO⁴ from low temperature calorimetry and fluorescence measurements

J. D. Bauer[1], A. Hirsch[2], L. Bayarjargal[1], L. Peters[2], G. Roth[2], B. Winkler[1]
[1]*Goethe-Universität Frankfurt am Main, Frankfurt am Main, Germany*
[2]*RWTH Aachen, Institut für Kristallographie, Aachen, Germany*

The orthophosphates of the lighter lanthanides (*Ln*PO_4) crystallize in the monoclinic monazite structure type. Ceramics of these compounds are of high interest for spent nuclear fuel conditioning [1]. In an assessment of the suitability for that purpose, it is crucial to know fundamental thermodynamic properties, such as the heat capacities (C_p). However, these properties have only been measured for some *Ln*-monazites.

The total heat capacity of ordered insulating crystalline solids is the sum of several contributions. The main contribution results from lattice vibrations. The thermal excitation of 4f electrons in the *Ln* gives rise to a contribution that is called the Schottky anomaly[2].

The low temperature C_p of flux grown single crystals (fig. 1) of the (La,Pr)PO_4 solid solution was measured. The Pr-containing samples show a Schottky anomaly at low *T* (maximum ~ 60 K), that can be separated from the lattice contribution by subtracting C_p of $LaPO_4$. The Schottky anomaly is due to the thermal population of the 9 levels of the $^3H_4$ electronic ground state of Pr^{3+}. The splitting of the energy levels was analysed by measuring the $^3P_0 \rightarrow {}^3H_4$ fluorescence transitions. 7 energy levels can be assigned unambiguously from the spectra.

An analysis of the Schottky anomaly was performed by a fit using an equation derived from statistical mechanics[3] including constraints for the levels obtained from fluorescence spectra (fig. 2). This fit yielded a complete set of energy levels of the $^3H_4$ ground state of Pr^{3+} in $PrPO_4$. These levels reproduce the temperature dependence of the experimentally determined Schottky anomaly much more accurately than the energy levels published earlier.

The present study demonstrates, that neither C_p measurements nor fluorescence spectroscopy alone provide sufficient information to obtain all energy levels, and that only the combination of the two techniques gives robust results.

Financial support from BMBF, Germany, grants 02NUK019E, 02NUK021E & 02NUK021F is gratefully acknowledged.

[1] H. Schlenz et al., *Z. Kristallogr.*, **2013**, *228*, 113-123

[2] E. Westrum Jr., *J. Therm. Anal.*, **1985**, *30*, 1209-1215

[3] B.H. Justice and E. Westrum Jr., *J. Phys. Chem.*, **1963**, *67*, 339-345

Fig. 1: $La_{0.2}Pr_{0.8}PO_4$ single crystal (the bar denotes 1 mm).

Fig. 2: The Schottky anomaly of $PrPO_4$: experimental (symbols) and calculated (line).

Figure 1

Figure 2

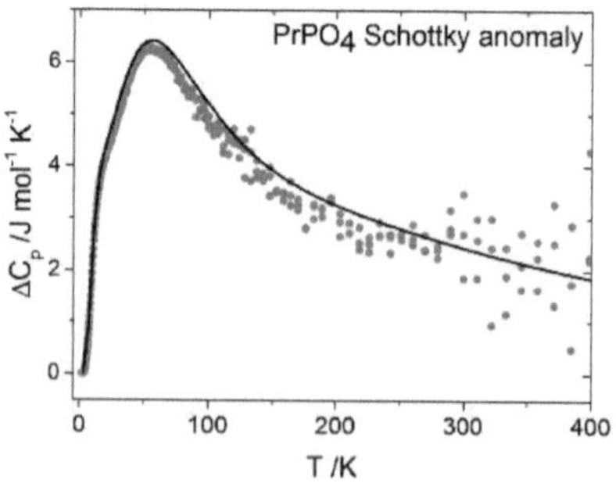

P151

Crystal Growth and Characterization of Mn-substituted Barium Hexaferrites $BaFe_{12-x}Mn_xO_{19}$

S. Nemrava[1], D. Vinnik[2], I. Zakharchuk[3], E. Lähderanta[3], R. Niewa[1]
[1]*University of Stuttgart, Institute of Inorganic Chemistry, Stuttgart, Germany*
[2]*South Ural State University, Chelyabinsk, Russian Federation*
[3]*Lappeenranta University of Technology, Lappeenranta, Finland*

Hexaferrites with the magnetoplumbite structure, e. g. $BaFe_{12}O_{19}$, are magnetic materials with a wide field of applications. Traditionally, they are used as permanent magnets, but have more recently gained importance as microwave absorbing materials. Partial substitution of Fe shows notable effects on the physical properties, like changes of the magnetization, the coercive field, or the Curie temperature, because of which this method is often used to study and modify the magnetic characteristics. [1 - 3]

Single crystals of $BaFe_{12-x}Mn_xO_{19}$ were obtained by a solid state reaction of Fe_2O_3, MnO and $BaCO_3$, while Na_2CO_3 was used as flux. All starting materials were ground in an agate mortar, filled in a platinum crucible and heated to 1260 °C for 3 h. After cooling to room temperature, single crystals of $BaFe_{12-x}Mn_xO_{19}$ with x up to 1.5 and sizes up to 8 mm were obtained after leaching in hot nitric acid.

Single crystal XRD measurements of the samples confirm the *M*-type structure with space group $P6_3/mmc$ (Fig. 1). PXRD refinements show that the substitution of Fe by Mn only has small effects on the lattice parameters compared with pure $BaFe_{12}O_{19}$.

Saturation magnetization of samples $BaFe_{12-x}Mn_xO_{19}$ increases with decreasing temperature (Fig. 2), which is typical for *M*-type ferrites, but the saturation magnetization is lower in comparison with pure $BaFe_{12}O_{19}$. This is caused by the weakening of the Fe–O–Fe superexchange interactions upon substitution. The Curie temperature is also lowered by the Mn substitution.

Assuming an isovalent substitution of Fe^{3+} by Mn^{3+}, the substitution range of about 12 at.-% for $BaFe_{10.5}Mn_{1.5}O_{19}$ seems rather narrow. However, due to the significant different coordination environments of the substitution sites (four-, five- and six-fold coordination) a partial charge disproportionation to Mn^{2+} and Mn^{4+} was earlier indicated, which may lead to a limited substitution level. [4]

Figure 1: Section of the crystal structure (*M* = Fe, Mn).

Figure 2: Hysteresis curves and saturation magnetization at different temperatures of $BaFe_{10.5}Mn_{1.5}O_{19}$. [3]

[1] J. M. D. Coey, *Scripta Mater.* **2012**, *67*, 524 - 529.
[2] A. Ghasemi, A. Hossienpour, A. Morisako, X. Liu, A. Ashrafizadeh, *Mater. Design* **2008**, *29*, 112 - 117.
[3] D. Vinnik, I. Zakharchuk, E. Lähderanta, *Mater. Sci. Forum* **2016**, *843*, 155 - 160.
[4] J. Fontcuberta, X. Obradors, J. B. Goodenough, *J. Phys. C* **1987**, *20*, 441 - 449.

Figure 1

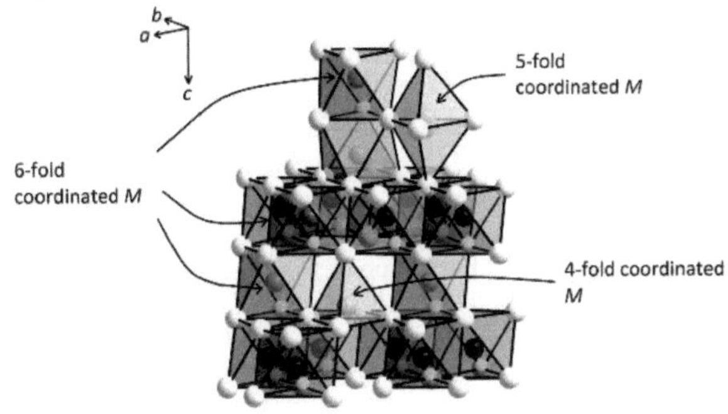

Figure 2

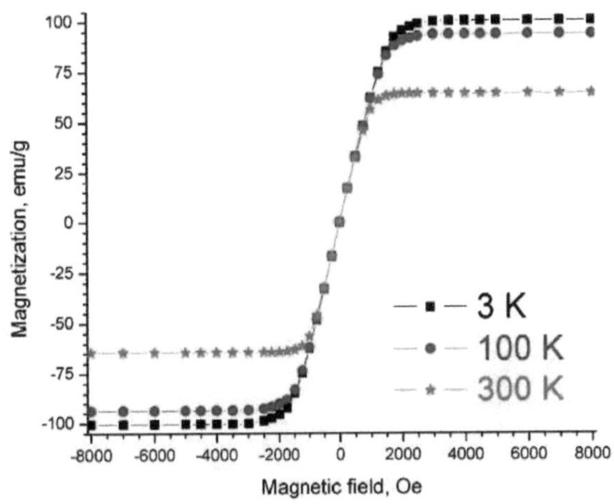

P152

Crystal Growth and Characterization of $BaMRu_5O_{11}$ (*M* = Li, Cu)

S. Nemrava[1], L. Link[1], B. Blaschkowski[1], R. Niewa[1]
[1]*University of Stuttgart, Institute of Inorganic Chemistry, Stuttgart, Germany*

For a long time, hexaferrites have attracted high interest due to their promising physical properties like high Curie temperatures or magnetic anisotropy; in combination with their high thermal stability they are good candidates for different applications in electronic devices. A special family of these hexaferrites are the so-called "*R*-type" ferrites, particularly the ruthenates, whose physical properties are sensitive to their composition. For example, $BaM'_{2\pm x}Ru_{4\mp x}O_{11}$ (*M'* = Fe, Co, Mn) are soft ferromagnetic materials, while $BaMRu_5O_{11}$ (*M* = Li, Cu) show paramagnetic behaviour. [1, 2]

Here we present the crystal growth of the quaternary *R*-type ferrites $BaMRu_5O_{11}$ with *M* = Li, Cu. Single crystals of $BaLiRu_5O_{11}$ were grown from a LiCl flux with $BaCO_3$ and RuO_2 in appropriate amounts and a quadruple amount of Li_2CO_3 as educts, while single crystals of $BaCuRu_5O_{11}$ were obtained from a KCl flux with a triple amount of CuO as starting materials. The initial batches were ground in an agate mortar, pressed into pellets and heated to 1130 °C in a corundum crucible, respectively. After cooling to room temperature, small crystals with sizes of about 0.2 mm were separated from the flux by treatment with nitric acid.

Single crystal X-ray diffraction reveal $BaMRu_5O_{11}$ (*M* = Li, Cu) to crystallize in the hexagonal space group $P6_3/m$. The structures can be described based on layers of edge-connected RuO_6 octahedra within (001), whilst the ruthenium atoms form a Kagome-net that is slightly distorted due to Ru–Ru bonding. These layers are interconnected via face-sharing octahedra-doubles Ru_2O_9 and trigonal bipyramids MO_5 (M = Li, Cu). Ba is surrounded anticuboctahedrally by oxygen (Fig. 1). [3]

Measurements of the magnetic suszeptibility of $BaCuRu_5O_{11}$ confirmed its formerly revealed paramagnetic behaviour with a temperature independent contribution of $\chi_{0,mol} = 1.8 \cdot 10^{-3}$ cm³/mol. Above 50 K the suszeptibility can be sufficiently fitted applying a Curie-Weiss law with $C \approx 0.4665$ cm³ · K · mol^{-1} and $\Theta = -191.3$ K. The negative Weiss constant hints at a ferro- or ferrimagnetic order at low temperatures.

Figure 1: Section of the crystal structure (M = Li, Cu).

[1] M. L. Foo, Q. Huang, J. W. Lynn, W.-L. Lee, T. Klimczuk, I. S. Hagemann, N. P. Ong, R. J. Cava, *J. Solid State Chem.* **2006**, *179*, 563 - 572.
[2] L. Shlyk, S. Kryukov, B. Schüpp-Niewa, R. Niewa, L. E. De Long, *Adv. Mater.* **2008**, *20*, 1315 - 1320.
[3] L. Shlyk, S. Strobel, T. Schleid, R. Niewa, *Z. Kristallogr.* **2012**, *227*, 545 - 515.

Figure 1

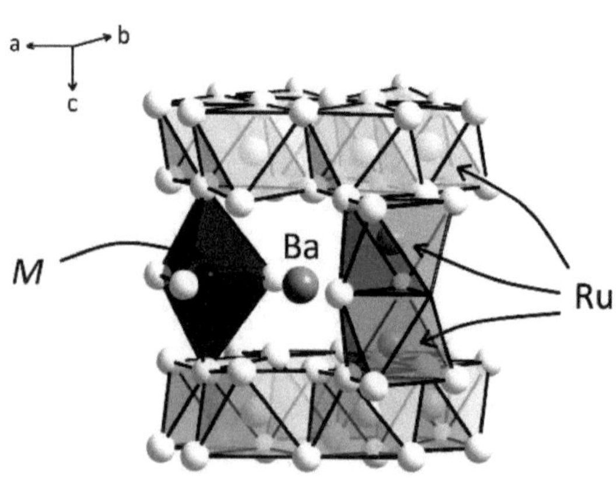

P153

Crystal growth, linear optical properties, pyroelectricity and vibrational spectroscopy of barium antimony tartrate, $Ba[Sb_2((+)C_4H_2O_6)_2]\cdot 3H_2O$

L. Bohatý[1], I. Nemec[2], H. Schneeberger[3], P. Becker[1]
[1]*Universität zu Köln, Institut für Geologie und Mineralogie, Abteilung Kristallographie, Köln, Germany*
[2]*Charles University in Prague, Faculty of Science, Department of Inorganic Chemistry, Prague, Czech Republic*
[3]*Landensberg/, Glöttweng, Germany*

Among the known antimony tartrates of divalent cations the hexagonal compounds $M^{II}[Sb_2((+)C_4H_2O_6)_2]\cdot 2H_2O$ (M^{II} = Ca, Sr) possess attractive optical [1] and pyroelectric [2] properties and allow phase matching for collinear second harmonic generation (SHG) [3]. Concerning the polar tetragonal (s.g. $P4_1$) barium compound, $Ba[Sb_2((+)C_4H_2O_6)_2]\cdot 3H_2O$, however, so far only the crystal structure is known. It consists of anionic tartrato(4-)-bridged binuclear complex groups $[Sb_2((+)C_4H_2O_6)_2]^{2-}$, tenfold coordinated barium atoms $[BaO_6(H_2O)_4]$ and additional H_2O molecules [4]. In the present work we report on the growth of large single crystals of $Ba[Sb_2((+)C_4H_2O_6)_2]\cdot 3H_2O$, the temperature dependence of thermal expansion coefficients and of the pyroelectric effect. In addition, precision refractive indices and their dispersion in the wavelength range 365 - 1083 nm are given and analyzed concerning possibilities for phase matching for second harmonic generation (SHG). A Raman and IR spectroscopic investigation of $Ba[Sb_2((+)C_4H_2O_6)_2]\cdot 3H_2O$ at room temperature exhibits the crucial role of the binuclear complex group $[Sb_2((+)C_4H_2O_6)_2]^{2-}$ in the character of the recorded vibrational spectra. The quantum-chemical calculations of vibrational modes of this complex group were used as a fundament for the presented spectral assignment. $Ba[Sb_2((+)C_4H_2O_6)_2]\cdot 3H_2O$ shows a marked anisotropy of thermal expansion ($\alpha_{11} = 17.8 \cdot 10^{-6}$ K^{-1}, $\alpha_{33} = 45.0 \cdot 10^{-6}$ K^{-1} at 293 K), and the pyroelectric coefficient (at constant stress) at 293 K is Cm^{-2}K^{-1}. The refractive indices and their dispersion allow for both, type I (ss-f, s = slow wave, f = fast wave) and type II (sf-f) SHG phase matching.

[1] L. Bohatý, Z. Kristallogr. **163** (1983) 255.

[2] L. Bohatý, P. Held, H. Schneeberger, T. Zheng, and P. Becker, Cryst. Res. Technol. **50** (2015) 482.

[3] L. Bayarjargal, P. Becker, and L. Bohatý, Cryst. Res. Technol. **43** (2008) 508.

[4] L. Bohatý, R. Fröhlich, and K.F. Tebbe, Acta Cryst. C **39** (1983) 59.

P154

Crystal growth, crystal structure, linear optical properties and vibrational spectroscopy of guanidinium orthophosphates

I. Nemec[1], I. Matulková[1], P. Held[2], D. Li[2], L. Bohatý[2], P. Becker[2]
[1]*Charles University in Prague, Faculty of Science, Department of Inorganic Chemistry, Prague, Czech Republic*
[2]*Universität zu Köln, Institut für Geologie und Mineralogie, Abteilung Kristallographie, Köln, Germany*

Among the salts of the guanidinium cation $[N(CH_2)_3]^+$ ("Gu") examples of crystals with promising $\chi^{(2)}$- or $\chi^{(3)}$-based nonlinear optical properties are reported, e.g. [1,2]. In the present work a study of guanidinum phosphates is presented. Starting from aqueous solutions of the educts guanidinium carbonate (Gu_2CO_3) and phosphoric acid (H_3PO_4) we could identify three different crystalline product phases formed though evaporation of the solvent in the temperature range between room temperature and 60°C: GuH_2PO_4, $Gu_2HPO_4\cdot H_2O$ and $Gu_3PO_4\cdot 3/2\ H_2O$. For all three phases the crystal structure was determined by single crystal X-ray diffraction.

	space group	*a* [Å]	*b* [Å]	*c* [Å]	*β* [°]
GuH_2PO_4	*P* $2_1/c$	12.204(2)	7.984(1)	12.976(2)	101.42(1)

$Gu_2HPO_4 \cdot H_2O$	$P\bar{4}2_1c$	16.8502(6)	16.8502(6)	7.2529(4)	90
$Gu_3PO_4 \cdot 3/2\ H_2O$	Cc	6.561(5)	17.601(5)	22.969(5)	92.138(5)

The results for $Gu_2HPO_4 \cdot H_2O$ corroborate earlier structure data [2,3]. Large single crystals of GuH_2PO_4 and $Gu_2HPO_4 \cdot H_2O$ of dimensions up to 45 x 35 x 20 mm^3 and 20 x 20 x 60 mm^3, respectively, were grown from aqueous solution by controlled evaporation of the solvent at 38°C. Refractive indices and their dispersion of these crystals were measured in the visible and near IR wavelength region. FTIR and Raman spectra of polycrystalline samples were recorded at room temperature and their assignment is discussed especially with respect to particular symmetry of the crystals and involved anions.

[1] L. Bohatý, P. Becker, E. Haussühl, et al., Z. Kristallogr. **230** (2015) 639.
[2] X. Liu, X. Wang, X. Yin, et al., CrystEngComm **16** (2014) 930.
[3] J.M. Adams and R.W.H. Small, Acta Cryst. B **32** (1976) 832.

P155

Hydride fluoride analogy and the design of Eu^{2+} emission colours

N. Kunkel[1], H. Kohlmann[2]
[1]*IRCP - Chimie ParisTech , Materials for Photonics and Opto-Electronics, Paris, France*
[2]*Universität Leipzig, Institut für Anorganische Chemie, Leipzig, Germany*

Since the ionic radii of hydride and fluoride may be very similar in some structures, ionic compounds of hydride and fluoride often show structural analogies and partial or complete solid solution series exist [1]. However, hydride shows a much higher polarizability than fluoride; a fact, that can also be observed from the variations in its effective radius from 127 to 153 pm [2]. Comparing Eu^{2+} -doped $LiBaH_3$ and $LiBaF_3$, it was shown that this large difference in the polarizability can strongly influence the emission energies of the d-f transition of Eu^{2+} [3].

In principle, different compositions in mixed hydride fluoride solid solution series should allow for tailoring different polarizabilities and therefore, different emission energies. The solid solution series EuH_xF_{2-x} was studied for the fluoride rich side on which the compounds crystallize in the fluorite structure type and luminescence spectra were recorded. Furthermore, we studied the hydride fluoride solid solution series $LiSrH_{3-x}F_x$, $LiBaH_{3-x}F_x$, $KMgH_{3-x}F_x$ and $EAH_{2-x}F_x$ (EA = Ca, Sr, Ba) by first principle calculations.

In EuH_xF_{2-x} an increase of the hydride content leads to a shift in the emission energies to longer wavelength. This can be explained by the strong polarizability and the larger nephelauxetic effect of hydride compared to fluoride [4]. The theoretical calculations suggest that in case of $LiBaH_{3-x}F_x$ and $KMgH_{3-x}F_x$ a complete solid solution exists, but in case of $LiSrH_{3-x}F_x$ only a partial solid solution series.

The present results show that substitution of hydride and fluoride in mixed systems is a tool for tuning emission wavelengths of Eu^{2+} and that a number of candidates for host lattices with mixed anion occupation exists.

References

[1] A. J. Maeland, W. D. Lahar, Z. Phys. Chem. 1993, 179, 181-185.

[2] T. R. P. Gibb, Jr., D. P. Schumacher, J. Phys. Chem. 1969, 64, 1407-1409.

[3] N. Kunkel, A. Meijerink, H. Kohlmann, Phys. Chem. Chem. Phys., 2014, 16, 4807-4813

[4] N. Kunkel, A. Meijerink, H. Kohlmann, Inorg. Chem. 2014, 53, 4800-4802.

P156

Plasma electrolytic oxidation of Titanium to form multifunctional porous Titaniumdioxide layers

A. Friedemann[1,2], P. Plagemann[1], T. M. Gesing[2,3]
[1]*Fraunhofer-Institut für Fertigungstechnik und Angewandte Materialforschung IFAM, Bremen, Germany*
[2]*Chemische Kristallographie fester Stoffe, Institut für Anorganische Chemie und Kristallographie, Leobener Straße /NW2, Universität Bremen, 28359 Bremen, Germany*
[3]*MAPEX Center for Materials and Processes, Bibliothekstraße 1, Universität Bremen, 28359 Bremen, Germany*

There is an ongoing search for titanium to improve its beneficial properties like its biocompatibility to strengthen its wide range of applications with an adapted chemical surface treatment. The high corrosion resistance and strength compared to other metals as well as a lower module of elasticity can be increasingly used in chemical industry, energy, automotive and medical technology [1, 2]. Anodization is one of the typical methods to modify the oxide layer on the surface of titanium and its alloys. The plasma electrolytic oxidation (PEO) (micro-arc oxidation MAO, also: plasma chemical oxidation, anodic spark deposition) is a type of high-voltage anodic oxidation capable to produce a stable oxide layer on valve metals (Al, Zr, Mg, Ti etc.). It is very suitable as a surface treatment due to the combined adjustment of morphology and chemical composition, while offering the possibility of generating a uniform crystalline surface [3]. In the present study the TiO_2-surfaces were prepared in acidic or alkaline electrolyte in a voltage range between 100 V and 300 V. During the electrochemical process, a high amount of characteristically micro-discharges occurred. These discharges are due to a local breakdown of the growing oxide layer which forms a porous, "crater-like" structure [4] and a partly crystalline surface. In comparison to other chemical surface treatments the plasma electrolytic process provides a high layer thickness and surface roughness linearly to the applied voltage. The surface structure was investigated using a scanning electron microscope (SEM). According to the different electrolytes the SEM images give an overview about the size of the pores at the variable applied voltages. In this case it is possible to control the size of the topography of the surface with changing the applied voltage and the behavior of the electrolyte. During the process the micro discharges on the whole surface intensify with the increasing voltage, resulting in bigger pores. XRD-measurements of the surfaces show the crystallinity of the TiO_2 phases (Anatase and Rutile) which increase with the applied voltage because of high local energy input of the discharges into the oxide layer. In this case it is possible to control the topography and the crystallinity of the surface with changing behavior of the electrolyte and the applied voltage.

References

[1] M. Peters, H. Clemens; *BHM*; 155 (9) **(2010)** 402-408

[2] E. Wintermantel, S.-W. Ha; *Medizintechnik-Life Science Engineering*; **(2009)** 191-217; 551-556

[3] S. Stojadinovi´c , R. Vasili , M. Petkovi, B. Kasalica, I. Belˇca, A. ˇZeki´c, Lj. Zekovi; *Applied Surface Science 265 (2013) 226-233*

[4] H.-J. Oh, J.-H. Leeb, Y. Jeongc, Y.-J. Kimd, C.-S. Chi; *Surface & Coatings Technology*; 198 **(2005)** 247-252.

P157

Comparative band gap determination of photocatalytic active $Bi_2Fe_4O_9$

A. Kirsch[1], M. M. Murshed[1], M. Schowalter[2], M. Curti[3], C. B. Mendive[3], T. M. Gesing[1]

[1]*Universität Bremen, Chemische Kristallographie fester Stoffe, Institut für Anorganische Chemie und Kristallographie, Leobener Straße /NW2, und MAPEX Center for Materials and Processes, Bremen, Germany*

[2]*Universität Bremen, Institute of Solid State Physics, Otto-Hahn-Allee /NW1, Bremen, Germany*

[3]*Universidad Nacional de Mar del Plata, Departamento de Química, Facultad de Ciencas Exactas y Naturales, Dean Funes3350, B7600AYL, Mar del Plata, Argentina*

Bismuth ferrates, $BiFeO_3$ and $Bi_2Fe_4O_9$, have attracted great attention due to their interesting physical and catalytic properties [1]. The synthesis of these compounds in the nano-size regime is of special interest due to enhanced or modified properties. In this study, a mullite-type $Bi_2Fe_4O_9$ precursor was synthesized using a polyol-mediated method and post-heated under different conditions. Analysis of X-Ray powder diffraction (XRD) data indicates that the as-synthesized sample is nano-crystalline. During heating, the X-ray amorphous powder transformed into a rhombohedral perovskite-type $BiFeO_3$ followed by a second transformation into mullite-type $Bi_2Fe_4O_9$. *In-situ* XRD measurements demonstrate that the nucleation of both $BiFeO_3$ and $Bi_2Fe_4O_9$ might simultaneously commence, however, their growth and ratios depend on temperature. Temperature-dependent Fourier transform infrared spectra helped understand some local features of the samples. The locally-resolved high-resolution transmission electron micrographs revealed that the surface of some heated samples is covered by 4 - 13 nm sized particles which were identified (from fringe distances) as crystalline $Bi_2Fe_4O_9$. Diffuse UV/Vis reflectance spectra showed fundamental absorption edges between 1.80 eV and 2.75 eV. A comparative study between the recently developed "derivation of absorption spectrum fitting " (DASF) [2] and the mostly used Tauc method was performed for the calculation of both direct and indirect band gaps. Using the Tauc method it is indispensable to know the nature of transition before the calculation whereas the DASF method provides the opportunity to overcome this limitation. The band gaps obtained by DASF method are in good agreement to the values obtained by Tauc's method (2.13(3) eV) considering the direct transitions for the samples calcined above 873 K. The obtained band gaps have been compared to those calculated by the density functional theory. This study on a mullite type compound bears particular importance to calculate the band gap using both the DASF and Tauc's method, which help determine the band gap as well as the nature of transition.

[1] Sun, S.; Wang, W.; Zhang, L.; Shang, M. *J. Phys. Chem. C* **2009**, *113*, 12826-12831.

[2] Souri, D.; Tahan, Z. E. *Appl. Phys. B* **2015**, *119*, 273-279.

P158

Structure-property relations in chalcopyrite based intermediate band solar absorber materials

J. Marquardt[1], A. Franz[1], C. Stephan[2,3], S. Schorr[1,2]

[1]*Helmholtz-Zentrum Berlin für Materialien und Energie, Berlin, Germany*

[2]*Freie Universität Berlin, Berlin, Germany*

[3]*BAM Bundesanstalt für Materialforschung und -prüfung, Berlin, Germany*

By now, the progress in manufacturing $Cu(Ga,In)Se_2$ absorbers used for thin film solar cells led to conversion efficiencies of more than 22% [1]. In general, compound semiconductors own the advantage of adjusting the band gap by changing the composition of the solid solution, as in $CuIn_{1-x}Ga_xSe_2$. To optimize the utilization of solar energy we try to establish an intermediate band [2]. The Shockley-Queisser limit for chalcopyrite semiconductors is at 32% [3], with an intermediate band gap it is proposed, that the efficiency can be raised up to 63.3%, in ideal conditions [4]. Therefore we use pure $CuGaS_2$ which has the widest band gap in the solid solution series of these two ternary system. Martì et al. proposed different transition elements, such as $Ti^{4+/3+}$ and $Fe^{3+/2+}$ which may cause an intermediate band within the energy band gap of $CuGaS_2$ [5]. In this study we focused at first on Cr^{3+} instead of Ti, because of better solubility proposed from thermodynamic calculations and Mn^{2+} as analogue [6]. The initial composition of our samples follows the pseudo-binary section of $CuGaS_2$ and Cr_2S_3 or MnS. Thus $(CuGaS_2)_{x-1}$ + $(Cr_2S_3)_x/(MnS)_x$ powder samples have been synthesized by solid state reaction at 900°C. The products of this synthesis were grounded and annealed two times for 300h at 900°C with a heating rate of 10K/h. After that the powder samples were analyzed in terms of crystal structure as well as phase content by X-ray and neutron diffraction and chemical composition by electron microprobe (WDX). We will show the first structural trends of $(CuGaS_2)_{x-1}$ + $(Cr_2S_3)_x/(MnS)_x$ sample. Furthermore we will show cation distributions of a previous $CuGa_{1-x}Mn_xS_2$ sample series.

[1]www.businesswire.com/news/home/20151207006574/en/Solar-Frontier-Achieves-World-Record-Thin-Film-Solar

[2] Luque A. and Martì A., Prog Photovolt: Res. Appl., 2001, Vol. 9, pp. 73-86

[3] Scheer, R. Schock, H. W., Chalcogenide photovoltaics: physics, technologies, and thin film devices (Wiley-VCH, Weinheim, 2011)

[4] Luque A. et al., Physica B, 2006, Vol. 382, pp. 320-327

[5] Martì A. et al., J. Appl. Phys., 2008, Vol. 103

[6] Palacios P. et al., J. Phys. Chem. C, 2008, Vol. 112, pp. 9525-9529

P159

New Scandium (III) Coordination Polymers at Bulk and Nano-Scale, Synthesis, Characterization, Thermal and Gas sensing Properties

E. SAHIN[1]
[1]*Ataturk University, Chemistry, Erzurum, Turkey*

New Scandium(III) coordination polymer at bulk and nanopowders as Scandium 1,4-benzene dicarboxylic acid hydroxide $[Sc(OH)(BDCA)]_n$, (BDCA = 1,4-Benzene Dicarboxylic Acid) has been synthesized by the reaction of a mixture Sc(III) Nitrate and BDCA in MeOH by simple Branched tube and sonochemical method. The nanopowders of Sc_2O_3 was prepared from the calcinations of the NCP at air atmosphere. The structure of the CP and NCP (CP = Coordination Polymers and NCP = Nano Coordination Polymers) were determined by X-ray crystallography, while nano-structural materials were characterized by X-ray powder diffraction (XRPD), Thermal Gravimetric Analysis (TGA) and Scanning Electron Microscopy (SEM). These NCP and Sc_2O_3 nanostructures have been tested for CO_2 gas monitoring by depositing them as thick films on an interdigitated alumina substrate and evaluating the surface resistance of the deposited layer as a function of operating temperature and CO_2 concentrations. The gas sensitivity tests have demonstrated that the Sc_2O_3 nanopowders, exhibit high sensitivity to CO_2 proving their applicability in gas sensors. The role of the nanopowders on the sensing properties of NCP and Sc_2O_3 are also discussed.

Keywords: Coordination Polymers, Sc_2O_3, Nanopowders, CO_2, gas sensing.

Figure. 1. Projection down [001] of structure CP showing chains of scandium octahedral, joined by hydroxyl groups.

References:

[1] J. C. Bailar, Preparative Inorganic Reaction. Vol1, Interscience, York, 1964.

[2] C. Janiak. Dalton Trans, (2003) 2781.

[3] C.B. Murray, C.R. Kagan, M.G. Bawendi, Annu. Rev. Matter Sci. 30 (2006) 545

Figure 1

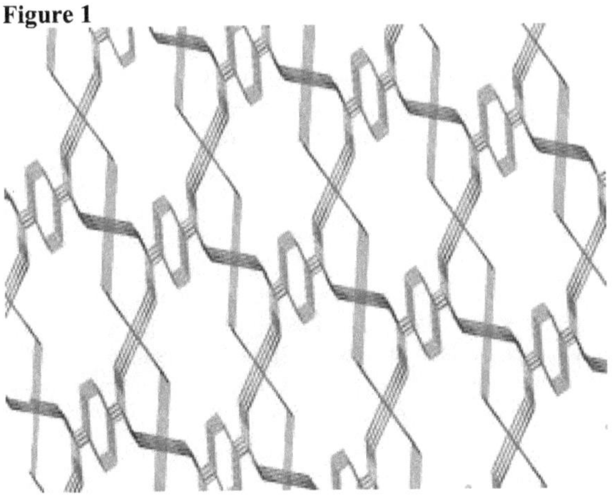

P160

Microscopic understanding of axial negative thermal expansion: a study on potassium arsenotungstate using neutron diffraction and DFT

M. M. Murshed[1], P. Zhao[1], M. Fischer[2], A. Huq[3], E. V. Alekseev[4], T. M. Gesing[1]
[1]*University of Bremen, Chemische Kristallographie fester Stoffe, Bremen, Germany*
[2]*University of Bremen, Kristallographie, Geowissenschaften, Bremen, Germany*
[3]*Oak Ridge National Laboratory, Chemical and Engineering Materials Division, Oak Ridge, United States*
[4]*Forschungszentrum Jülich and Institut für Kristallographie, RWTH Aachen, Institute of Energy and Climate Research (IEK-6), Jülich, Germany*

Negative thermal expansion (NTE) indicates strong anisotropy either in the Grüneisen parameters or in the elastic properties. However, in many cases neither of them can explain the axial NTE [1]. The temperature-dependent neutron time-of-flight diffraction data Rietveld refinement reveal that the framework type $K[AsW_2O_9]$ [2] exhibit positive thermal expansion in the **a**- and **c**-directions, and NTE in the **b**-direction between 10 K and 900 K. Using the quasi-harmonic approximation the DFT calculations demonstrate that both acoustic and optic mode negative Grüneisen parameters contribute to the phonon density of states (PDOS). These modes are associated with the transverse vibration of oxygen atoms normal to As-O-W linkages as observed from the phonon wavevector analyses, which are not parallel to any crystallographic axis-direction. Temperature-dependent Raman spectra do not show hardening of any phonon modes. That is, the isochoric anharmonicity seems to be dominating over the isothermal anharmonicity in the system. The lattice thermal expansion was fitted using Grüneisen first order approximation for the zero pressure equation of state at 0 K, where the vibrational energy was calculated using the Debye-Einstein-Anharmonicity model [3]. In this model the thermoelastic constants determine the isothermal anharmonicity while the anharmonicity constant measures the isochoric one. The PDOS spectrum helped guide to set the respective characteristic frequencies during the simulation. A negative axial Grüneisen parameter was set in the **b**-direction for mechanical stability, which is however not the necessary condition for the associated NTE. The cross-linking values (elastic compliance multiplied by corresponding Grüneisen parameter) ultimately determine the sign of the axial coefficients of the expansion [1]. This study on a complex framework oxide demonstrates that a slight anisotropy in the Grüneisen parameter can change the elastic cross-linking value large enough, leading to the axial expansion sign negative.

References

[1] R.W. Munn, J. Phys. C: Solid State Phys., 5 (1972) 535.

[2] E.V. Alekseev, O. Felbinger, S. Wu, T. Malcherek, W. Depmeier, G. Modolo, Th.M. Gesing, S.V. Krivovichev, E.V. Suleimanov, T.A. Gavrilova, L.D. Pokrovsky, A.M. Pugachev, N.V. Surovtsev, V.V. Atuchin, J. Solid State Chem. 204 (2013) 59.

[3] M.M. Murshed, C.B. Mendive, M. Curti, M. Sehovic, A. Friedrich, M. Fischer, Th.M. Gesing, J. Solid State Chem. 229 (2015) 87.

P161

Modified Novel Perovskite-type Oxides and Oxyfluoronitrides for Thermolelectric and Solar Water Splitting Application

M. Widenmeyer[1], X. Xaio[1], J. Häcker[1], C. M. Bubeck[1], W. Xie[1], A. Weidenkaff[1]

[1] *University of Stuttgart, Institute for Materials Science, Stuttgart, Germany*

During the last decade perovskite-type oxides (ABO_3) and perovskite-type oxynitrides ($AB(O,N)_3$) have attracted great interest due their possibilities of being used in various fields of application such as high-temperature thermoelectrics (TE, T > 1000 K) and photoanodes for solar water splitting (SWS).[1] The high stability and tolerance towards substitution on *A*-, *B*- and O-site favors materials development. The successful substitution on any of these 3 positions leads to a significant manipulation of the electronic band structure and hence the physical properties.

Eu^{2+} substituted $Ba_{1-x}Eu_xTiO_{3-\delta}$ precursors were synthesized by Pechini method and afterwards sintered at 1473-1673 K in reducing atmosphere. Rietveld refinements of the PXRD data revealed a perovskite structure and that all $Ba_{1-x}Eu_xTiO_{3-\delta}$ ($0 < x < 1$) samples are single phase (s. Figure 1). The electrical conductivity increases significantly with rising Eu^{2+} substitution. The Seebeck coefficient is negative (*n*-type semiconductor) and exhibits an opposite trend. Eu^{2+} substitution induced point defects and activates a mass fluctuation phonon scattering mechanism. The minimum lattice thermal conductivity was achieved at $x = 0.5(1)$ (s. Figure 2). The significant enhancement of electrical conductivity and reduction of lattice thermal conductivity makes $Ba_{1-x}Eu_xTiO_{3-\delta}$ ($0 < x < 1$) promising for HT-TE usage. The lower reaction temperature during the Pechini process allowed an enhanced sintering grade above 90 % significantly higher than in conventional synthesis.

Partial anionic substitution of O^{2-} by N^{3-} leading to perovskite-type $AB(O,N)_3$ (A = Mg, Ca, Sr, Y, La; B = Ti, Zr) was proven to reduce the optical band gap (oBG) by up to 2 eV compared to ABO_3. About 10 at.% N are required for this purpose. However, simultaneously applied *A*-site substitution for charge compensation induced an opening of the oBG by an enhanced tilting of the octahedral network.[2] An alternative charge compensation is achieved by double anionic substitution of O by N and F. Therefore, oxyfluorides $AB(O,F)_3$ were formed by solid state reactions between ACO_3, BO_2 and AF_2 at $T \approx 1300$ K in air and subsequently transferred in the respective oxyfluoronitrides $AB(O,N,F)_3$ by thermal ammonolysis. For a single phase synthesis of $AB(O,N,F)_3$ a precise adjustment of the T program and regrinding steps are required. The presence of approximately 1 at.% of N and F resulted in a reduction of the oBG by up to 1.7 eV and lower anionic vacancy concentrations as observed for cationic substituted oxynitrides $AB(O,N)_3$.

Figure 1. Rietveld refinements of the crystal structure of $Eu_{0.4}Ba_{0.6}TiO_{3-d}$.

Figure 2. Transport properties (electrical *s* and thermal conductivity k_L, Seebeck coefficient *S*) of sintered $Eu_{0.4}Ba_{0.6}TiO_{3-d}$ measured in forming gas.

[1] S. Yoon, A. Maegli, L. Karvonen, A. Shkabko, S. Populoh, K. Gałązka, L. Sagarna, M. Aguirre, P. Jakes, R. Eichel, S. Ebbinghaus, S. Pokrant, A. Weidenkaff., *Z. Anorg. Allg. Chem.* **2014**, *640*, 797.

[2] M. Widenmeyer, C. Peng, A. Baki, R. Niewa, A. Weidenkaff, *Solid Stat. Sci.* **2015**, in press.

Figure 1

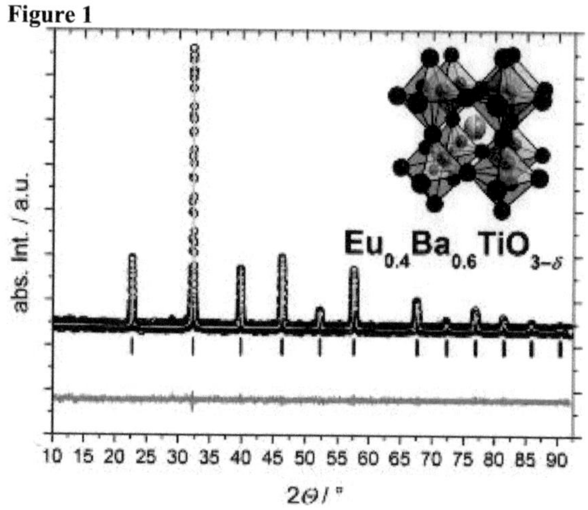

Figure 2

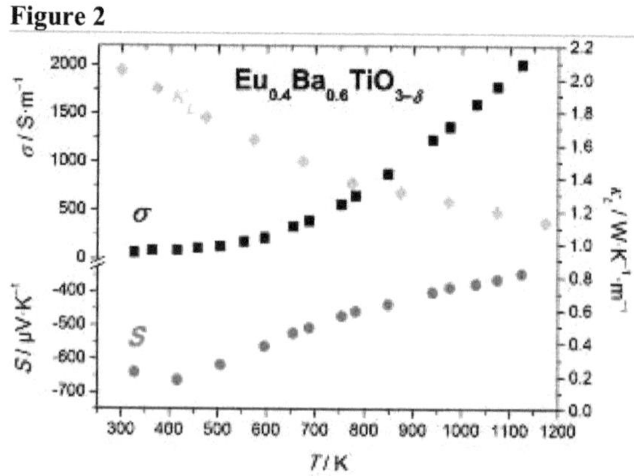

P162

Re-evaluation of the compressibility of the rhenium boride phases Re_3B and Re_7B_3

C. Neun[1], B. Petermüller[2], L. Bayarjargal[1], W. Morgenroth[3,1], H. Huppertz[2], B. Winkler[1]

[1] *Institut für Geowissenschaften, Abteilung für Kristallographie, Frankfurt, Germany*

[2] *Institut für generelle, anorganische und theoretische Chemie, Innsbruck, Austria*

[3] *Deutsches Elektronensynchrotron, Hamburg, Germany*

Numerous transition metal borides are high-performance materials. In this context, the system Re-B is of special interest, as Re has a very high concentration of valence electrons. While ReB_2 has been studied extensively due to its outstanding properties [1], Re_7B_3 and Re_3B have attracted less interest, even though they might also exhibit unusual properties.

The object of this study is to re-evaluate the high-pressure behavior of Re_7B_3 and Re_3B on phase pure samples. In an earlier study [2], Re_7B_3 was found to be very incompressible with a bulk modulus of 435(16) GPa. On the other hand the experimentally determined compressibility of Re_3B in that study [2] with a B_0 of 320(15) GPa was much lower than expected and not in a good agreement with the DFT calculations performed.

In the earlier study [2], the Re-borides had been synthesized in a laser-heated DAC and were present as parts of complex phase mixtures only, which prevented an accurate determination of their compressibility. Here, samples were synthesized in an HF furnace and via arc-melting. X-Ray diffraction experiments were performed at the beamline P02.2 at PETRA III using an energy of 43 keV. In order to better understand the structure-property relations of the synthesized we performed density functional theory (DFT) calculations.

The newly found value for B_0 of 409(4) GPa for Re_3B is now consistent with the calculated value of 400 GPa (figure 1). The bulk modulus of Re_7B_3 was found to be 390(11) GPa, which is considerably lower than the value found earlier and much closer to the theoretically derived value of 394 GPa (figure 2).

The discrepancies between the experimental and theoretical compressibilities of Re_3B and Re_7B_3 have been significantly reduced by using single phase samples. Whereas Re_7B_3 could not be verified as being ultra-incompressible, Re_3B was found to have a very high bulk modulus.

The DFG and the FWF are gratefully acknowledged for funding this project in the framework of an ERA-chemistry project (WI 1232 and I 1636-N19), and so is the BMBF project 05K13RF1.

Figure 1: Compressibility of Re_3B

Figure 2: Compressibility of Re_7B_3

[1] M. Frotscher, M. Hölzel, and B. Albert. *Z. Anorg. Allg. Chem.* **636** (2010),1783-1786.

[2] E. A. Juarez-Arellano, B. Winkler, A. Friedrich, L. Bayarjargal, W. Morgenroth, M. Kunz, and V. Milman. *Solid State Sci.* **25** (2013), 85 - 92.

Figure 1

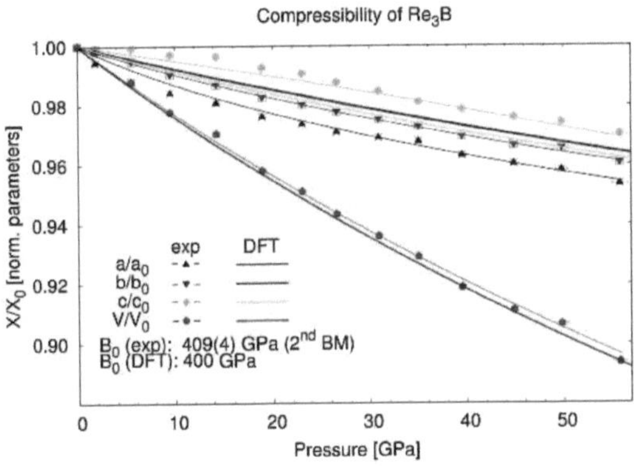

Figure 2

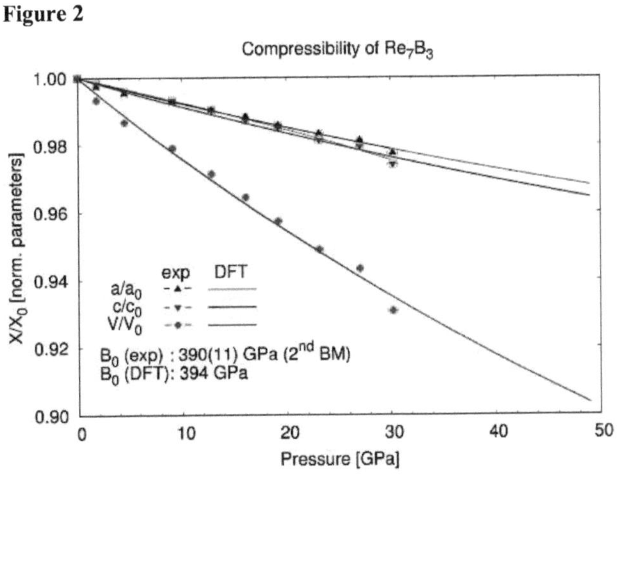

P163

Texture and anisotropic elastic behavior of the mineral phase of the charonia lampas lampas shell

S. OUHENIA[1], D. Chateigner[1]

[1]*University of Bejaia, physics, Bejaia, France*

Introduction

Molluscan shells are fascinating examples of high performance organic/inorganic biocomposites materials. Organic components represent only about 1-5% the weight of the shell and they are responsible for the remarkable enhancement of the strength and elasticity of the material as compared to geological mineral. Molluscan shells are mainly built of two polymorphs of calcium carbonate: calcite and aragonite. Most of the organic phase is located between crystallites (intercrystalline), but some organic molecules (intracrystalline) are also intercalated within the crystalline lattice.

Objectives

In this work, the preferred crystallographic orientation of the aragonitic shell of the Gastropod *Charonia lampas lampas* is examined using X-ray and combined analysis. This approach allows to work on the real layers of the shell without necessity of powderisation, and provides the orientation distributions of the three layers of the shell, together with their structure refinement. We examine the correspondence between textures and morphology (as seen by scanning electron microscopy) of the shell structures. The macroscopic elastic behaviour of the mineral parts of the layers is simulated and discussed.

Materials and Methods

We collected a 20 cm large shell (Fig. 1a) from the coast of Algeria. The inorganic part of the shell is only composed of aragonite. A SEM image at low magnification (Fig. 1b) shows three distinct layers which we will refer to as outer, intermediate and inner layers. We first measured the outer layer, then removed this layer with a diluted solution of HCl to analyse the intermediate layer and the inners layers.

We examined the shell microstructures of gold-sputtered fractured cross-sections using a SEM microscope . X-ray diffraction measurements were carried out using a four-circle goniometer and curved position-sensitive detector (CPS-120, Inel). Grid measurement in tilt and azimuthal angles was carried out to cover the whole pole figure, resulting in 936 diagrams measured for each layer. All data were analysed within the so-called ''Combined Analysis" formalism (Chateigner, 2004) using the MAUD software (Lutterotti et al., 1999).

Results

The SEM images reveal that Charonia shell is composed of three crossed lamellar layers of biogenic aragonite. The X-ray diffraction combined analysis of preferred orientations shows that the outer layer exhibits a fibre texture, the intermediate crossed lamellar layer is radial with a split of its c-axis and single twin pattern of its a-axis, and the inner layer is comarginale with split c-axis and double twinning. A lost of texture strength is quantified from the inner layer outward. The simulation of macroscopic elastic stiffness cij for the mineral part of the three layers of Charonia, calculated from the OD, using the geometric mean method shows that the strong orientations present in the successive layers render maximum benefits to the shell in terms of rigidity and shear resistance.

5. Conclusion

The simulation of the macroscopic elastic tensors of the mineral part of the three layers, from texture data, reveals an optimisation of the elastic coefficient of compression and shear in all directions of the shell as an overall view.

References

Chateigner, D. (Ed.), 2004. Combined Analysis: Structure-Texture-Microstructure-Phase-Stresses-Reflectivity Analysis by X-ray and Neutron Scattering.

Lutterotti, L., Matthies, S., Wenk, H.-R., 1999. MAUD. National Research Council of Canada, Ottawa, 1999, pp. 1599-1604.

figure legends

Fig. 1.(a) The Charonia lampas lampas shell studied in this work. (b) Cross-section SEM image of the fractured shell at the location indicated in (a).G, M and N indicate the Growth, Margin and Normal directions, respectively

Figure 1

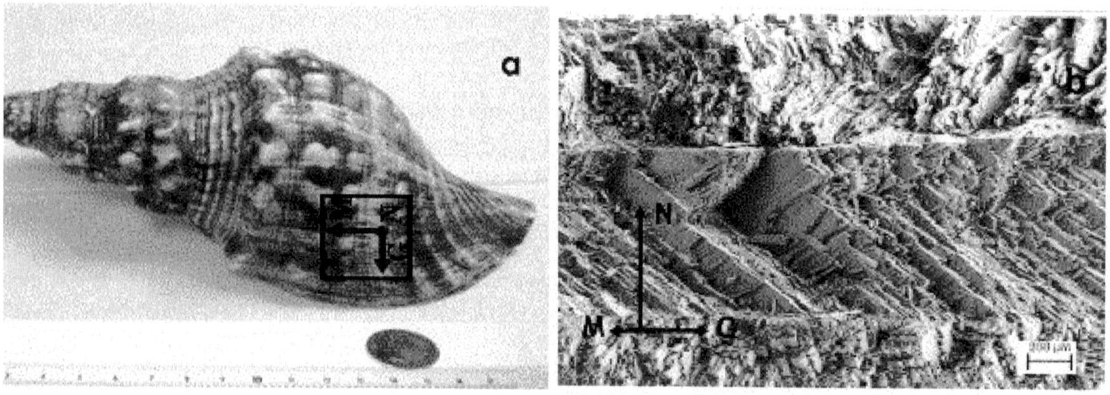

P164
ELIMINATION OF METHYLENE BLUE BY ADSORPTION ON RAW ORE GRAPHITE (DEPOSIT POINT 214-AREA SIDI BOUOTHMANE- MARRAKECH)

S. sabir[1]
[1]*Faculty of Science Ben M'sik University Hassan II, chemistry, Casablanca, Morocco*

This work has allowed us to study the adsorption phenomenon of BM methylene blue on raw ore graphite (deposit point 214 sidi bouothmane region Marrakech).

At the beginning of this work, we conducted a series of experiments to study this phenomenon methylene blue playing on a number of parameters, namely: The pH of the medium, the particle size, the mass of graphite, the concentration of methylene blue, the main results of this study are:

The adsorption is done in a first time in a monolayer when the initial concentration of methylene blue does not exceed 10mg / l.

Once this concentration limit is outdated, a new multilayer adsorption phenomenon began to appear according to the concentration put out.

Otherwise, the saturation of a monolayer of 1 g of GP214 with 10mg of (BM) has confirmed by two models (Langmuir, Freundlich)

We are planning into perspective to continue our trials multilayer adsorption subject to try to discover the farthest upper boundary layer of the support and which could result in improvement, the adsorption capacity of the GP214 natural material.

Keywords: Graphite; Adsorption; modeling; Methylene blue; Langmuir; Freundlich

Spectroscopic methods in crystallography

P165
Kinetics of the hydrogen defect in lithium niobate and lithium tantalate

T. Köhler[1], E. Mehner[1], J. Hanzig[1], G. Gärtner[1], H. Stöcker[1], D. C. Meyer[1]
[1]*Technische Universität Bergakademie Freiberg, Institut für Experimentelle Physik, Freiberg, Germany*

$LiNbO_3$ and $LiTaO_3$ crystals are used in many optical and electronic devices, therefore, understanding the structural defects is helpful to optimize related material parameters. One important defect is hydrogen, which is incorporated during the growth process and forms OH^- ions. The aim of the present study is the development of a structural model for the diffusion behavior of the OH^- defect.

The migration of H^+ is investigated by FT-IR and UV/VIS spectroscopy using different crystallographic orientations and excitation polarisations. Depending on the Li content, the hydrogen defect causes an OH^- band with two (congruent material) or four sub-bands (near stoichiometric material). The formation or decline of the sub-bands is examined under reducing and oxidizing conditions.

It was found that the H^+ out-diffusion is anisotropic in C-$LiNbO_3$, while C-$LiTaO_3$ has an isotropic behavior. The [0001] direction seemingly is the preferred diffusion path. The coloration of the reduced crystals can be related to the incorporation of oxygen vacancies, allowing new diffusion paths.

For the protonation two setups are used: hot water bath and oxidizing in water vapor-rich atmosphere. Only the hot water bath treatment shows a formation of new sub-bands. Here, it is assumed, that the incorporation is coupled with the out-diffusion of Li^+.

The reprotonation proves that due to the preceding incorporation of oxygen vacancies an open structure is present that allows an in-diffusion in water vapor-rich atmosphere.

Regarding the structural model, it can be concluded that the OH^- band is formed by four components, wherein the hydrogen is present disordered on interstitial lattice sites in the materials. An accelerated out-diffusion is possible by the incorporation of additional defects, as well as an incorporation of H^+ on additional binding sites, which are not occupied in the untreated material.

P166
Thermal annealing of radiation damaged minerals

P. Zietlow[1], T. Beirau[1], B. Mihailova[1], J. Schlüter[1], C. Paulmann[1], R. Škoda[2], L. A. Groat[3], U. Bismayer[1]
[1]*Universität Hamburg, Hamburg, Germany*
[2]*Masaryk University, Brno, Czech Republic*
[3]*University of British Columbia, Vancouver, Canada*

Minerals with radiation damage are structurally disordered on different length scales -resulting from the decay processes of built-in uranium and thorium nuclei. These structural damages are metastable and influence the macroscopic properties. Thermal annealing reorganizes the metastable structure. This is of great interest for material sciences, for instance for the better understanding of embedding materials for actinides in radioactive waste. [1], [2]

The analysis of pyrochlores and titanites exposed to different radiation doses on annealing using X-ray diffraction, Raman and IR spectroscopy, and NMR spectroscopy, respectively, gives insight into their structural damage and their recrystallization behaviour on different length scales. Recrystallization temperatures are sensitive to crystal chemistry, structural topology and disorder phenomena. In our study we present the thermal annealing behaviour of three pyrochlores and one titanite of different degrees of initial structural damage with respect to low actinide reference samples. [3]

[1] Ewing, R. C., Meldrum, A., Wang, L.M., Wang, S.X. (2000): Radiation-Induced Amorphization, Reviews in Mineralogy and Geochemistry, Ed. P.H. Ribbe, Min. Soc. America **39**, 319-361.

[2] Hawthorne, F.C., Groat, L.A., Raudsepp, M., Ball, N.A., Kimata, M., Spike, F., Gaba, R., Halden, N.M., Lumpkin, G.R., Ewing, R.C., and others (1991): Alpha-decay damage in titanite. American Mineralogist, **76**, 370-396.

[3] Zietlow, P., Beirau, T., Groat, L. A., Paulmann, C., Bismayer, U. (2014): ^{29}Si MAS NMR and synchrotron XRD study of metamict Cardiff titanite. Zeitschrift für Kristallographie, **229**, 551-554.

P167
Mechanical properties of multiferroic $Bi_2Mn_4O_{10}$: Full set of elastic constants determined by inelastic neutron scattering

F. Ziegler[1], M. Mangir Murshed[2], H. Gibhardt[1], O. Sobolev[1], T. M. Gesing[2], G. Eckold[1]
[1]*Georg August Universität Göttingen, Institut für Physikalische Chemie, Göttingen, Germany*
[2]*Universität Bremen, Institute of Inorganic Chemistry and Crystallography and MAPEX Center for Materials and Processes, Bremen, Germany*

1. Introduction

Mullite-type $Bi_2Mn_4O_{10}$ [1] is an example for a multiferroic compound with a Néel temperature of 39 K, although its crystal structure [2] is not typical for allowing ferroelectricity. The dipoles of the constituent $Mn^{4+}O_6$ (along *c*-axis, bond valence sum BVS = 3.90(2) v.u.) and $Mn^{3+}O_5$ (along *b*-axis, BVS = 3.04(2) v.u.) polyhedra are oriented in different directions [3]. Both its nuclear and magnetic structural features differ from those of other rare-earth members of the $R_2Mn_4O_{10}$ family. For instance, all of the coordination polyhedra ($Mn^{4+}O_6$, $Mn^{3+}O_5$ and $Bi^{3+}O_8$ (BVS = 3.18(2) v.u.) show distinctive distortions due to slight but significant lone electron pair activity of the Bi^{3+} cation [4].

2. Objective

Mechanical distortions are frequently found to play important roles during ferroelectric phase transitions. Therefore, the knowledge of the elastic behavior seems crucial for a profound understanding of the interplay of magnetic and electric ordering behavior. Hence, we have undertaken an inelastic neutron scattering study on single crystals of $Bi_2Mn_4O_{10}$ aiming at the determination of the full set of nine elastic constants c_{ij} provided by the orthorhombic system (*Pbam*).

3. Methods

The inelastic neutron scattering experiments have been performed at the three-axes spectrometer PUMA@FRM II in Garching. The dispersion curves of the acoustic phonon branches in the Brillouin zone have been measured at room-temperature for different propagation directions close to the zone center (Γ-point) and for different polarization vectors. The linear slopes of the dispersion curves are connected with a specific combination of the c_{ij}

depending on the propagation directions and the polarization vectors of the modes.

4. Results

We were able to determine all nine elastic constants c_{ij} provided by the orthorhombic crystal (see table). These results show a clear mechanical anisotropy indicating that the crystallographic *a*-axis is softer than the other principal directions. This behavior is associated with the mutual linking of the individual building blocks of the unit cell. Since the magnetization in the low-temperature phase is also oriented along the *a*-axis, this can be regarded as an indication for the important role of the magnetoelastic coupling for the existence of multiferroicity.

Furthermore, some interesting resolution effects were observed allowing the detection of a phonon that does not obey the polarization selection rule (see figure).

5. Conclusion

In conclusion, using inelastic neutron scattering we were able to determine the complete set of single crystal elastic constants of $Bi_2Mn_4O_{10}$ at ambient condition with reasonable accuracy. These data provide a useful basis for modelling the microscopic behavior of this multiferroic system.

References

[1] H. Schneider, R. X. Fischer, Th. M. Gesing, J. Schreuer, M. Mühlberg, Int. J. Mat. Res. 103 (2012) 422 - 429.

[2] M. Burianek, Th. F. Krenzel, M. Schmittner, J. Schreuer, R. X. Fischer, M. Mühlberg, G. Nénert, H. Schneider, Th. M. Gesing, Int. J. Mat. Res. 103 (2012) 449 - 455.

[3] L. M. Volkova, D.V. Marinin, J. Phys.: Condens. Matter 21 (2009) 015903.

[4] M. Curti, Th. M. Gesing, M. Mangir Murshed, Thomas Bredow, Cecilia B. Mendive, Z. Kristallogr. 228 (2013) 629-634.

Figure 1: All nine elastic constants

ij	c_{ij} / GPa
11	160 ± 16
22	233 ± 23
33	257 ± 26
44	83 ± 8
55	56 ± 6
66	58 ± 6
12	134 ± 18
13	157 ± 14
23	187 ± 16

Figure 2: Constant energy scan at 1 THz. All three acoustic phonons were measured in this scan due to an unusal resolution effect although the phonon T2A should have no intensity

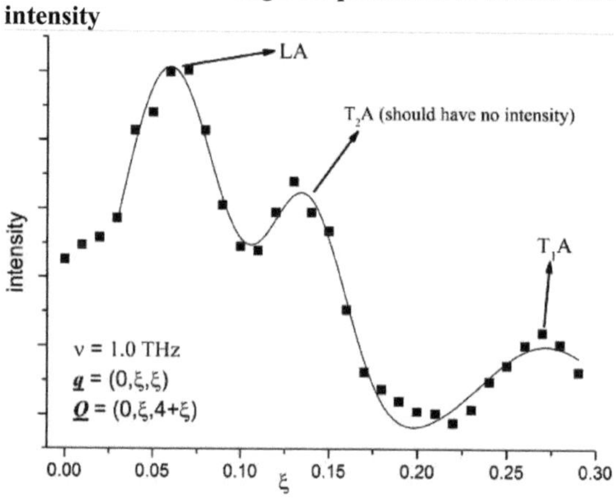

P168

Local Atomic Ordering of Ga in $Ba(Al_{1-x}Ga_x)_4$ with $0.025 \leq x \leq 0.075$ Investigated by Quantum Mechanical Calculations and Solid State NMR Spectroscopy

V. Peters[1], B. Mausolf[1], F. Haarmann[1]
[1] *Institute for inorganic Chemistry, RWTH Aachen University, Aachen, Germany*

Improved properties of new materials are often based on structural variations of existing materials.[1] For further improvement of these materials it seems to be necessary to understand the structure-property-bonding relation. A powerful tool to investigate the structure and bonding of intermetallic phases is the NMR spectroscopy.

The aim of this study is the determination of local atomic environments the atoms in $Ba(Al_{1-x}Ga_x)_4$ with solid state NMR spectroscopy and by quantum mechanical (QM) calculations.[2]

$Ba(Al_{1-x}Ga_x)_4$ can be understood as $BaAl_4$ were Al is substituted with Ga. It crystalizes in the $BaAl_4$ type (*I4/mmm*) with Al atoms occupying the Wyckoff positions 4d and 4e with respect to their coordination number referred to as 4b and 5b, respectively. The 4b site is tetrahedron-like coordinated by four 5b atoms. The 5b site is quadratic pyramidal coordinated by one 5b atom and four 4b atoms.

Substituting small amounts of Al with Ga varying local arrangements are realized in the crystal structure. By modelling the local environments using super lattice structures and calculating their coupling parameters the NMR signals can be simulated and compared with experimental data.

Due to similar bonding situations the corresponding NMR signals are likely to overlap and distinguishing those can be challenging. To increase the experimental resolution the crystallites can be aligned in the magnetic field. The alignment is achieved by exposing grinded powder suspended in glue to the orienting magnetic field B_{or}. During the hardening of the glue the crystallite orientation freezes. The sample can be rotated about *c*, which describes the angle between B_{or} and magnetic field used for the measurement B_0.[3] The orientation dependent NMR signals differ for the various possible atomic arrangements. Especially the so called satellite transitions are strongly dependent on the sample orientation and therefore well-defined indicators for different environments even with similar NMR parameters.

The ^{69,71}Ga-NMR measurements of the central transition of $Ba(Al_{1-x}Ga_x)_4$ with x = 0.05 indicate only isolated Ga-atoms occupying the 5b site (Fig.1). Considering this model QM calculations predict twelve different Al(5b) environments. The ^{27}Al experiments on the aligned powder sample at c=0° are in good agreement with the simulated signals (Fig.2). This supports the model of Ga atoms being coordinated exclusively by Al atoms.

Using the orientation dependence of the satellite transitions of $Ba(Al_{1-x}Ga_x)_4$ with x = 0.05 contains valuable information which are well accessible by measuring aligned powders in combination with QM calculations.

Fig1: ^{71}Ga-NMR signals of regular (RP) and aligned powders at c=90° and c=0° (top) and ^{69}Ga aligned powder at c=90° and c=0°(bottom). Measured and simulated signals are depicted in grey and black, respectively. Simulations are based on a single isolated Ga(5b) site.

Fig2: ^{27}Al measurements of the central and satellite transitions (grey). The sum of the twelve considered Al(5b) environments are shown in black.

[1] G. J. Snyder, E. S. Toberer, Nature Materials 7, 105 - 114 (**2008**).

[2] P. Blaha, K. Schwarz, G. K. H. Madsen, D. Kvanicka und J. Luitz, *WIEN2k, An Augmented Plane Wave Plus Local Oribitals Program for Calculating Crystal Properties,* Vienna University of Technology , **2001**.

[3] F. Haarmann, *Quadrupolar NMR of Intermetallic Compounds*, in: Enzyclopedia of Magnetic Resonance (Eds.: R. K. Harris, R. E. Wasylishen), John Wiley & Sons, Ltd, Chichester, **2011**.

Figure 1

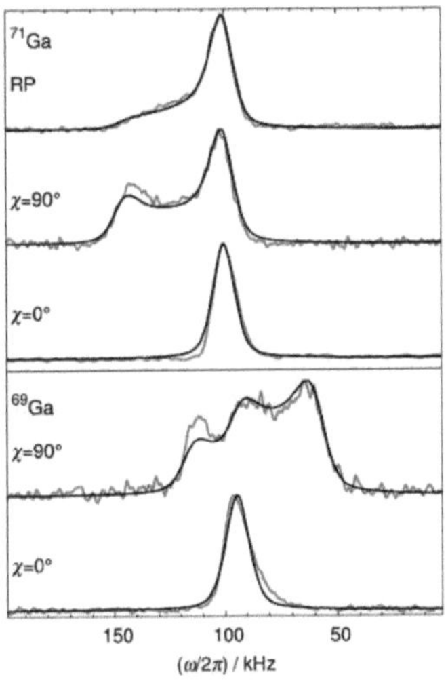

Figure 2

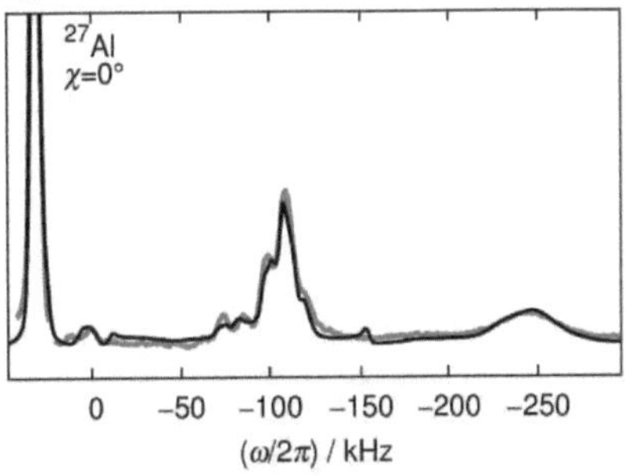

P169

Structural transformations in the ferroelectric (1-x)$PbTiO_3$-$x$$BiMg_{0.5}Ti_{0.5}O_3$ solid solution studied by Raman spectroscopy

I. Margaritescu[1], K. Datta[1], J. Chen[2], B. Mihailova[1]
[1]*University of Hamburg, Earth Sciences, Hamburg, Germany*
[2]*University of Science and Technology Beijing, Physical Chemistry, Beijing, China*

Lead-based ferroelectric solid solutions with the perovskite structure type (ABO_3) exhibit excellent dielectric and piezoelectric properties and have been widely used for a long time in different fields such as microelectronics, optoelectronics, medical imaging etc. However, the restrictions on the use of Pb have increased over the past years, due to its potential environmental toxicity during disposure. As a consequence, the structure and properties of lead-free or lead-poor ferroelectrics are gaining continuously growing attention and are becoming subjects to extensive investigations. Solid solutions of type (1-x)$PbTiO_3$-$x$$BiMeO_3$, Me = Sc, In, $Mg_{0.5}Ti_{0.5}$, $Ni_{0.5}Ti_{0.5}$, $Co_{1/2}Ti_{0.5}$, $Mg_{0.5}Zr_{0.5}$, $Ni_{0.5}Zr_{0.5}$, have been recently spotted as a very good compromise between the relatively low content of Pb and excellent piezoelectric and electromechanical responses combined with a high Curie temperature. Similar to Pb-pure ferroelectric solid solutions, (1-x)$PbTiO_3$-$x$$BiMeO_3$ exhibit morphotropic phase boundary (MPB), at which the long-range ferroelectric order changes from tetragonal to rhombohedral/monoclinic, and the response functions are strongly enhanced near x_{MPB}. However, the relation chemistry - structure - properties in (1-x)$PbTiO_3$-$x$$BiMeO_3$ solid solutions is still poorly understood and studies dedicated to the mesoscopic-scale structural transformations are scarce. The aim of this contribution is to study the local structure and atomic dynamics in (1-x)$PbTiO_3$-x$BiMg_{0.5}Ti_{0.5}O_3$ (PT-xBMT) solid solution by Raman spectroscopy. For the purpose, a series of PT-xBMT ceramics with x varying between 0.10 and 0.70 across the MPB composition x_{MPB} = 0.63 were analyzed at room temperature. Four selected compounds with x = 0.20, 0.50, 0.63, and 0.70 were further analyzed at different temperatures on cooling from 870 K to 100 K. The results show that the composition-driven phase transition at room temperature is clearly mirrored by the phonon mode near 120 cm^{-1} involving A-BO_3 translations. However the Raman scattering near 50 and 270 cm^{-1}, arising from off-centred A-site and B-site cations respectively, indicates composition-induced structural rearrangements already at x = 0.50. The high-temperature Raman spectra of PT-xBMT clearly reveal doubling of the perovskite structure on the mesoscopic length scale as well as A-site and B-site cationic off-centre displacements existing at least up to 870 K. The temperature-induced structural transformations for different x aside and near the MPB composition will be compared and discussed.

P170

Toplogical studies in bivalve shells

J. He[1], S. Zhao[2], U. Bismayer[1]
[1]*Universität Hamburg, Mineralogisch-Petrographisches Institut, Hamburg, Germany*
[2]*China University of Geosciences, School of Earth Sciences, Wuhan, China*

Shells of bivalves and gastropods are in the focus of biomaterial research; usually they are inorganic-organic composite materials with hierarchical topological arrangements. The combination of a small amount (~5%) of organic macromolecules and a major part of mineral substance (generally polymorphs of calcium carbonate) in an interlaced way on different length scales provides superb mechanical function that natural minerals don't achieve. For example, the difference of hardness between shells and calcite/aragonite can be significant. The key for this excellent mechanical performance lies in the microstructure and texture, and most importantly it is influenced by the formation of such textures.

In this study, Raman spectroscopy and electron backscatter diffraction(EBSD) were applied to obtain the mineral components and local crystal orientation of bivalve shells. Polarized Raman spectra show that the nacre layer and the prismatic layers are mainly composed by aragonite and calcite, respectively. The polarized spectra also indicate preferential orientation compared with non-polarized spectra. EBSD results show that the *c*-axes of most aragonite nano-particles is perpendicular to the shell body, while *a* and *b*-axes are organized in two level domain structure. In the primary domain structure, there is about a 64° misorientation of the *a*-axes between different domains. In the secondary domain structure, the misorientation angles of two different aragonite sheets are around 10° or 20°. This kind of domain structure is important to better understand the growth mechanism of aragonite in nacre.

P171

Low-temperature structure of Pb-lawsonite, $PbAl_2[(OH)_2|Si_2O_7]\cdot H_2O$

M. Ende[1], G. Giester[1], B. Wunder[2], M. Koch-Müller[2], E. Libowitzky[1]
[1]*University of Vienna, Department of Mineralogy and Crystallography, Vienna, Austria*
[2]*German Research Centre for Geosciences, Potsdam, Germany*

1. Introduction

A phase transition, which changes the space group of Pb-lawsonite from *Cmcm* to *Pbnm*, has been detected around 450 K [1], well above the comparable 273 K transition observed for lawsonite [2] (*Cmcm* - *Pmcn*). New temperature-dependent Raman spectroscopic measurements in the region between 400 cm^{-1} - 700 cm^{-1} indicate a further phase transition at low temperatures, as it has been observed for lawsonite too.

2. Objectives

A comparison of Pb-lawsonite crystal structures refined from single-crystal X-ray diffraction at RT and 100 K in combination with detailed Raman band (shift) analyses should reveal the reason for the non-uniform changes around 150 K. Using the d-ν correlation of Libowitzky [3] allows the estimation of hydrogen bond lengths in the form of d(O···O) and d(H···O) distances from Raman O-H stretching band positions (ν - wavenumber) and makes it possible to compare those distances with the results of crystal structure analyses.

3. Materials & Methods

Pb-lawsonite samples were synthesised by a piston cylinder technique at pressures around 4 GPa at a temperature of 873 ± 10 K for around 120 h at the GFZ, Potsdam, Germany. As starting material a glass with stoichiometric Pb-lawsonite cation composition ($PbAl_2Si_2O_{x(7-8)}$) and water was used [1]. Raman spectroscopic data were collected on powder samples with a notch filter-based micro-Raman system Horiba Jobin Yvon LabRAM-HR Evolution equipped with an Olympus BX41 microscope. Crystallographic data were acquired using a BRUKER Kappa Apex II single-crystal X-ray diffractometer with a BRUKER APEX CCD detector and a Mo Kα Incoatec Microfocus Source IμS.

4. Results

The crystal structure determination at 100 K results in the same space group (*Pbnm*) as it was determined for the RT structure of Pb-lawsonite. No further mirror plane is lost for Pb-lawsonite. Only the strong hydrogen bond O4h···O6 seems to vanish, resulting in a non-uniform change of temperature-dependent Raman shifts around 150 K. In addition, the hydrogen atoms of the H_2O molecule apparently build up stronger dynamical hydrogen bonds. Using the d-ν correlation of Libowitzky [3], from band positions estimated distances could be compared to the observed bonds (Figure 1).

5. Conclusion

The phase transition of lawsonite located around 125 K is accompanied with the loss of the "m.." mirror plane [2], while Pb-lawsonite already lost it around 450 K [1]. Due to the missing splitting of the O4h atom position, possibly caused by stabilising of the mirror plane "..m" by the Pb-cation, those strong hydrogen bonds are not exactly the same as those observed for lawsonite (Figure 1). The hydrogen bonding of the hydroxyl group is different, comprising Hha···O6 and Hhb···O4h bonds, which by their disappearance possibly evoke the non-uniform framework changes around 150 K.

[1] Ende, M., Wunder, B., Koch-Müller, M., Pippinger, T., Buth, G., Giester, G., Lengauer, C. L., Libowitzky, E. (2016): T-induced displacive phase transition of Pb-lawsonite end member. Mineral. Mag., DOI: 10.1180/minmag.2016.080.015

[2] Libowitzky, E., Armbruster, T. (1995): Low-temperature phase transitions and the role of hydrogen bonds in lawsonite. Am. Mineral., 80, 1277-1285

[3] Libowitzky, E. (1999): Correlation of O-H stretching frequencies and O-H···O hydrogen bond lengths in minerals. Mh. Chem., 130, 1047-1059

Figure 1: Temperature-dependent (83 - 573 K) Raman band shifts of hydroxyl and H_2O stretching vibrations.

Figure 1

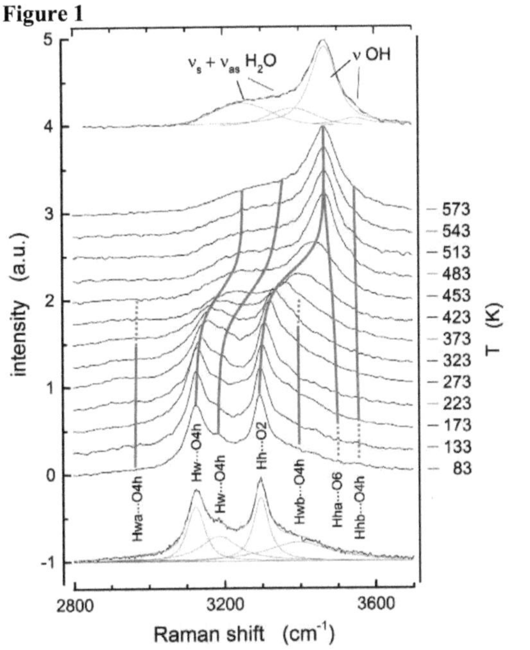

P172

Utilization of diaminobenzidine oxidation for the generation of specific contrast in X-Ray Zernike phase-contrast microscopy

O. Lorbeer[1,2], M. Warmer[1], C. Schneider[2], V. Mordhorst[2], I. Vartiainen[3], N. Stübe[1], B. Reime[1], J. Meyer[1], P. Fischer[1], R. Reimer[2], A. Meents[1]
[1]Deutsches Electronen-Synchrotron DESY, Hamburg, Germany
[2]Heinrich-Pette Institute, Leibniz Institute for Experimental Virology, Hamburg, Germany
[3]Paul Scherrer Institute PSI, Villigen, Switzerland

X-ray microscopy of biological samples with sub-cellular resolution has become an established method especially in the energy range of the so called water window (280 to 530 eV), where the absorption of biological matter is high compared to cellular water. Biological samples mainly consist of light elements such as carbon, oxygen and nitrogen providing only poor absorption contrast. X-ray Zernike phase-contrast microscopy is a promising approach to increase the contrast with biological samples and to maintain the required penetration depth. At X-ray full-field microscope at beamline P11 PETRA III a resolution of 50 nm can be achieved with biological specimens. A major step forward in observing cellular function on a microscopic level has been the introduction of fluorescent proteins. To obtain high resolution images of specifically labelled sub-cellular protein structures in whole cells with X-ray Zernike phase-contrast microscopy we adopted staining techniques from light and electron microscopy utilizing enzymatic oxidation and photooxidation of 3,3'-diaminobenzidine (DAB) in human foreskin fibroblasts (HFF) followed by subsequent osmophication.

Crystal physics

P173

Thermodynamic properties of Ra sulfate estimated from calorimetric data on isostructural crystals

L. Bayarjargal[1], V. Vinograd[2], J. Bauer[1], B. Winkler[1], M. Luchitskaia[1], C. M. Pina[3], K. Refson[4]
[1]*Institute of Geosciences, Goethe University, Frankfurt am Main, Germany*
[2]*Institute of Energy and Climate Research (IEK6) Nuclear Waste Management and Reactor Safety, Forschungszentrum Jülich , Jülich, Germany*
[3]*Facultad de Ciencias Geológicas, Universidad Complutense de Madrid. Instituto de Geociencias IGEO (UCM – CSIC), Departamento de Cristalografía y Mineralogía, Madrid, Spain*
[4]*Department of Physics, Royal Holloway, University of London, London, Great Britain*

Safety assessments of the direct disposal of spent nuclear fuel in deep geological formations include investigations of the unlikely event of container failure in which ground water could be contaminated by radionuclides. Ra^{2+} is likely to be able to easily migrate from the waste into the host rock. The potential danger is greatly reduced, however, if the host rock is saturated with respect to barite, $BaSO_4$. Therefore the thermodynamic characterization of $RaSO_4$ and Ra containing solid solutions is important for quantitative assessment of safety aspects [1].

So far, there is no complete set of experimental data of $RaSO_4$, due to difficulties in direct experimental handling. The available thermodynamic properties of $RaSO_4$ are estimated from indirect approaches or from atomistic simulations. Here, we present an approach based on the Debye model for $RaSO_4$.

Heat capacities of Sr-, Ba- and Pb-sulfates were measured by relaxation calorimetry (PPMS) between 2 K and 395 K. For $SrSO_4$ and $BaSO_4$ our values agree well with previous measurements [2-4]. However, there were no experimental data concerning $PbSO_4$. Our heat capacity data of $PbSO_4$ does not differ much from that of $BaSO_4$ or $SrSO_4$ above 300 K. However, at lower temperatures, the heat capacity of $PbSO_4$ is markedly larger. The standard entropy of $PbSO_4$ calculated from the measured heat capacity is 147.1(1.5) J K^{-1} mol^{-1}.

The Debye model suggests that the heat capacity should vary linearly with the parameter $M^{3/2}V^{-1/2}K^{-3/2}$ where M is weight, V is volume and K is bulk modulus. A linear interpolation of the measured C_p values for $SrSO_4$, $BaSO_4$ and $PbSO_4$ against this parameter is used here to predict the heat capacity of $RaSO_4$. This interpolation suggests that the standard entropy of $RaSO_4$ should be approximately similar or slightly larger than that of $PbSO_4$ and thus about 11 J/K/mol larger than the currently published value of 137.8 J/K/mol [1]. This conclusion could be modified, however, if the entropy of $PbSO_4$ has an additional contribution presumably arising from the positional disorder of the Pb^{++} cation.

The study has been funded in part by BMBF through grants (02NUK019E and 02NUK021F).

[1] V. Vinograd et al., Geochim. Cosmochim. Acta 122, 398 (2013).
[2] V. Gurevich et al., Geokhimiya 35 (3), 305311 (1997).
[3] W. M. Latimer et al., J. Phys. Chem. 1, 620 (1933).
[4] J. Majzlan et al., Geochim. Cosmochim. Acta 66, 1839 (2002).

P174

A Low-Cost Laue Method for the Orientation of Single Crystals (Part II)

W. Lottermoser[1], A. Jakhu[2], G. Amthauer[1]
[1]*University of Salzburg, Chemistry and Physics of Materials, Salzburg, Austria*
[2]*Indian Institute of Technology, Delhi, India*

Introduction
With the availability of low-cost powerful computers, the Laue technique has experienced a splendid resurrection during the last decades. Main application fields are the orientation of inorganic and organic crystals. However, the mechanical and recording conditions of Laue patterns were not as encouraging, in particular from an economic point of view: Films are too cumbersome to handle and chemically hazardous, electronic films or image plates together with their scanners too expensive for the budget of many smaller research units.

Objectives
Our aim was to develop a low-cost method for the recording of reflections and for the crystal handling within a common X-ray device using the Laue technique

Materials and Methods

On the basis of a standard PHYWE XRE 4.0 X-ray expert system [1] we developed a fully functional crystal orientation device comprising of a HUBER big arc goniometer head [2], in-house equipped by 3-axis motor drives and webcam-controlled, a recording fluorescent foil and a DSLR with connection to an external computer for purposes of storing and evaluating the emerging Laue patterns. The system was tested in transmission geometry with several single crystals of different absorption coefficients detecting no further obstacles or disadvantages compared to conventional systems. The evaluation of the Laue patterns is done by comparison with a common software generating theoretical Laue reflections [3] and a crystallographic structure display program [4].

Results
Due to the close distances between collimator, crystal and foil the very low power of the X-ray tube (35 W) does not appear to be disadvantageous. The recording times are low, the reflections comparably bright and can easily be after-processed by a suitable image enhancing program. Since the PHYWE system mentioned above is a full protection device with a very low radiation, the operating staff need not be supervised radiometrically. In the whole, the costs of the whole system could be reduced to almost half of the common value of the cheapest commercial system with digital recording.

Conclusions
Even research units with a very low budget can now afford a digital crystal orientation system which is comparable to much more expensive devices

[1] PHYWE XRE 4.0 X-ray expert, Phywe-Lehrmittel Göttingen

[2] HUBER X-ray Diffraction Equipment Ltd., Rimsting

[3] JAVA Lauegram (c) Steffen Weber, July 1997 update 10/27/01

[4] CaRIne Crystallography 3.1, C.Boudias & D.Monceau 1989

P175

Mechanical properties of natural radiation damaged minerals and temperature-induced structural reorganization

T. Beirau[1,2], W. D. Nix[3], R. C. Ewing[1], G. A. Schneider[4], L. A. Groat[5], U. Bismayer[2]
[1]*Stanford University, Department of Geological Sciences, Stanford, United States*
[2]*University of Hamburg, Department of Earth Sciences, Hamburg, Germany*
[3]*Stanford University, Department of Materials Science and Engineering, Stanford, United States*
[4]*Hamburg University of Technology, Institute of Advanced Ceramics, Hamburg, Germany*
[5]*University of British Columbia, Department of Earth, Ocean and Atmospheric Sciences, Vancouver, Canada*

We present new insights into the relation between thermally induced structural reorganization and the macroscopic mechanical properties of radiation-damaged titanite (Beirau et al. 2016). Low-temperature annealing affects only slightly the sample stiffness and leads to a softening resulting from the defect annihilation in crystalline regions. In the high-temperature annealing regime, amorphous domains recrystallize and this leads to further recovery of defects, reduction of interfaces, grain growth, and, in general, an increase in the long-range order. The thermally induced recrystallization is accompanied by massive dehydration leading to considerable stiffening and hardening. This interpretation of the recrystallization process in titanite based on the correlation of new results from nanoindentation and Raman-spectroscopic measurements complementing previous investigations using thermogravimetric and gas analyses by Hawthorne et al. (1991) and infrared spectroscopy by Zhang et al. (2001). Additionally, we present the results of recent alpha-radiation dose dependent nanoindentation measurements of zircon. With increasing radiation dose, the defect concentration and the amount of amorphous fraction increase and the density decreases which has a complex influence on the materials elastic modulus and hardness.

Beirau, T., Nix, W.D., Ewing, R.C., Schneider, G.A., Groat, L.A. and Bismayer, U. (2016) *Mechanical properties of natural radiation damaged titanite and temperature induced structural reorganization: A Nanoindentation and Raman spectroscopic study*. American Mineralogist, in press.
Hawthorne, F.C., Groat, L.A., Raudsepp, M., Ball, N.A., Kimata, M., Spike, F., Gaba, R., Halden, N.M., Lumpkin, G.R., Ewing, R.C., and others. (1991) *Alpha-decay damage in titanite*. American Mineralogist, 76, 370-396.
Zhang, M., Groat, L.A., Salje, E.K.H., and Beran, A. (2001) *Hydrous species in crystalline and metamict titanites*. American Mineralogist, 86, 904-909.

P176

Single crystal elastic properties of $Bi_{12}SiO_{20}$ as a function of pressure: Brillouin spectroscopy and atomistic simulations

H.- J. Reichmann[1], E. Haussühl[2], A. Friedrich[2,3]
[1]*Deutsches GeoForschungsZentrum GFZ, Sektion 3.3, Potsdam, Germany*
[2]*Goethe Universität Frankfurt, Institut für Geowissenschaften, Abt. Kristallographie, Frankfurt am Main, Germany*
[3]*Universität Würzburg, Institut für Anorganische Chemie, Würzburg, Germany*

$Bi_{12}MO_{20}$ with M = Si, Ti, and Ge are members of the sillenite group and are relevant for industrial applications due to their prominent optic, elasto-optic, electro-optic and piezoelectric applications. For example, they serve as devices utilized in optical computing, optical information processing, and high sensitive photo detectors [1]. The origin of these interesting properties can be found in the acentric crystal structure (space group *I*23) [2]. Sillenite crystals in general exhibit low acoustic wave velocities. Therefore they are ideally suited for delay lines in electronic circuits. Up to now $Bi_{12}SiO_{20}$ (BSO) was investigated in respect to its elastic behavior in a low pressure range from ambient conditions to 0.16 GPa only [3]. In our earlier studies we have investigated the structural compression and vibrational properties of BSO up to 39 GPa [2,4]. Anomalies of the BSO lattice compression from an idealized behavior were attributed to a high sensitivity of the sillenite structure to slight deviatoric stress, which is induced when using neon instead of helium as a pressure medium above 15 GPa, rather than to the occurrence of a phase transition [4]. In order to confirm the absence of an anomaly in helium pressure medium, to better understand structure-property relations in BSO and to complement our earlier results, we have determined the complete sets of elastic coefficients by means of Brillouin interferometry up to 27 GPa and by density functional theory (DFT) based model calculations. We found that the moduli c_{11} and c_{12} continuously increase with pressure while c_{44} only slightly increases and remains nearly constant within standard deviation above 15 GPa. This flattening could be an indication for a phase transition at higher pressures (mode softening).

The authors gratefully acknowledge Prof. B. Winkler for the DFT calculations and the Deutsche Forschungsgemeinschaft DFG for financial support of this investigation (HA 5137/3). AF thanks the DFG for financial support within SPP1236 (FR 2491/2-1).

[1] Mel'nikova, T.I., Kuz'micheva, G.M., Rybakov, V.B., Bolotina, N.B., and Dubovsky, A.B. (2011) Synthesis, compositon, and structure of sillenite-type solid solutions in the Bi_2O_3 - SiO_2 - MnO_2 system. Inorg. Chem., 50, 2002-2009.
[2] Wiehl, L., Friedrich, A., Haussühl, E., Morgenroth, W., Grzechnik, A., Friese, K., Winkler, B., Refson, K. and Milman, V. (2010) Structural compression and vibrational properties of $Bi_{12}SiO_{20}$ sillenite from experiment and theory. J. Phys.: Cond. Matter, 22, 505401.
[3] Gospodinov, M., Haussühl, S., Sveshtarov, P., Tassev, V., and Petkov, N. (1988) Physical Properties of Cubic $Bi_{12}SiO_{20}$. Bulg. J. Phys., 15, 140-143.
[4] Wiehl, L., Friedrich, A., Haussühl, E., Morgenroth, W., Biehler, J., Winkler, B., and Hanfland, M. (2013) High pressure powder X-ray diffraction of sillenites $Bi_{12}MO_{20}$ (M = Si, Ge, Ti) and $Bi_4Ti_3O_{12}$. J. Solid State Chem., 208, 35-42.

P177

Order parameter behaviour of Pb-lawsonite

M. Ende[1], W. Schranz[2], E. Libowitzky[1]
[1]*University of Vienna, Department of Mineralogy and Crystallography, Vienna, Austria*
[2]*University of Vienna, Faculty of Physics, Vienna, Austria*

1. Introduction

Lawsonite ($CaAl_2[(OH)_2|Si_2O_7]\cdot H_2O$) has a phase transition at 273 K [1], which is related to framework displacements as well as proton ordering and which is thermodynamically described as tricritical [e.g. 2]. Recently, it was shown that Pb-lawsonite ($PbAl_2[(OH)_2|Si_2O_7]\cdot H_2O$) has a similar phase transition around 450 K, which is accompanied by a coordination change of the Pb^{2+}-cation and consequent framework displacements as well as an additional huge contribution of hydrogen bonds as well [3].

2. Objectives

Investigations of the thermodynamic behaviour of the structural phase transition in Pb-lawsonite around 450 K should characterise it energetically and reveal similarities and differences to lawsonite.

3. Materials & Methods

Infrared spectroscopic experiments on a synthetic Pb-lawsonite/KBr powder-pellet were done using a Bruker Tensor 27 IR spectrometer with Hyperion microscope and Linkam FTIR600 cool./heat. stage.

Landau theory describes phase transitions and expands the Gibbs free energy (G) in terms of an order parameter (Q). For isobaric thermal expansion it is expressed: $G(T) = G_0(T) + 1/2A(T-T_c)Q^2 + 1/4BQ^4 + 1/6CQ^6$, where G_0 is the Gibbs free energy of the high-temperature phase. Usually three types of energetic behaviour can be distinguished, i.e. 2nd-order, tricritical, and 1st-order transitions. From low to high temperature, for tricritical phase transitions (2-6 potential, $B \approx 0$) Q approaches zero as T approaches T_c (transition temperature) rather faster than in 2nd-order transitions (2-4 potential, B>>C). In 1st-order transitions (2-4-6 potential, B is negative), the order parameter jumps discontinuously to zero near T_c.

Based on earlier studies band position (ω), intensity (I) ≙ absorbance (A) and width (FWHM) most commonly depend for this zone boundary transition on Q by the simple relationship [e.g. 4, 5]: $\Delta\omega \sim \Delta I \sim \Delta FWHM \sim AQ^2 + BQ^4 + CQ^6$

4. Results

In the temperature-dependent IR measurements of Pb-lawsonite non-uniform changes of the spectra are observed around 90 K below T_c. Similar anomalies were observed for lawsonite too, but they are very small [6]. Since these anomalies in lawsonite were considered as separate contributions of proton ordering and framework distortions, the same is anticipated for Pb-lawsonite. However, both a tricritical 2-4-6 potential and a model with order parameter coupling fit the experimental data equally well.

5. Conclusion

Both transitions, that of lawsonite around 273 K and that of Pb-lawsonite around 450 K, have tricritical order character and both comprise framework- and hydrogen-related order parameters. Furthermore, both phases seem to exhibit anomalies possibly revealing complex order parameter coupling, which should be discussed in more detail.

[1] Libowitzky, E., Rossman G.R. (1996) Am. Mineral., 81, 1080-1091
[2] Sondergeld, P., Schranz, W., Kityk, A.V., Carpenter, M.A., Libowitzky, E. (2000) Phase Transit., 71, 189-203
[3] Ende, M., Wunder, B., Koch-Müller, M., Pippinger, T., Buth, G., Giester, G., Lengauer, C.L., Libowitzky, E. (2016) Mineral. Mag., DOI: 10.1180/minmag.2016.080.015
[4] Schranz, W., Tröster, A., Gardon, M., Krexner, G., Prem, M., Carpenter, M.A., Sondergeld, P., Armbruster, T. (2005) Z. Kristallogr., 220, 704-711
[5] Salje, E.K.H., Bismayer, U. (1997) Phase Transit., 63, 1-75
[6] McKnight, R.E.A, Carpenter, M.A., Darling, T.W., Buckley, A., Taylor, P.A. (2007) Am. Mineral., 92, 1665-1672

P178

Bi lone-pair induced unusual charge disproportionation of mixed-valent chromate $BiCu_3Cr_4O_{12}$

M. Isobe[1], M. Etter[1], H. Sakurai[2], R. E. Dinnebier[1], H. Takagi[1]
[1]*Max Planck Institute for Solid State Research, Stuttgart, Germany*
[2]*National Institute for Materials Science (NIMS), Tsukuba, Ibaraki, Japan*

Many 3d transition metal perovskites exhibit various interesting phenomena. The $AA'_3B_4O_{12}$ belong to wide family of *A*-site-ordered perovskite, which crystallize with an *Im*3 cubic lattice. They provide an opportunity for the mixed valence state of *B* cations. *A'* sites are of a pseudosquare planar coordination suited to Jahn-Teller active ions such as Cu^{2+} and Mn^{3+}, while *A* sites are occupied by typical large *A*-site ions such as alkaline earth and rare earth metal ions.

Recently A new ordered perovskite $BiCu_3Cr_4O_{12}$ has been synthesized under high-pressure and high-temperature conditions. $BiCu_3Cr_4O_{12}$ is a mixed-valent chromate. In this compound, a ferrimagnetic transition with a monoclinic structural distortion occurs at around 190 K [1]. At the transition the resistivity exhibits an anomaly; however it has a metallic conductivity down to 2 K. From the structural analysis of the low-temperature phase, we conclude that it is an unusual charge disproportionation [1], whose trigger is the local distortion of the BiO_{12} polyhedron for Bi 6s-O 2p hybridization.

In some ferroelectric perovskites including Bi^{3+} in the *A*-site, the Bi cation has a stereochemically active $6s^2$ lone-pair which causes the Bi 6p (empty) orbital to come closer in energy to the O 2p orbitals. This leads to hybridization between the Bi 6s and O 2p orbitals and drives the off-centering of the cation towards the neighboring anion in the low-temperature phase. In general, this can be the origin of the ferroelectric transition in the perovskites including Bi^{3+}.

In $BiCu_3Cr_4O_{12}$, the off-centering of Bi^{3+} in figure 1(b) induces the distortion of the neighboring CrO_6 octahedra. The distortion stabilized the Cr-O hybridization of the $Cr^{4+}O_6$ octahedra. As the result, the two kinds of $Cr^{3.5+}O_6$ and $Cr^{4+}O_6$ chains along the b direction ordered in a rocksalt-type substructure in ac plane in the low-temperature phase. This is the first observation of the charge disproportionation in chromium oxides induced by Bi lone-pair.

Figure 1: (a) Cubic crystal structure of $BiCu_3Cr_4O_{12}$ at 300K. (b) Monoclinic crystal structure of $BiCu_3Cr_4O_{12}$ at 100K. The charge disproportionation and charge ordering of the Cr ions are indicated by the coloring of the CrO_6 octahedra (yellow : $Cr(1)^{4+}$, blue : $Cr(2)^{3.5+}$, purple : $Cr(3)^{3.5+}$).

[1] M. Etter, M. Isobe, H. Sakurai, R. E. Dinnebier and H. Takagi, A new monoclinic crystal structure type for *A*-site ordered quadruple perovskites. *23th Annual Meeting of the German Crystallographic Society* (2015).

Figure 1

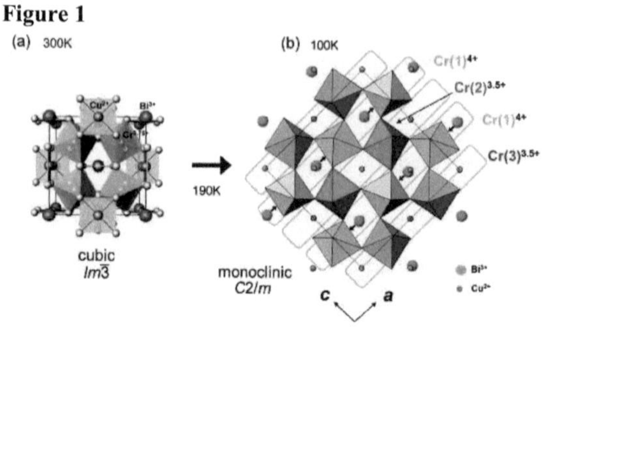

Supplement Issue No. 36
Zeitschrift für Kristallographie
ISBN 978-3-11-044490-2

Bei Fragen zur Produktsicherheit wenden Sie sich bitte an:
If you have any questions regarding product safety,
please contact:

Walter de Gruyter GmbH
Genthiner Straße 13
10785 Berlin
productsafety@degruyterbrill.com